Springer Series on Polymer and Composite Materials

Series Editor

Susheel Kalia, Army Cadet College Wing, Indian Military Academy, Dehradun, India

More information about this series at http://www.springer.com/series/13173

Swati Sharma · Ashok Kumar
Editors

Lignin

Biosynthesis and Transformation for Industrial Applications

Springer

Editors
Swati Sharma
University Institute of Biotechnology (UIBT)
Chandigarh University
Mohali, Punjab, India

Ashok Kumar
Department of Biotechnology
and Bioinformatics
Jaypee University of Information
Technology
Waknaghat, Himachal Pradesh, India

ISSN 2364-1878 ISSN 2364-1886 (electronic)
Springer Series on Polymer and Composite Materials
ISBN 978-3-030-40662-2 ISBN 978-3-030-40663-9 (eBook)
https://doi.org/10.1007/978-3-030-40663-9

This Springer imprint is published by the registered company Springer Nature Switzerland AG
The registered company address is: Gewerbestrasse 11, 6330 Cham, Switzerland

Preface

This book had been composed to provide concrete information about the structural properties and various industrial applications of lignin. As a well-known biopolymer on the earth's surface still, lignin is underutilized and not well explored. The complexities in the molecular structure and properties and poor solubility of the lignin pose some challenges for utilization of lignin at industrial scale. Moreover, this is one of the by-products of two major industries bioethanol and paper/pulp manufacturing units. In North American and European countries, the excessive production of lignin from these industries also added the lignin as a pollutant to the environment. Thus, it becomes very necessary to find out the possible methods to minimize its production or further utilization to convert it into value-added products. Thus, the present book was encouraged by the ambition that lignin can be used as powerful raw material for the synthesis of various products of commercial value. This book will provide detailed information about the sources, techniques of lignin treatment origin, structure, physicochemical properties, and various applications in agriculture, food, pharmaceutical, and nanotechnology industries. We hope that information provided in the chapters written by various experienced researchers will surely helpful for the researchers, industrialists, and young students working in the field of biomass utilization, biofuels production, and waste management. In this book, we compiled the chapters contributed by well-experienced academicians and scientists working in the bio-based products and green chemistry. The fundamental rationale of this book is to provide important information about the various topics on lignin on a single platform. The editors have a keen interest in the lignin conversion and its utilization which encouraged them to compose this book and provide in the public domain. One does not need to spend many hours to gather information on a single topic and get puzzled by various newly published research articles and old data which sometime become contradictory to each other. Thus, in the book, we mentioned the chapter contents on all the relevant topics on lignin which become the area of thrust for various research groups. Our chapter "Lignin as Potent Industrial Biopolymer: An Introduction" described the basic introduction characteristics and properties of lignin. Our chapter "Structure and Characteristics of Lignin" provided a detailed description of structure and properties of the lignin. In this continuation, chapters "Lignin Synthesis and Degradation" and "Analysis of Lignin Using Qualitative and Quantitative Methods" were dealing with the synthesis and degradation of lignin and its analysis of lignin using qualitative as well as quantitative methods. Chapter "Chemical Modification of Lignin by Polymerization and Depolymerization" had been emphasized on the chemical modification of lignin by polymerization and depolymerization. Also, chapters "Lignin and Its Composites" and "Synthesis of Micro- and Nanoparticles of Lignin" give detailed information on composites of lignin and synthesis of nanoparticles from lignin. Chapters "Synthesis and Applications of Lignin-Derived Hydrogels" and "Lignin Composites for Biomedical Applications: Status, Challenges and Perspectives" described the application of lignin in the synthesis of hydrogel of pharmaceutical importance and various biomedical applications of lignin composites. Chapter "Applications of Lignin in the Agri-Food Industry" gives important information about the application of lignin in the agricultural and food industry. We are firmly hopeful that this book will be helpful for all biotechnologists, academicians, and industrialists.

Mohali, India — Swati Sharma
Waknaghat, India — Ashok Kumar

Contents

About the Editors

Dr. Swati Sharma is working as Assistant Professor in the University Institute of Biotechnology, Chandigarh University Mohali, Punjab, India. Dr. Sharma has completed her Ph.D. from University Malaysia Pahang, Malaysia. She worked as a visiting researcher in the college of life and environmental sciences at Konkuk University, Seoul, South Korea. Dr. Sharma has completed her masters (M.Sc.) from Dr. Yashwant Singh Parmar University of Horticulture and Forestry, Nauni Solan HP. India. She has also worked as program co-coordinator at Himalayan Action Research Center, Dehradun, and senior research fellow at Indian Agricultural Research Institute in 2013–2014. Dr. Sharma has published her research papers in reputed international journals. Presently, Dr. Sharma's research is in the field of bioplastics, hydrogels, keratin nanofibers and nanoparticles, biodegradable polymers and polymers with antioxidant and anti-cancerous activities and sponges. Dr. Swati has published 16 research papers in various internationally reputed journals, two books and a couple of book chapters.

Dr. Ashok Kumar is working as an Assistant Professor in the Department of Biotechnology and Bioinformatics, Jaypee University of Information Technology, Waknaghat, Solan, Himachal Pradesh, India. He has completed his Ph.D. in biotechnology from the Department of Biotechnology, Himachal Pradesh University, Summer Hill, Shimla, India, in 2013. He worked as a post-doctoral fellow in State Key Laboratory of Agricultural Microbiology, Huazhong Agricultural University, Wuhan, China for two years, 2014–2016. He also worked as Brain Pool researcher/ Assistant Professor at Konkuk University, Seoul, South Korea for 2016–2017. Dr. Ashok has a keen interest in enzyme immobilization, biomaterials, biopolymers, nanobiotechnology, biocatalysis, waste management, biomass degradation, biofuel synthesis, and biotransformation. Dr. Ashok has published more than 50 research papers in various internationally reputed journals, two books and 17 book chapters. He also contributes as editor and reviewer in various international journals. He is teaching industrial enzyme technology, metabolic engineering, and environmental biotechnology to degree students. He is also supervising the doctorate, master, and bachelor degree students in his laboratory.

Lignin as Potent Industrial Biopolymer: An Introduction

Swati Sharma, Abhishek Sharma, Sikandar I. Mulla, Deepak Pant, Tanvi Sharma and Ashok Kumar

Abstract Lignin is a complex biological molecule and well known as a second abundant biopolymer on the planet. It also produced as a by-product of lignocellulosic biomass, paper and pulp industry. It has not been well explored for industrial applications due to its poor solubility and complex molecular structure. Hence, since last two decades, lignin has gained the attention of researchers as a raw material for the synthesis of value-added products of commercial importance. However, the major part of the lignin derived from the pulp industry is being utilized as boiler fuel. The efficacy of lignin produced from lignocellulosic-based bio-ethanol industry to synthesize the renewable polymeric material is much higher as compared to the lignin obtained from pulp industry. These days lignin has also been exploited to synthesize various chemicals like esters, ethers, bio-adhesives, biolubricants, foams, nano-particles and nano-composites. The excessive production of lignin from paper or bio-ethanol industry has also appeared as critical environmental problem. The conversion of huge amount of lignin will not only help to solve the environmental threat but also boost the bio-economy of bio-ethanol and paper industry. In this chapter, we highlight the sources, extraction of lignin and its immense industrial importance.

Keywords Ethanol · Lignocellulosic biomass · Biofuel · Esters · Ethers · Adhesives · Foam

S. Sharma
University Institute of Biotechnology, Chandigarh University, Mohali, Punjab 140413, India

A. Sharma
Department of Biotechnology, Himachal Pradesh University, Summer Hill, Shimla 171005, India

S. I. Mulla
Department of Biochemistry, School of Applied Sciences, REVA University, Bangalore, India

D. Pant
Department of Chemistry, School of Chemical Sciences, Central University of Haryana, Mahendragarh, Haryana 123031, India

T. Sharma · A. Kumar (✉)
Department of Biotechnology and Bioinformatics, Jaypee University of Information Technology, Waknaghat, Solan 173 234, India
e-mail: ashok.nadda@juit.ac.in; ashok.nadda09@gmail.com

S. Sharma and A. Kumar (eds.), *Lignin*, Springer Series on Polymer and Composite Materials, https://doi.org/10.1007/978-3-030-40663-9_1

1 Introduction

Lignin is the most abundant in plants and an important component of the cell wall. It is highly hydrophobic and insoluble in water and alcohol. Lignin is a complex organic polymer formed of monolignol units. Lignin waste is also appeared as an environmental issue as this waste is dumped or burned as low-grade fuel causes wastage of natural resources. Hence, upgradation of lignin valourization through technological improvements for the realization of sustainable resource management is very necessary. Due to the technological improvements like thermochemical tailor technologies of lignin, numerous value-added chemicals can be synthesized, e.g. ethylbenzene (vital intermediate for organic synthesis) (Regmi et al. 2017), guaiacol (phenolic resin making) (vom Stein et al. 2015) and *p*-hydroxyl acetophenone (cholagogue drug making and perfume synthesis) (Luo et al. 2018). Currently, more than 80 million tons of lignin is being produced as a by-product from lignocellulosic biorefinery and paper/pulp industry. On the other hand, only a small fraction of this lignin is used as fuel or energy source in the industry. Thus, researchers are intended to synthesize various fine chemicals of higher commercial value from lignin to make the biorefinery process more sustainable. However, the direct conversion of lignin into valuable chemicals is also challenging due to its poor solubility and complex phenolic ring structures with high bond energy. Lignin falls within important biomass resources due to the depletion of fossil resources and increasing environmental pollution by the excessive use of fuels (Peng et al. 2014).

Literature showed that lignin is a good source for manufacturing the phenolics as it is one of the potent sources of aromatic chemical compounds (Alonso et al. 2010). As stated in previous reports, lignin can be hydrolysed into phenolic resins and alkylphenols which can act as highly efficient fuels and energy sources (Graglia et al. 2015; Li et al. 2015; Xu et al. 2014; Zakzeski et al. 2010). The lignin-derived derivatives can also be used in the synthesis of polyurethane resins, polyols, foams, antifungal agents and various products of industrial use (Dos Santos et al. 2015; Mahmood et al. 2016; Rößiger et al. 2018). The major challenges in the lignin conversion are formation of char as a by-product, complexity in the purification and separation of products and also poor selectivity for product synthesis which are needed to be considered during operation of lignin-based biorefinery.

2 Sources and Structure of Lignin

2.1 Major Sources of Lignin

Lignin was primarily studied for its significance in the pulp and paper industry, where fibrous product rich in cellulose was produced by eliminating lignin from the wood. Therefore, the major studies were focused on the composition and structure of lignin (Chen et al. 2018; Fernández-Rodríguez et al. 2019; Glasser 2019; Rahimi et al. 2014). Some studies of lignin were also related to herbaceous plants (Buranov and Mazza 2008; del Río et al. 2012a, b; Marques et al. 2010) where lignin partly triggered by digestibility and dietary conversion issues in animal feed (Campbell and Sederoff 1996). But nowadays, these lignocellulosic wood species and wastes provide opportunities for the production of green chemicals, bioproducts and energy within the biorefinery (de Gonzalo et al. 2016; Graglia et al. 2015; Keegstra 2010; Zakzeski et al. 2010). Lignin is mostly bound to polysaccharides in close association with cellulose and hemicelluloses in plants. But the distribution of lignin, cellulose and hemicellulose is not uniform throughout the plant body. Vascular plants contain lignin mainly between the cells, with second largest in quantity to cellulose among natural polymers (Gellerstedt 2015; Gierer 1985; Graglia et al. 2015; Keegstra 2010; Moura et al. 2010; Zakzeski et al. 2010). Lignin is proven fundamental source of complex organic polymers for the chemical industry in a biorefinery. More specifically in woody plants, the lignin content is approximately 15–40% (Fengel and Wegener 1984), whereas in herbs or annual plants, the lignin content is less than 15% (Table 1). So, the paper and pulp sectors yield a large quantity of lignin with greater potential in biorefineries. Negligible lignin is recovered from solid waste, where their content (lignin) is assessed to be less than 15%.

Table 1 Quantity of lignin in various classes of plants

Plant	Lignin contents (%)	Reference
Piceaabies sp.	20–40	Önnerud (2003), Önnerud and Gellerstedt (2003)
Pinus pinaster	25–35	Baptista et al. (2008)
Pinus taeda	28–45	Huang et al. (2011)
Pinus radiata	25–30	Sjöström (1993)
Acasia mollissima	20–25	Sjöström (1993)
Acacia mangium	27–35	Pinto et al. (2005)
Cynara cardunculus	16–25	Gominho et al. (2001)
Betula verrucosa	15–25	Önnerud (2003)
Eucalyptus globulus	20–25	Sjöström (1993)
Populus tremula	18–22	Önnerud and Gellerstedt (2003)
Quercus suber	25–40	Lourenço et al. (2016)
Tectona grandis	30–45	Lourenço et al. (2015)
Arundo donax	15–20	Shatalov and Pereira (2005)
Miscanthus x *giganteus*	15–20	Pascoal Neto et al. (1997)
Musa acuminate	10–25	Oliveira et al. (2007)

2.2 Structural Composition of Lignin

In order to understand the properties of lignin and to propose a lignin-based pathway, its structural aspects need to be understand deeply. Lignin is primarily constituted of three monolignol precursors: coniferyl alcohol (guaiacyl), *p*-coumaryl alcohol (hydroxyphenyl) and sinapyl alcohol (syringyl) (Fig. 1) which synthesizes lignin polymers in the cytoplasm via shikimate pathway (Boudet 2000; Glasser 2019; Rahimi et al. 2014; Rößiger et al. 2018). The chemical structure of lignin possessed abundance of the chemical sites which indicated that lignin might be a key precursor in the synthesis of aromatic chemicals. Few other non-conventional monolignol precursors are also identified in negligible amounts (Gellerstedt and Henriksson 2008). Also, these polyphenolic rings provide hydrophobicity in aqueous systems which ultimately gave resistance to chemical and biological degradation of plants (Keegstra 2010). The composition of these monolignol precursors in different types of plants is represented in Table 2. In angiosperms, syringyl–guaiacyl-type lignin is

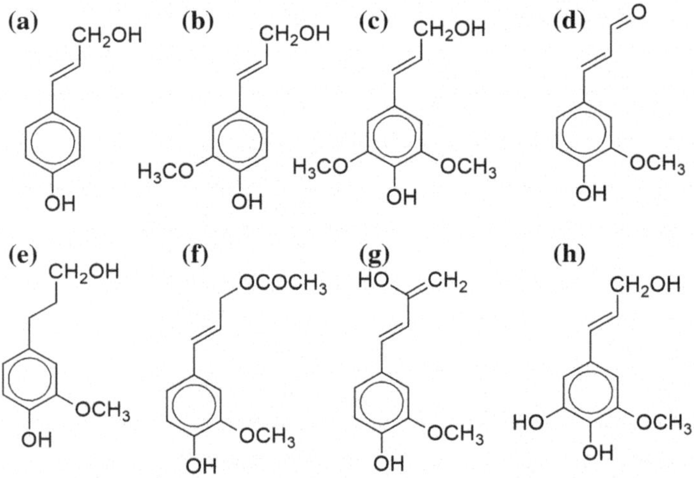

Fig. 1 Types of monolignol precursors for lignin syntheses: **a** *p*-coumaryl alcohol, **b** coniferyl alcohol, **c** sinapyl alcohol, **d** coniferaldehyde, **e** dihydroconiferyl alcohol, **f** coniferyl alcohol-9-acetate, **g** ferulic acid and **h** 5-hydroxyconiferyl alcohol

Table 2 Amount of monolignol precursors is in different types of plants (Gellerstedt and Henriksson 2008)

Plant	Hydroxyphenyl (H) (%)	Guaicayl (G) (%)	Syringyl (S) (%)
Softwoods	5	95	0
Hardwoods	0–8	25–50	45–75
Grasses	5–35	35–80	20–55

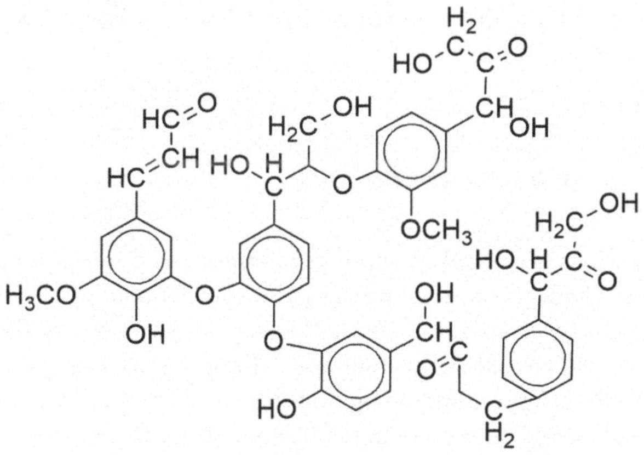

Fig. 2 Structure of lignin showing G, S and H units

present which makes angiosperm (hardwood) lignin more readily detached during pulping process, while in gymnosperm, only guaiacyl lignin is present (Fig. 2).

3 Lignin Extraction from Different Types of Biomass

The extraction of lignin from different types of biomass usually depends on a starting material or the wood to obtain the pulp (Chen et al. 2018; Radotić and Mićić 2016). The yield of the lignin is affected by various parameters such as the method of extraction, reaction time, type of reaction medium and temperature. There are several methods which have been developed for the extraction of lignin from cellulosic pulp which involves enzymatic, chemical and physical.

3.1 Biological or Enzymatic Methods

Enzymatic methods involve the use of cellulolytic enzymes for carbohydrate's hydrolysis which produce the insoluble fraction of lignin (Radotić and Mićić 2016). Successive enzymatic treatments of the pulps ensure absolute hydrolysis of carbohydrates to enhance the recovery of lignin with good yield as compared to other methods like unbleached kraft pulps. These enzymes are quite effective to hydrolyse the entire pulp polysaccharides of the wood. The obtained fractions contain ~65 to 80% lignin with ~7 to 8% carbohydrates and some fraction of other impurities (Radotić and Mićić 2016). These impurities are difficult to remove by enzymatic methods. However, the residual lignin obtained from enzymatic hydrolysis or dissolution of carbohydrates is considered to be effective and economic as lignin structures remain unchanged with a higher yield. In order to remove the impurities and ensure the high purity of lignin, we need to apply an integrated approach which involves the use of chemical agents and mechanical amenities along with enzymatic method.

3.2 Chemical Methods

Various chemical methods for extraction of lignin are employed; some of them are acid hydrolysis, kraft process, lignosulphonate process and organosolvolysis. In acid hydrolysis, the extracted pulp from the wood is refluxed under N_2 or Ar (argon) and HCl to solubilize the lignin (Fernández-Rodríguez et al. 2019; Gellerstedt et al. 1994). The solubilized lignin is free from contaminants with minimum impurities and makes this method better and rapid as compared to enzymatic hydrolysis. However, strong acidic environment results in structural changes in the lignin like the cleavage of aryl and alkyl ethers in benzyl alcohol moiety, alterations of molecular structure, which could be the possible drawbacks of this method.

The extraction of lignin through lignosulphonate process involves the treatment of wood at high temperatures with sodium sulphite (aqueous) under acidic conditions. The functional moieties, e.g. carboxylic, phenolic and –SH, allow their use as surfactants, additives, dispersants and flocculants (Aro and Fatehi 2017). Approximately, 1 million tons of lignosulphonates are produced every year as dry solids. Nowadays, kraft process is more preferred as compared to sulphite pulping because of its better product quality and fibre strength obtained in lignin extracts (Zhu et al. 2009).

Presently, the usage of volatile organic solvents (VOS) is growing to prevent environmental pollution and minimization of the greenhouse effect (Dos Santos et al. 2015; Mahmood et al. 2016). Also, ionic liquids are in trend for biomass treatment due to their low volatility as well as reusable behaviour. Lignin obtained by organosolvents like ethanol or alkaline NaOH is assisted with the combined treatment of ionic liquid 1-butyl-3-methylimidazolium methylsulphate [Bmim][$MeSO_4$] (Meng et al. 2012). Overall, when compared to lignosulphonates or kraft lignins, the alkaline and organosolvolignins are sulphur-free and most suitable for the synthesis of various value-added products (Zhang 2011).

3.3 Physical Methods

Various physical or mechanical methods like pyrolysis and heat steams are used in lignin isolation from wood. Pyrolysis is known to be a thermochemical process as it can convert 60–75% of the original biomass into crude oil (Garcia-Perez et al. 2007; Jiang et al. 2010). This crude oil of biomass can replace the petroleum-based fuels (Czernik and Bridgwater 2004). Pyrolysis is generally operated at relatively high heating rate, ambient temperature range and short residence time of gas in order to obtain more liquid products (Demirbaş 2001). There are only a few reports on extracting pyrolytic lignin from bio-oils (Jiang et al. 2010; Scholze et al. 2001). It may be separated by adding water to precipitate lignin-derived hydrophobic components.

Another method for lignin extraction is steam-assisted process. In this process, wood is subjected to high temperature (200 °C) and quick release of pressure which ultimately liberates bundles of fibres. Nowadays, this process has gained more attention due to its low-cost ease of polymer separation. However, chances of homolytic cleavage reactions of lignin are higher due to the high temperature (200 °C) which could be the major drawbacks of this method. After condensation reaction, this method results in a highly heterogeneous lignin-containing discrete fragments as well as recombined bundles (Tian et al. 2017).

3.4 Miscellaneous Processes of Lignin Extraction and Conversion

Lignin content is isolated from plant tissues and purified cell wall regions. Cell breakage is mostly used technique for lignin isolation. These techniques enable to achieve pure cell wall material. FastPrep method for isolation of lignin helps to preserve its native structure which is free from proteins and other contaminants. This type of lignin is principally used for accurate determination of structure and sensitive assays of its functions (Chen et al. 2000). The major advantage of using FastPrep method is improved homogenization which accelerates the purification and product recovery and also making this process faster than other methods. Similarly, Saisu et al. (2003) conducted a study on conversion of lignin using supercritical H_2O and phenol. They evaluated that biomass of plants consists of cellulose (50 wt%), hemicellulose (20wt%) and lignin (30wt%). In a previous study, the gasification of lignin with a nickel catalyst in supercritical water was carried out at 673 K (Hagaman and Lee 1999). The conversion of lignin by hydrothermal reaction using alkali has also been used to convert lignin into chemicals/fuels. Hydrothermal degradation has also been reported to perform under subcritical or supercritical conditions for the improved production of phenolics (Peng et al. 2014).

4 Challenges in Lignin Conversion

Presently, the main challenge is to convert lignin into valuable chemicals by improving the overall economics of production to make it suitable for commercial scale. The development of competent methods for product separation, complexities in purification and minimizing the char formation are major challenges of lignin processing (Rößiger et al. 2018). Thus, the valourization of lignin is one of the most important challenges in the development of sustainable and economic biorefinery processes. However, the biological strategies for lignin valourization hold the stance by overcoming the complexity of lignin in the improvement of bioprocesses. The polyalkylphenolic lignin molecule can be cleaved into low-molecular-weight products like phenols, alkylphenols and phenol resins using enzymes or chemical catalysts (Graglia et al. 2015; Zakzeski et al. 2010). These products are extremely efficient fuels, fuel additives (Li et al. 2015; Xu et al. 2014) antifungal components (Dos Santos et al. 2015) and can synthesize polyols or polyurethane resins (Mahmood et al. 2016). Till date, the highly stable intramolecular bonds in the lignin make its depolymerization as a difficult task. In addition to this, efficient technologies for the selectivity of defined products, downstreaming process, product recovery and their purification from the reaction cocktail are still the major subjects of concern related to lignin conversion.

5 Lignin-Derived Products of Industrial Importance

5.1 Lignin-Based Foams

Foam is generally formed by mixing suspension of lignin and carbohydrates which results in the formation of small particles, long fibres, sheets and fibreboards. A variety of foams have been developed from lignin such as polyurethane lignin foams (PUL foam), phenolic resin lignin foams (PRL foam), silicone–lignin foams (SiL foam), lignin–furan foams (LF foam) starch–lignin foams (SL foams), polystyrene-lignin foams (PS foams) and tannin–lignin foams (TL foam). These foams have immense applications in goods packaging and foundries' cushioning of utensils and machinery. PU foams are synthesized from isocyanate and polyol components of lignin-containing various free OH groups (Cateto et al. 2014; Obaid et al. 2016). The PU foams can be soft or hard types. The synthesis of hard or rigid PU foam was reported by Li and Ragauskas (2012). PRL foams are generally synthesized from phenol urea-formaldehyde resin, and these are widely used as insulating material (Ghorbani et al. 2017). SL foams have been prepared by moulding and compression of the lignin mixture and widely employed for the packing of goods and machinery (Obaid et al. 2016). Another important class of foam is lignin furan foam which can be synthesized from used spent liquor of pulp industry. The industrial liquor produced in magnesium bisulphite pulping reaction was precipitated using 30–40% of surfactant (Tondi and Pizzi 2009). Similarly, PS–lignin foams are the polymeric foams produced from styrene and lignin which have various applications in transportation industries for packaging.

Silicone–lignin foams are prepared by hydrosilanes and lignin polymerization as reported previously (Brook et al. 2011; Zhang et al. 2015). Tannin–lignin foams are well known for their high flame resistance and low chemical reactivity (Sánchez-Martín et al. 2013; Szczurek et al. 2014). The synthesis of TL foams was carried out by a mechanical frothing process as reported for PU foam preparation. Thus, all these lignin-based foams have high commercial value and numerous applications in day-to-day life.

5.2 *Epoxy Resins from Lignins*

The petroleum-based epoxy resins are synthesized using bisphenol A and epoxy resin in the presence of thiols, phenols or amines. The bisphenol A component is not environmentally friendly. Lignin is one of the biological components which can replace the bisphenol A and can be used to synthesize the epoxy resin with better stability and physicochemical properties (Delmas et al. 2013). To prepare the epoxy resins, the liquefied wood was subjected to glycidyl etherification using NaOH and epichlorohydrin under boiling conditions (Kishi and Fujita 2008; Kishi et al. 2006).

5.3 *Lignin-Based Ethers*

Ethers are fragrance and flavouring agents used in the cosmetics, food and confectionary industry. The ethers are usually produced by chemical reactions, but these can be synthesized from the lignin too. A polyoxyethylene is industrially important ether which can be synthesized from kraft lignin (KL) and polyethylene glycol (PEG). In a recent study, alkaline lignin derivatives were synthesized by etherification and epoxidation reactions. These can be used to increase enzymatic hydrolytic reaction in detergents and fermentation industry (Chen et al. 2016). In a previous study, the addition of hydrophilic and lipophilic groups in the lignin using dodecyl glycidyl ether in the presence of dimethyl benzylamine and followed by sulphonation using chlorosulphonic acid yields the ethers of commercial value.

5.4 *Lignin-Based Esters*

Esters are also one of the important organic compounds synthesized by the chemical reaction of acid and alcohol in the presence of catalyst such as alkali (NaOH). The esters of fatty acids are essential oils and many naturally occurring flavours. In the last decades, enzymatic synthesis of esters also becomes popular due to environmentally friendly biocatalytic reaction (Chandel et al. 2011; Sharma et al. 2017). The microbial lipolytic enzymes and esterases have also been reported to catalyse the conversion of fatty acid in the presence of alcohol into their respective esters (Kumar and Kanwar 2011, 2012; Kumar et al. 2013, 2015; Ramesh et al. 2014; Sharma et al. 2018). Zhao et al. (2017) have proposed the mechanism of esterification of lignin which involved the mechanical activation of solid-phase synthesis technology. In this reaction, sodium acetate and sulphuric acid were used as the catalyst. Generally, the esterification was carried out using acid chloride, acid anhydride, imidazole, pyridine or dimethylaminopyridine (DMAP) under controlled conditions (Nevárez et al. 2011; Thielemans and Wool 2005). In the literature, the esterification of hardwood and softwood was performed using various anhydrides. These conventional processes of esterification are more time-consuming and laborious. Thus, with the advancement of technology, the rate of reaction and recovery of product can be enhanced using supercritical CO_2 and ionic liquid as solvent (Cachet et al. 2014). The rate of reaction can also be enhanced using microwaves and mixture of ionic liquids (Monteil-Rivera and Paquet 2015; Wang et al. 2018). In a recent study, the esters of *p*-coumaric acid and ferulic acid were obtained by esterification of lignin (Tana et al. 2019). Conclusively, the esterification of lignin can reduce the amount of waste product and improve the bio-economy of the process.

5.5 Lignin-Based Paints and Coatings

Lignin-based paints can be prepared by mixing the lignin liquor with paints and treatment with phenol oxidizing agents such as enzymes (Sena-Martins et al. 2008). The enzyme-treated mixture will be incubated to increase the viscosity of paint. The lignin has also been used to coat the fertilizers such as urea for its controlled release in the soil and increase its availability to the plants. Urea has been partly soluble in aqueous lignin dispersion, and the lignin layer results in fine coatings around the urea particles. Lignin has abroad range of molecular mass and various free functional groups which make it easier to modify with various polymeric materials to form insulation coating (Kai et al. 2016).

6 Conclusion

Despite various challenges in lignin valorisation and conversion into products of commercial value, it still holds the potential of alternative of various harsh chemicals and fossil fuels. Continuous efforts have been made by the researchers to scale up the lignin-based biorefineries and replace the chemical products by these biobased products. The question arises whether the lignin-based industries are sustainable and economical or not? How the formation of by-products such as char, phenols and alcohols can be controlled and materialized for mankind use? The presence of oxygen-rich contents in the lignin and its by-products hinders the easy implementation in industries. New strategies for the easy recovery of lignin with high purity level are the need of hour which should be scaled up for industrial usage.

Acknowledgements The financial support from the Jaypee University of Information Technology, Waknaghat, to undertake this study is thankfully acknowledged. The author [SS] is also thankful to UIBT Chandigarh University, Mohali, Punjab, for providing necessary facilities. Further, the authors have no conflict of interest either among themselves or with the parent institution.

References

Alonso DM, Bond JQ, Dumesic JA (2010) Catalytic conversion of biomass to biofuels. Green Chem 12(9):1493–1513. https://doi.org/10.1039/C004654J

Aro T, Fatehi P (2017) Production and application of lignosulfonates and sulfonated lignin. Chemsuschem 10(9):1861–1877. https://doi.org/10.1002/cssc.201700082

Baptista C, Robert D, Duarte AP (2008) Relationship between lignin structure and delignification degree in *Pinus pinaster* kraft pulps. Bioresour Technol 99(7):2349–2356. https://doi.org/10.1016/j.biortech.2007.05.012

Boudet A-M (2000) Lignins and lignification: selected issues. Plant Physiol Biochem 38:81–96. https://doi.org/10.1016/S0981-9428(00)00166-2

Brook MA, Grande JB, Ganachaud F (2011) New synthetic strategies for structured silicones using B(C6F5)3. In: Muzafarov AM (ed) Silicon polymers. Springer, Berlin, Heidelberg, pp 161–183

Buranov AU, Mazza G (2008) Lignin in straw of herbaceous crops. Ind Crops Prod 28(3):237–259. https://doi.org/10.1016/j.indcrop.2008.03.008

Cachet N, Camy S, Benjelloun-Mlayah B, Condoret J-S, Delmas M (2014) Esterification of organosolv lignin under supercritical conditions. Ind Crops Prod 58:287–297. https://doi.org/10.1016/j.indcrop.2014.03.039

Campbell MM, Sederoff RR (1996) Variation in lignin content and composition (mechanisms of control and implications for the genetic improvement of plants). Plant Physiol 110(1):3–13. https://doi.org/10.1104/pp.110.1.3

Cateto CA, Barreiro MF, Ottati C, Lopretti M, Rodrigues AE, Belgacem MN (2014) Lignin-based rigid polyurethane foams with improved biodegradation. J Cell Plast 50(1):81–95. https://doi.org/10.1177/0021955x13504774

Chandel C, Nadda A, Kanwar S (2011) Enzymatic synthesis of butyl ferulate by silica-immobilized lipase in a non-aqueous medium. J Biomater Nanobiotechnol 2:400–408

Chen M, Sommer AJ, McClure JW (2000) Fourier transform–IR determination of protein contamination in thioglycolic acid lignin from radish seedlings, and improved methods for extractive-free cell wall preparation. Phytochem Anal 11(3):153–159. https://doi.org/10.1002/(sici)1099-1565(200005/06)11:3%3c153:aid-pca502%3e3.0.co;2-z

Chen C, Zhu M, Li M, Fan Y, Sun R-C (2016) Epoxidation and etherification of alkaline lignin to prepare water-soluble derivatives and its performance in improvement of enzymatic hydrolysis efficiency. Biotechnol Biofuels 9(1):87. https://doi.org/10.1186/s13068-016-0499-9

Chen J, Dong J, Yang G, He M, Xu F, Fatehi P (2018) A process for purifying xylosugars of pre-hydrolysis liquor from kraft-based dissolving pulp production process. Biotechnol Biofuels 11(1):337. https://doi.org/10.1186/s13068-018-1336-0

Czernik S, Bridgwater AV (2004) Overview of applications of biomass fast pyrolysis oil. Energy Fuels 18(2):590–598. https://doi.org/10.1021/ef034067u

de Gonzalo G, Colpa DI, Habib MH, Fraaije MW (2016) Bacterial enzymes involved in lignin degradation. J Biotechnol 236:110–119. https://doi.org/10.1016/j.jbiotec.2016.08.011

del Río JC, Prinsen P, Rencoret J, Nieto L, Jiménez-Barbero J, Ralph J, Gutiérrez A (2012a) Structural characterization of the lignin in the cortex and pith of elephant grass (*Pennisetum purpureum*) stems. J Agric Food Chem 60(14):3619–3634. https://doi.org/10.1021/jf300099g

del Río JC, Rencoret J, Prinsen P, Martínez ÁT, Ralph J, Gutiérrez A (2012b) Structural characterization of wheat straw lignin as revealed by analytical Pyrolysis, 2D-NMR, and reductive cleavage methods. J Agric Food Chem 60(23):5922–5935. https://doi.org/10.1021/jf301002n

Delmas G-H, Benjelloun-Mlayah B, Bigot YL, Delmas M (2013) Biolignin™ based epoxy resins. J Appl Polym Sci 127(3):1863–1872. https://doi.org/10.1002/app.37921

Demirbaş A (2001) Biomass resource facilities and biomass conversion processing for fuels and chemicals. Energy Convers Manag 42(11):1357–1378. https://doi.org/10.1016/S0196-8904(00)00137-0

Dos Santos P, Erdocia X, Gatto DA, Labidi J (2015) Bio-oil from base-catalyzed depolymerization of organosolv lignin as an antifungal agent for wood. Wood Sci Technol 50. https://doi.org/10.1007/s00226-015-0795-8

Fengel D, Wegener G (1984) Reactions in alkaline medium. Wood: chemistry, ultrastructure, reaction, Walter de Gruyter, Berlin, New York, pp 296–318

Fernández-Rodríguez J, Erdocia X, Hernández-Ramos F, Alriols MG, Labidi J (2019) Lignin separation and fractionation by ultrafiltration. In: Separation of functional molecules in food by membrane technology, Academic Press, pp 229–265. https://doi.org/10.1016/C2017-0-02808-2

Garcia-Perez M, Chaala A, Pakdel H, Kretschmer D, Roy C (2007) Characterization of bio-oils in chemical families. Biomass Bioenerg 31(4):222–242. https://doi.org/10.1016/j.biombioe.2006.02.006

Gellerstedt G (2015) Softwood kraft lignin: raw material for the future. Ind Crops Prod 77:845–854. https://doi.org/10.1016/j.indcrop.2015.09.040

Gellerstedt G, Henriksson G (2008) Lignins: major sources, structure and properties. In: Belgacem MN, Gandini A (eds) Monomers, polymers and composites from renewable resources. Elsevier, Amsterdam, pp 201–224 (Chap. 9)

Gellerstedt G, Pranda J, Lindfors EL (1994) Structural and molecular properties of residual birch kraft lignins. J Wood Chem Technol 14(4):467–482. https://doi.org/10.1080/02773819408003108

Ghorbani M, Konnerth J, Budjav E, Silva AR, Zinovyev G, Van Herwijnen HWG, Liebner F (2017) Ammoxidized fenton-activated pine kraft lignin accelerates synthesis and curing of resole resins. Polymers 9(2):43. https://doi.org/10.3390/polym9020043

Gierer J (1985) Chemistry of delignification. Wood Sci Technol 19(4):289–312. https://doi.org/10.1007/bf00350807

Glasser WG (2019) About making lignin great again—some lessons from the past. Front Chem 7(565). https://doi.org/10.3389/fchem.2019.00565

Gominho J, Fernandez J, Pereira H (2001) *Cynara cardunculus* L.—a new fibre crop for pulp and paper production. Ind Crops Prod 13:1–10. https://doi.org/10.1016/S0926-6690(00)00044-3

Graglia M, Kanna N, Esposito D (2015) Lignin refinery: towards the preparation of renewable aromatic building blocks. ChemBioEng Rev 2(6):377–392. https://doi.org/10.1002/cben.201500019

Hagaman EW, Lee SK (1999) Acid-catalyzed cross-linking reactions at benzylic sites in fluorene monomers, polymers, and lignin[1]. Energy Fuels 13(5):1006–1014. https://doi.org/10.1021/ef980269s

Huang F, Singh PM, Ragauskas AJ (2011) Characterization of milled wood lignin (MWL) in loblolly pine stem wood, residue, and bark. J Agric Food Chem 59(24):12910–12916. https://doi.org/10.1021/jf202701b

Jiang X, Ellis N, Zhong Z (2010) Characterization of pyrolytic lignin extracted from bio-oil. Chin J Chem Eng 18:1018–1022. https://doi.org/10.1016/S1004-9541(09)60162-2

Kai D, Tan MJ, Chee PL, Chua YK, Yap YL, Loh XJ (2016) Towards lignin-based functional materials in a sustainable world. Green Chem 18(5):1175–1200. https://doi.org/10.1039/C5GC02616D

Keegstra K (2010) Plant cell walls. Plant Physiol 154(2):483–486. https://doi.org/10.1104/pp.110.161240

Kishi H, Fujita A (2008) Wood-based epoxy resins and the ramie fiber reinforced composites. Environ Eng Manag J 7:517–523. https://doi.org/10.30638/eemj.2008.074

Kishi H, Fujita A, Miyazaki H, Matsuda S, Murakami A (2006) Synthesis of wood-based epoxy resins and their mechanical and adhesive properties. J Appl Polym Sci 102(3):2285–2292. https://doi.org/10.1002/app.24433

Kumar A, Kanwar SS (2011) Synthesis of ethyl ferulate in organic medium using celite-immobilized lipase. Bioresour Technol 102(3):2162–2167. https://doi.org/10.1016/j.biortech.2010.10.027

Kumar A, Kanwar S (2012) An efficient immobilization of *Streptomyces* sp. STL-D8 lipase onto photo-chemically modified cellulose-based natural fibers and its application in ethyl ferulate synthesis. Trends Carbohydr Res 4:13–23
Kumar A, Sharma V, Sharma P, Kanwar SS (2013) Effective immobilisation of lipase to enhance esterification potential and reusability. Chem Pap 67(7):696–702. https://doi.org/10.2478/s11696-013-0377-x
Kumar A, Zhang S, Wu G, Wu CC, Chen J, Baskaran R, Liu Z (2015) Cellulose binding domain assisted immobilization of lipase (GSlip-CBD) onto cellulosic nanogel: characterization and application in organic medium. Colloids Surf B Biointerfaces 136:1042–1050. https://doi.org/10.1016/j.colsurfb.2015.11.006
Li Y, Ragauskas AJ (2012) Kraft lignin-based rigid polyurethane foam. J Wood Chem Technol 32(3):210–224. https://doi.org/10.1080/02773813.2011.652795
Li C, Zhao X, Wang A, Huber GW, Zhang T (2015) Catalytic transformation of lignin for the production of chemicals and fuels. Chem Rev 115(21):11559–11624. https://doi.org/10.1021/acs.chemrev.5b00155
Lourenço A, Neiva D, Gominho J, Marques A, Pereira H (2015) Characterization of lignin in heartwood, sapwood and bark from *Tectona grandis* using Py–GC–MS/FID. Wood Sci Technol 49:159–175. https://doi.org/10.1007/s00226-014-0684-6
Lourenço A, Rencoret J, Chemetova C, Gominho J, Gutiérrez A, del Río JC, Pereira H (2016) Lignin composition and structure differs between Xylem, Phloem and Phellem in *Quercus suber* L. Front Plant Sci 7(1612). https://doi.org/10.3389/fpls.2016.01612
Luo M, Lin H, Li B, Dong Y, He Y, Wang L (2018) A novel modification of lignin on corncob-based biochar to enhance removal of cadmium from water. Bioresour Technol 259:312–318. https://doi.org/10.1016/j.biortech.2018.03.075
Mahmood N, Yuan Z, Schmidt J, Tymchyshyn M, Xu C (2016) Hydrolytic liquefaction of hydrolysis lignin for the preparation of bio-based rigid polyurethane foam. Green Chem 18(8):2385–2398. https://doi.org/10.1039/C5GC02876K
Marques G, Rencoret J, Gutiérrez A, del Río J (2010) Evaluation of the chemical composition of different non-woody plant fibers used for pulp and paper manufacturing. Open Agric J 4:93–101. https://doi.org/10.2174/1874331501004010093
Meng LY, Kang SUM, Zhang X, Yu-Ying WU, Sun RC (2012) Isolation and physico-chemical characterization of lignin from hybrid poplar in dmso/licl system induced by microwave-assisted irradiation. Cellul Chem Technol 46:409–418
Monteil-Rivera F, Paquet L (2015) Solvent-free catalyst-free microwave-assisted acylation of lignin. Ind Crops Prod 65:446–453. https://doi.org/10.1016/j.indcrop.2014.10.060
Moura JC, Bonine CA, de Oliveira Fernandes Viana J, Dornelas MC, Mazzafera P (2010) Abiotic and biotic stresses and changes in the lignin content and composition in plants. J Integr Plant Biol 52(4):360–376. https://doi.org/10.1111/j.1744-7909.2010.00892.x
Nevárez LAM, Casarrubias LB, Celzard A, Fierro V, Muñoz VT, Davila AC, Sánchez GG (2011) Biopolymer-based nanocomposites: effect of lignin acetylation in cellulose triacetate films. Sci Technol Adv Mater 12(4):045006. https://doi.org/10.1088/1468-6996/12/4/045006
Obaid N, Kortschot MT, Sain M (2016) 12—Lignin-based foaming materials. In: Faruk O, Sain M (eds) Lignin in polymer composites. William Andrew Publishing, pp 217–232
Oliveira L, Cordeiro N, Evtuguin D, Torres IC, Silvestre A (2007) Chemical composition of different morphological parts from 'Dwarf Cavendish' banana plant and their potential as a non-wood renewable source of natural product. Ind Crops Prod 26:163. https://doi.org/10.1016/j.indcrop.2007.03.002
Önnerud H (2003) Lignin structures in normal and compression wood. Evaluation by thioacidolysis using ethanethiol and methanethiol. Holzforschung 57:377–384. https://doi.org/10.1515/HF.2003.056
Önnerud H, Gellerstedt G (2003) Inhomogeneities in the chemical structure of spruce lignin. Holzforschung 57:165–170. https://doi.org/10.1515/HF.2003.025

Peng C, Zhang G, Yue J, Xu G (2014) Pyrolysis of lignin for phenols with alkaline additive. Fuel Process Technol 124:212–221. https://doi.org/10.1016/j.fuproc.2014.02.025

Pascoal Neto C, Seca A, Nunes AM, Coimbra MA, Domingues F, Evtuguin D, Silvestre A, Cavaleiro JAS (1997) Variations in chemical composition and structure of macromolecular components in different morphological regions and maturity stages of *Arundo donax*. Ind Crops Prod 6(1):51–58. https://doi.org/10.1016/S0926-6690(96)00205-1

Pinto PC, Evtuguin DV, Pascoal Neto C (2005) Chemical composition and structural features of the macromolecular components of plantation *Acacia mangium* wood. J Agric Food Chem 53(20):7856–7862. https://doi.org/10.1021/jf058081b

Radotić K, Mićić M (2016) Methods for extraction and purification of lignin and cellulose from plant tissues. Sample preparation techniques for soil, plant, and animal samples, Humana Press, New York, NY, pp 365–376. https://doi.org/10.1007/978-1-4939-3185-9_26

Rahimi A, Ulbrich A, Coon JJ, Stahl SS (2014) Formic-acid-induced depolymerization of oxidized lignin to aromatics. Nature 515(7526):249–252. https://doi.org/10.1038/nature13867

Ramesh K, Ashok K, Shamsher SK (2014). Synthesis of methyl succinate by natural-fibre immobilized lipase of *Streptomyces* sp. STL-D8. Curr Biotechnol 3(2):152–156. https://doi.org/10.2174/2211550102666140107232547

Regmi YN, Mann JK, McBride JR, Tao J, Barnes CE, Labbé N, Chmely SC (2017) Catalytic transfer hydrogenolysis of organosolv lignin using B-containing FeNi alloyed catalysts. Catal Today. https://doi.org/10.1016/j.cattod.2017.05.051

Rößiger B, Unkelbach G, Pufky-Heinrich D (2018) Base-catalyzed depolymerization of lignin: history, challenges and perspectives. Lignin Trends Appl 21:99–120. https://doi.org/10.5772/intechopen.72964

Saisu M, Sato T, Watanabe M, Adschiri T, Arai K (2003) Conversion of lignin with supercritical water-phenol mixtures. Energy Fuels 17(4):922–928. https://doi.org/10.1021/ef0202844

Sánchez-Martín J, Beltrán-Heredia J, Delgado-Regaña A, Rodríguez-González MA, Rubio-Alonso F (2013) Optimization of tannin rigid foam as adsorbents for wastewater treatment. Ind Crops Prod 49:507–514. https://doi.org/10.1016/j.indcrop.2013.05.029

Scholze B, Hanser C, Meier D (2001) Characterization of the water-insoluble fraction from fast pyrolysis liquids (pyrolytic lignin): Part II. GPC, carbonyl goups, and 13C-NMR. J Anal Appl Pyrol 58–59:387–400. https://doi.org/10.1016/S0165-2370(00)00173-X

Sena-Martins G, Almeida-Vara E, Duarte JC (2008) Eco-friendly new products from enzymatically modified industrial lignins. Ind Crops Prod 27:189–195. https://doi.org/10.1016/j.indcrop.2007.07.016

Sharma A, Kumar A, Meena K, Rana S, Singh M, Kanwar S (2017) Fabrication and functionalization of magnesium nanoparticle for lipase immobilization in n-propyl gallate synthesis. J King Saud University Sci 29. https://doi.org/10.1016/j.jksus.2017.08.005

Sharma A, Sharma T, Meena KR, Kumar A, Kanwar SS (2018) High throughput synthesis of ethyl pyruvate by employing superparamagnetic iron nanoparticles-bound esterase. Process Biochem 71:109–117. https://doi.org/10.1016/j.procbio.2018.05.004

Shatalov A, Pereira H (2005) Kinetics of organosolv delignification of fibre crop *Arundo donax* L. Ind Crops Prod 21:203–210. https://doi.org/10.1016/j.indcrop.2004.04.010

Sjöström E (1993) Wood chemistry: fundamentals and applications. Academic Press San Diego. https://doi.org/10.1016/C2009-0-03289-9

Szczurek A, Fierro V, Pizzi A, Stauber M, Celzard A (2014) A new method for preparing tannin-based foams. Ind Crops Prod 54:40–53. https://doi.org/10.1016/j.indcrop.2014.01.012

Tana T, Zhang Z, Beltramini J, Zhu H, Ostrikov K, Bartley J, Doherty W (2019) Valorization of native sugarcane bagasse lignin to bio-aromatic esters/monomers via a one pot oxidation–hydrogenation process. Green Chem 21(4):861–873. https://doi.org/10.1039/C8GC03655A

Thielemans W, Wool RP (2005) Lignin esters for use in unsaturated thermosets: lignin modification and solubility modeling. Biomacromol 6(4):1895–1905. https://doi.org/10.1021/bm0500345

Tian D, Chandra RP, Lee JS, Lu C, Saddler JN (2017) A comparison of various lignin-extraction methods to enhance the accessibility and ease of enzymatic hydrolysis of the cellulosic component of steam-pretreated poplar. Biotechnol Biofuels 10:157. https://doi.org/10.1186/s13068-017-0846-5

Tondi G, Pizzi AP (2009) Tannin-based rigid foams: Characterization and modification. Ind Crops Prod 29:356–363. https://doi.org/10.1016/j.indcrop.2008.07.003

vom Stein T, den Hartog T, Buendia J, Stoychev S, Mottweiler J, Bolm C, Klankermayer J, Leitner W (2015) Ruthenium-catalyzed C–C bond cleavage in lignin model substrates. Angew Chem Int Ed Engl 54(20):5859–5863. https://doi.org/10.1002/anie.201410620

Wang H, Chen W, Zhang X, Wei Y, Zhang A, Liu S, Liu C (2018) Structural changes of bagasse dusring the homogeneous esterification with maleic anhydride in ionic liquid 1-Allyl-3-methylimidazolium chloride. Polymers 10:433. https://doi.org/10.3390/polym10040433

Xu C, Arancon RAD, Labidi J, Luque R (2014) Lignin depolymerisation strategies: towards valuable chemicals and fuels. Chem Soc Rev 43(22):7485–7500. https://doi.org/10.1039/C4CS00235K

Zakzeski J, Bruijnincx PCA, Jongerius AL, Weckhuysen BM (2010) The catalytic valorization of lignin for the production of renewable chemicals. Chem Rev 110(6):3552–3599. https://doi.org/10.1021/cr900354u

Zhang Y-HP (2011) What is vital (and not vital) to advance economically-competitive biofuels production. Process Biochem 46:2091–2110. https://doi.org/10.1016/j.procbio.2011.08.005

Zhang J, Fleury E, Brook MA (2015) Foamed lignin–silicone bio-composites by extrusion and then compression molding. Green Chem 17(9):4647–4656. https://doi.org/10.1039/C5GC01418B

Zhao X, Zhang Y, Wei L, Hu H, Huang Z, Mei Y, Huang A, Wu J, Feng Z (2017) Esterification mechanism of lignin with different catalysts based on lignin model compounds by mechanical activation-assisted solid-phase synthesis. RSC Adv 7(83):52382–52390. https://doi.org/10.1039/C7RA10482K

Zhu JY, Pan XJ, Wang GS, Gleisner R (2009) Sulfite pretreatment (SPORL) for robust enzymatic saccharification of spruce and red pine. Bioresour Technol 100(8):2411–2418. https://doi.org/10.1016/j.biortech.2008.10.057

Structure and Characteristics of Lignin

Yong-Chao Lu, Yao Lu and Xing Fan

Abstract Lignin contributes up to 30% of lignocellulosic biomass by weight and provides an alternative renewable source of aromatic chemicals. However, it has not been exploited effectively due to its natural complexity and high stability. Therefore, continuous and vigorous efforts should be paid to achieve a better understanding of the chemical structure and composition of lignin in order to design more efficient and greener degradation processes, and further to explore new value-added applications. Lignin is biosynthesized by radical coupling of the basic building blocks by enzymatically catalyzed random polymerization reactions. The chemical structure of lignin depends not only on the type of biomass material but also on the part of the plant, location of tissue, age, the growing conditions, and other natural conditions of living plants. In addition, in the elucidation of structural characteristics of lignin, the degradation processes, extraction techniques, and even the analytical methods used for characterization can also influence the results. This chapter presents a comprehensive overview of structural characteristics of lignin and related developments in the studies including experimental and computational assays, physicochemical properties investigation, isolation and separation methods, and the instrumental technologies. Monolignols/units and the inter-unit linkages as well as their formation and transformation are summarized in the processes of radical reactions and biosynthesis. Macroscopic physicochemical properties of lignin together with parameters are also listed. Different methods for generating lignins are compared. Various techniques applying on characterization of lignin

Y.-C. Lu
School of Pharmacy, Xuzhou Medical University, Xuzhou 221004, China

Y. Lu
Advanced Analysis and Computation Center, China University of Mining and Technology, Xuzhou 221116, China

X. Fan (✉)
Key Laboratory of Coal Processing and Efficient Utilization, Ministry of Education, China University of Mining and Technology, Xuzhou 221116, China
e-mail: fanxing@cumt.edu.cn

Key Laboratory of Coal Cleaning Conversion and Chemical Engineering Process, Xinjiang Uyghur Autonomous Region, College of Chemistry and Chemical Engineering, Xinjiang University, Urumqi 830046, Xinjiang, China

S. Sharma and A. Kumar (eds.), *Lignin*, Springer Series on Polymer and Composite Materials, https://doi.org/10.1007/978-3-030-40663-9_2

structure are introduced and discussed, including spectroscopic, chromatographic, and spectrometric methods. New insight into structural characteristics of lignin and advanced separation methods and instrumental techniques is highlighted. Future investigations on the topic of lignin structure are also prospected and suggested.

Keywords Lignin · Structural features · Physicochemical properties · Degradation techniques · Analysis

1 Introduction

With the growing demands for energy and fine organic chemicals, as well as the global concerns about environmental problems induced by the uses of fossil fuels, renewable lignocellulosic biomass is regarded as a promising candidate to replace fossil fuels and to slow down environmental crisis basing on its abundant reserve and characteristic chemical structure. Cell walls of lignocellulosic biomass mainly consist of cellulose, hemicellulose, and lignin, with different ratios according to the origins of biomass (Chen 2014). Cellulose fibers bundle up into regular structures and determine the framework of cell walls. The intervals among fibers in compact plant cell walls are filled with hemicellulose and lignin molecules by chemical bonds and/or hydrogen bonds. The absolute separation of individual components of the three is significantly difficult due to various chemical bonds. The complex structure of plant cell walls in biomass and the difficulty in degradation of lignin as well as the oriented isolation of degraded species are the main reasons impeding the development of lignocellulosic biomass value-added application, such as production of advanced materials and fine chemicals.

Lignin is the most abundant natural source of aromatic moieties on the Earth (Luterbacher et al. 2014; Upton and Kasko 2015). However, besides its use as energy source, there are only very few industrial and commercial applications of lignin, such as construction materials, thermoplastics, and carbon-based value-added materials. The more efficient and comprehensive utilization of lignins is critically dependent on the in-depth knowledge of their chemical structural characteristics.

In lignocellulosic biomass, lignin contributes relative contents of weight and energy up to 30% and 40%, respectively. Lignin contains more than one-third of non-fossil organic carbon on the Earth. The amount of annually synthesized lignin in nature is estimated to 3×10^9 tons. Industrial lignins are often isolated from spent black liquor, agricultural wastes and woods. The annual industrial production of lignins is more than 70 million tons (Kumar et al. 2009). However, these lignins are often treated as waste or byproduct during pulping process and chemical production, and have not been exploited effectively. In fact, only about 2% of the available lignin is value-added used as industrial materials or chemicals (Stewart 2008). The main reason for this situation is the natural complexity and high stability of lignin. The effective, oriented, and commercial depolymerization of lignin is a huge challenging task (Lupoi et al. 2015). Hence, continuous and vigorous efforts should be paid to achieve a better understanding of the chemical structure and composition of lignin in order to design more efficient and greener degradation processes, and further to explore new value-added applications.

Lignin means wood and is derived from the Latin word *lignum* (Novo-Uzal et al. 2012). Lignin was first introduced by Candolle in 1819. Lignin was then described as "encrusting material" in wood by Payen in 1838. In 1865, Schulze defined chemical structure of lignin as a hydrophilic substance in plant cell walls (Schulze 1857). In 1875, Bente (Bente 1875) carried out quantitative determination of lignin by using color reactions and demonstrated that lignin was aromatic component. In 1890, methoxy groups were identified in lignin (Benedikt and Bamberger 1890). Klason (Klason 1920) suggested that lignin was a polymer that consisting of coniferyl alcohol. Karl Freudenberg (Freudenberg and Neish 1968) carried out fruitful works on the characterization of lignin structure. Now, it is accepted that lignin is a complex three-dimensional, cross-linked, and amorphous polymer consisting of various phenylpropanoid units combined with several types of linkages. The content and distribution of units and linkages vary according to the origin of lignin.

Lignin is biosynthesized by radical coupling of the basic building blocks, namely monolignols, by enzymatically catalyzed random polymerization reactions. The three main monolignols are *p*-coumaryl alcohol, coniferyl alcohol, and sinapyl alcohol, which correspond to the *p*-hydroxyphenyl (H unit), guaiacyl (G unit), and syringyl (S unit), respectively (Fig. 1) (CalvoFlores et al. 2015; Chen 2014; Harmatha 2005; Hatfield and Vermerris 2001; Kerr 1995; Lupoi et al. 2015). The amount and structural details of lignin depend not only on the type of biomass but also on the part of the plant, location of tissue, age, the growing conditions, and other natural conditions of living plants (Barros et al. 2015; Buranov and Mazza 2008; Chen et al. 2018; del Río et al. 2012; Lourenço et al. 2013, 2015b, 2016; Mansfield et al. 2012a; Marques et al. 2010; Rencoret et al. 2008; Río et al. 2007, 2012; Schuetz et al. 2013; Zhao et al. 2016). In addition, the degradation processes, extraction techniques, and even the analytical methods used for characterization can also influence the results in the

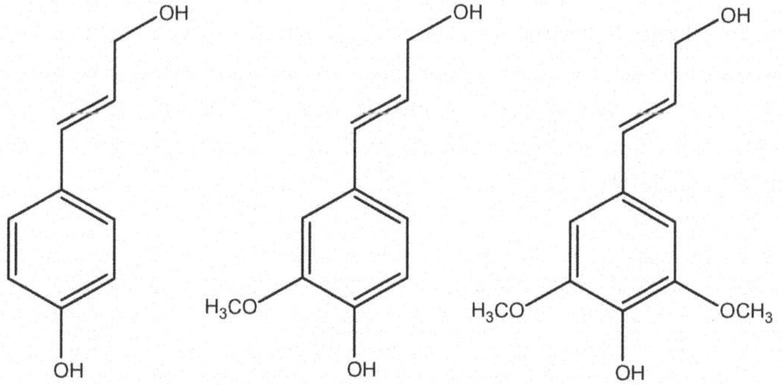

p-coumaryl alcohol (H unit) coniferyl alcohol (G unit) sinapyl alcohol (S unit)

Fig. 1 The three main monolignols of lignin

elucidation of structural characteristics of lignin (Blanch et al. 2011; Bule et al. 2013; CalvoFlores et al. 2015; Chen 2014; Chiaramonti et al. 2012; Li et al. 2016; Lupoi et al. 2015; Sathitsuksanoh et al. 2014; Shi et al. 2013; Zhen and Smith 2016).

Lignin is a major component in secondary cell walls (Müse et al. 1997), and it maintains integrity and rigidity of the plants by providing mechanical support, stress response, and pathogen resistance (Austin and Ballaré 2010). More importantly, it plays a crucial role in transportations of nutrients and water, and in the protection of plants against microbial attacks (Breznak and Brune 1994; Campbell and Sederoff 1996; Harmatha et al. 2011). The content of lignin varies significantly according to the kind of plant, e.g. 19–40% and 12–25% in woody plants and herbaceous plants, respectively, basing on dry mass fraction. The relative proportions of H/G/S units vary significantly among the types of biomass (Donaldson 2001; Fengel et al. 1984; Gellerstedt and Henriksson 2008b) (Table 1). In softwood lignins, such as pinus and spruce, more than 90% of the monolignols are G unit. However, in hardwood lignins, such as birch, beech, and aspen, the proportion of S unit ranges from 45 to 75%, which is larger than the amount of G unit (25–50%). Higher content of H unit (5–35%) exists in grasses than in softwood and hardwood lignins (CalvoFlores et al. 2015; Gellerstedt and Henriksson b; Li et al. 2016; Lupoi et al. 2015). More methoxyl groups on the aromatic rings in lignins prevent the formation of resistant C–C inter-links, i.e., 5–5′ linkage. Consequently, lignins in hardwoods and grasses are more linear and less condensed compared to softwoods.

Universal accepted opinion believed that natural lignin is synthesized via enzymatic oxidative processes by various monolignols, which are phenol species derived from saccharides (Abdelaziz et al. 2016). In natural lignin, more than two-thirds of the linkages between monolignols are C–O bonds, while the rest are C–C bonds. Hydroxyl, methoxyl, carbonyl, and carboxyl groups are the main functional groups in lignin. Among them, phenolic hydroxyl groups are most reactive and can significantly affect the chemical reactivity of lignin (Glasser et al. 1981; Hüttermann et al. 2001). Technical lignins can be separated from natural lignin in biomass by using industrial processes, such as Kraft lignin and lignosulfonates. The cleavage of covalent bonds in lignin is inevitable in the degradation processes. The polymeric structure of lignin is degraded or modified to various extent according to the conditions used in these processes. Hence, the structure of lignin isolated from these processes would have great differences. The characteristics of technical lignins include aromatic-rich, high carbon content, and high sulfur and ash contents (Sannigrahi et al. 2010).

Table 1 Relative amount of the different monolignols in lignins of different types of biomass (CalvoFlores et al. 2015; Gellerstedt and Henriksson b; Li et al. 2016; Lupoi et al. 2015)

Lignins	H unit/%	G unit/%	S unit/%
Hardwood	0–10	25–50	45–75
Softwood	<5	>95	~0
Grass	5–35	35–80	20–55

Although the clear and unique structural definition or specific formula of lignin has not been established due to its intrinsic molecular complexity and structural compositional diversity, extensive developments have been reported on the structural characteristics of lignin. In these investigations, various degradative methods were employed to decompose lignin macromolecules into fragments. Instrumental analyses are conducted on the depolymerized products to collect certain structural information of lignin. By using statistic methods, chemical structural features of lignin can be obtained including types of monolignols and their contents, distribution of inter-unit linkages and functional groups. Whichever method is used, fundamental understandings of the structural characteristics and the physicochemical properties of lignin are crucial important for advanced application of lignin.

In this chapter, we present the structural features of lignin from different aspects, compare different processes involved in the degradation of lignin, exhibit the physiochemical properties of lignin with different origins, and summarize the progress on the lignin structure elucidation by using various instrumental methods.

2 Structural Features of Lignin

Many natural polymers, such as proteins and polysaccharides, have specific structures with optical activity, and are synthesized/degraded in highly ordered pathways involved in specific enzymatic reactions. On the opposite, however, lignin is treated as the strange exception. It is racemic, and no living organism is able to degrade and use it as a sole source (Leisola et al. 2012). The following items present comprehensive information on elucidation of the detailed structure characteristics of lignin.

2.1 Monolignols/Units

The main structure of natural lignin is a three-dimensional heterogeneous polymer containing three types of hydroxycinnamyl alcohol subunits or monolignols, namely *p*-hydroxyphenyl (H unit), guaiacyl (G unit), and syringyl (S unit) phenylpropanoid units. They have different extension of methoxylation at the $3'$ or $3'$-$5'$ position of phenolic rings (Dence 1992; Ralph and Lu 1998) (Fig. 1). The diversity and complexity of lignin structure in different kinds of plants results from the combination of different amounts of monolignols with distinct distribution and substitution patterns (Mann et al. 2009) (Table 1).

The chemical structures of lignins in softwood and grass do not vary much (Akiyama et al. 2002), but hardwood lignins vary greatly among different plant species (Nimz et al. 1981). S/G ratio is often used to describe the structural differences of hardwood lignins, such as degree of condensation, amount of specific dimers, and contents of functional groups (Nimz et al. 1981). Accordingly, hardwood lignins with higher S/G ratio contain more methoxyl groups than softwood lignins.

Hardwood lignins present less condensed structures compared to softwood lignins (Nimz et al. 1981). Lignins from softwoods have more compact structures and have stronger recalcitrant resistance against degrading attacks compared to lignins from hardwood and grass (Agarwal and Atalla 1986; Jung et al. 1993; Morrow et al. 1997). Non-woody lignins are commonly acetylated, coumarylated, and/or *p*-hydroxybenzoylated, and grass lignins are often acylated, especially for S unit (Mann et al. 2009). The understanding of grass lignin structure is limited due to its more complexity and wider variability.

Besides the type of plants, the location of lignin in plants is a critical factor that determines the composition of monolignols and distribution of linkages (Rencoret et al. 2017; Santos et al. 2011b). For example, lignin content is higher at the top compared with other parts of a mature conifer (Zobel and Buijtenen 1989). In woody biomass, lignin mostly deposits in plant cell walls, especially in the mature xylem cell walls (Lourenço et al. 2016).

The relative contents of monolignols (H, G, and S units) affect digestibility of biomass significantly. The S/G ratio in cell walls reflects the performance of pretreatments on biomass and is another factor that affects the degradability of lignin, e.g., high S unit favors the removal of lignin (Min et al. 2013; Santos et al. 2011a), and the S/G ratio was negatively related to the enzymatic hydrolysis (Besombes and Mazeau 2005a, b; Min et al. 2013) and the hydrolyzability of biomass (Li et al. 2010; Mansfield et al. 2012b; Studer et al. 2011). Meanwhile, however, because of the complexities of biomass and the inevitable effects of the other components, inconsistent results suggesting S/G ratio positively related to the delignification were also reported during the pretreatment process (Li et al. 2016; Min et al. 2013; Santos et al. 2011a). In those cases, high S/G ratio benefited pretreatment. For example, S unit was preferentially removed compared to G unit in delignification of poplar (Trajano et al. 2013). S/G ratio also influenced the saccharification efficiency. For NaOH and H_2SO_4 pretreated *Miscanthus*, S/G ratio was negatively associated with the saccharification (Bo et al. 2016; Li et al. 2014; Xu 2012). However, the exact conclusion has not been established since inconsistent or unclear correlations on both sides are reported (Nimz et al. 1981; Papa et al. 2012; Wu et al. 2013; Yu et al. 2011). These results indicate that S/G ratio affects partially on biomass recalcitrance. Furthermore, the other complex components in biomass interfere with the degradation in pretreatments. Besides S/G ratio, high H unit content favors the degradation of biomass (Angela et al. 2010). This positive effect of H unit may be attributed to its reduced molecular weight, reduced H unit-glucan bonding, and high activity (Li et al. 2016). Consequently, the H/G ratio had a distinct positive correlation with glucose yield (Shi et al. 2016; Wu et al. 2013). Except for different types of species, the compositions of H/G/S units in lignin varied in different tissues (Lourenço et al. 2016; Normark et al. 2014; Takao et al. 2010; You et al. 2013). For example, xylem, phloem, and phellem in *Quercus suber* L. showed different composition of H/G/S units in lignin (Lourenço et al. 2016) with molar ratios of 1:45:55, 1:58:41, and 2:85:13, respectively.

New monolignols have been recognized including hydroxycinnamic acids/aldehydes, and coniferyl and sinapyl acetates or coumarates (Grabber

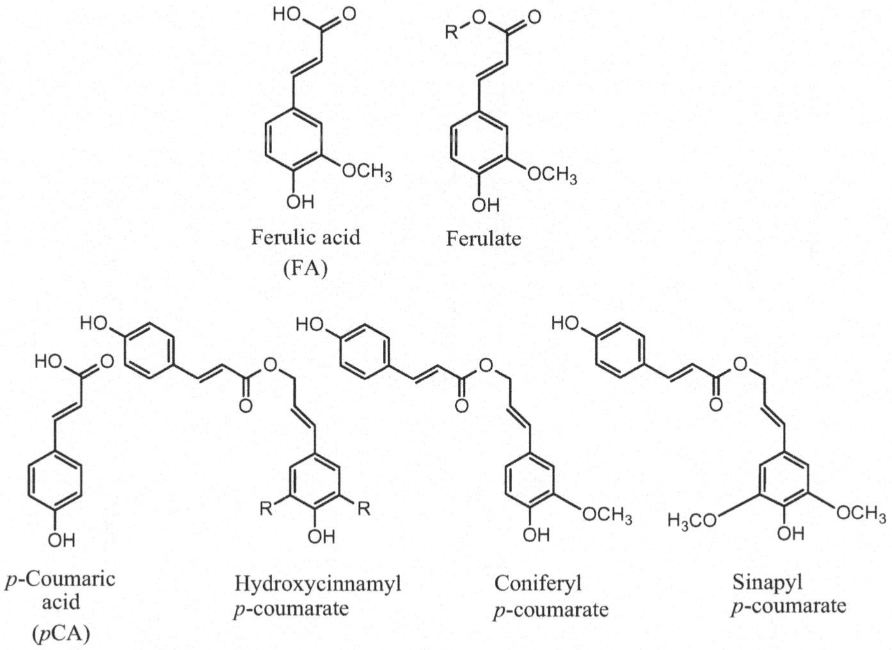

Fig. 2 Hydroxycinnamates derived from ferulic acid (FA) and *p*-coumaric acid (*p*CA)

et al. 2010; Ralph and Landucci 2010; Ralph et al. 2004). Ferulates (FA) and *p*-coumarates (*p*CA) (Fig. 2) are two important hydroxycinnamates detected in grass lignin (Buranov and Mazza 2008; Ralph 2010), which are closely related to biomass recalcitrance. FA facilitates the cross-linking reactions between polysaccharides and lignin molecules (Azarpira et al. 2011; Buanafina 2009) and increases the recalcitrance of biomass. The FA content in grass lignin is a good indicator for the digestibility of cell walls in plants (Grabber et al. 1995; Jung et al. 2011; Jung and Phillips 2010). Therefore, the FA-involved cross-linkages are evident targets in degradation of lignin and further enhance biomass decomposition in pretreatments (Chang et al. 2016; Martínez et al. 2016; Qin et al. 2015). *p*CA is usually linked on γ-position of S unit as esters and also exist in their acylated forms as polysaccharides (Petrik et al. 2014). The content of *p*CA is proportional to lignin content (Taboada et al. 2010). The degradation of biomass is negatively correlated with both contents of *p*CA and lignin (Angela et al. 2010; Li et al. 2012). Hydroxycinnamates connect to lignin with weak ester bonds (Ralph 2010) which improve degradation of biomass (Tobimatsu et al. 2012; Wilkerson et al. 2014). Some less abundant and non-conventional monolignols are identified in grassy and herbaceous biomass (CalvoFlores et al. 2015; Gandini and Belgacem 2008) (Fig. 3).

Most of these monolignols are phenylpropanoids (C_6–C_3 units), which are building blocks of lignin connected by various linkages forming dimmers or oligomers (Table 2). The relative proportion of monolignols varies according to

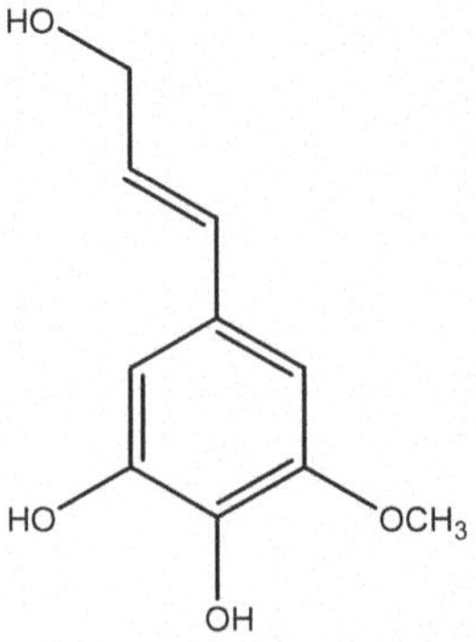

5-Hydroxyconiferyl alcohol

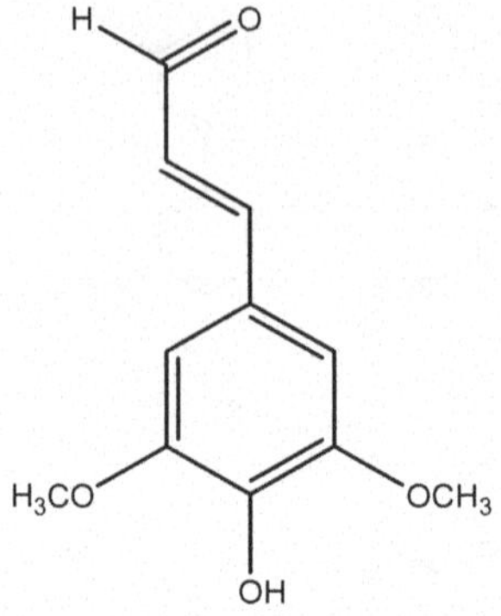

Synapaldehyde

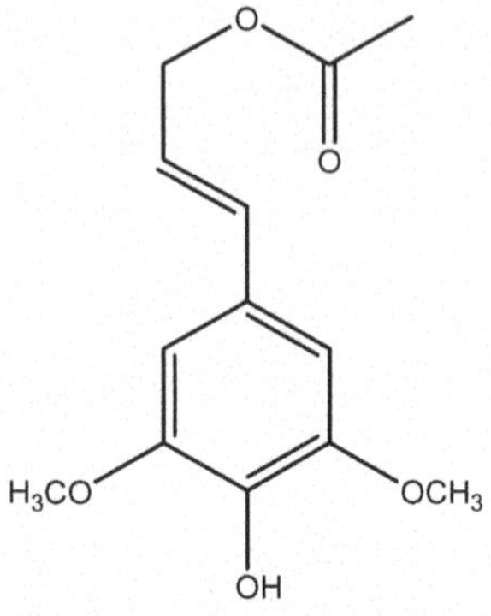

Sinapyl acetate

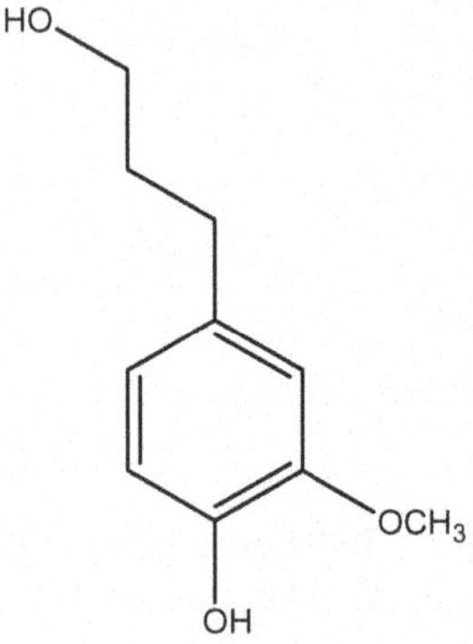

Dihydroconiferyl alcohol

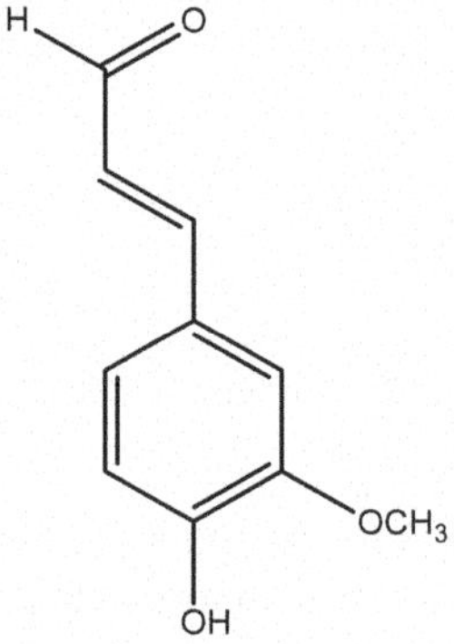

Coniferaldehyde

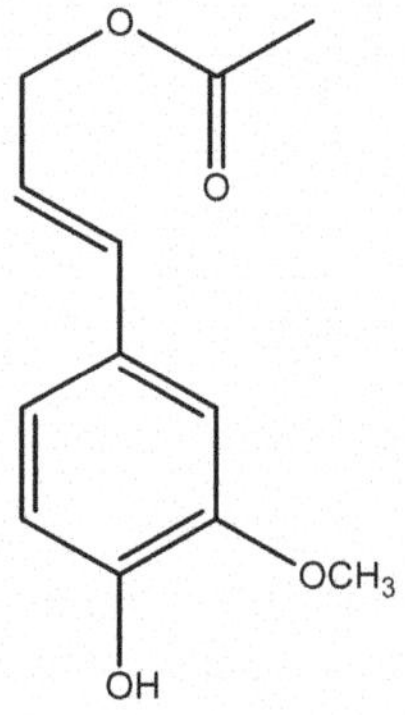

Coniferyl acetate

Fig. 3 Less abundant and non-conventional monolignols

the species, tissues of plant, cell location, and other natural conditions. These lignin monomers undergo a random and peroxidase-mediated free-radical coupling polymerization to form framework of cell walls in plants.

2.2 *Inter-Unit Linkages*

Natural lignin is a highly branched phenolic polymer produced from non-organized radical polymerization of monolignols (Buanafina and De 2009; Morreel et al. 2004). The structural diversity of lignin arises not only from the abundance and distribution of different monolignols, but also from the different ways in which these building blocks connect with each other. In the lignification, H/G/S units and other less abundant monomers conjugate variously to form the irregular and highly branched

Table 2 Phenylpropanoids derived species in lignins (with the form of compositions of carbon atoms) (Harmatha et al. 2011)

Composition of carbon atoms	Species
C_6	Benzene, phenols and benzoquinones
C_6–C_1	Phenolic aldehydes and acids
C_6–C_2	Acetophenones and benzofurans
C_6–C_3	Phenylpropanoids, benzopyranes and coumarins
C_6–C_4	Naphthoquinones
C_6–C_5	Ageratochromenes and prekocens
$(C_6)_2$	Dibenzofurans, dibenzoquinones and biphenyls
C_6–C_1–C_6	Dibenzopyranes, benzophenones and xanthones
C_6–C_2–C_6	Anthraquinones and phenanthrenes
C_6–C_3–C_6	Flavonoids, isoflavones, chalcones and aurones
C_6–C_4–C_6	Norlignans and diphenylbutadienes
C_6–C_5–C_6	Norlignans and conioids
$(C_6–C_3)_2$	Neolignans
$(C_6–C_3–C_6)_2$	Biflavonoids
$(C_6–C_3–C_6)_n$	Condensed tannins and flavolans
$(C_6–C_3)_n$	Lignins
$(C_6)_n$	Catecholmelanines

lignins. This process includes the transportation of monolignol precursors to the cell wall, the biosynthesis of monolignols, and the polymerization of the monolignols producing the final lignin macromolecules.

The structural features of lignin are generally elucidated by the relative abundance of H/G/S units and the distribution of various inter-unit linkages. Basing on the structural analysis of lignin, nine inter-unit linkages are identified (Fig. 4) (Lange et al. 2013). These inter-molecular linkages between different units contribute to the three-dimensional structure of lignin (Table 3). One should note that not all H/G/S units involved in all the linkage styles presented. Among these linkages, the coupling reaction generally favors at the β-position, leading to the formation of arylglycerol-β-aryl ethers (β-O-4), phenylcoumarans (β-5′), pinoresinols (β-β′), diphenylethane dimers (β-1′), and spirodienones (SD). The coupling reactions take place at positions 4 and 5 producing diaryl ethers (4-O-5′) and biphenyls (5-5′). The 5-5′ dimer undergoes α-β-O-4-4′ coupling mode to form dibenzodioxocine. The dibenzodioxocine and 4-O-5′ oligomers provide branching points and result in helical structure of lignin molecules (Faulon and Hatcher 1994).

The β-O-4 ether is the most frequent linkage accounting for 40–60% of all in lignin (Adler 1977; Sarkanen and Ludwig 1971), especially in dicots. The high abundance of this linkage may be attributed to the extraordinary reactivity of phenoxy oxygen and β-carbon. It is also the easiest one to be cleaved. It is the major target for tracking structural changes during lignin decomposition, fractionation, or depolymerization

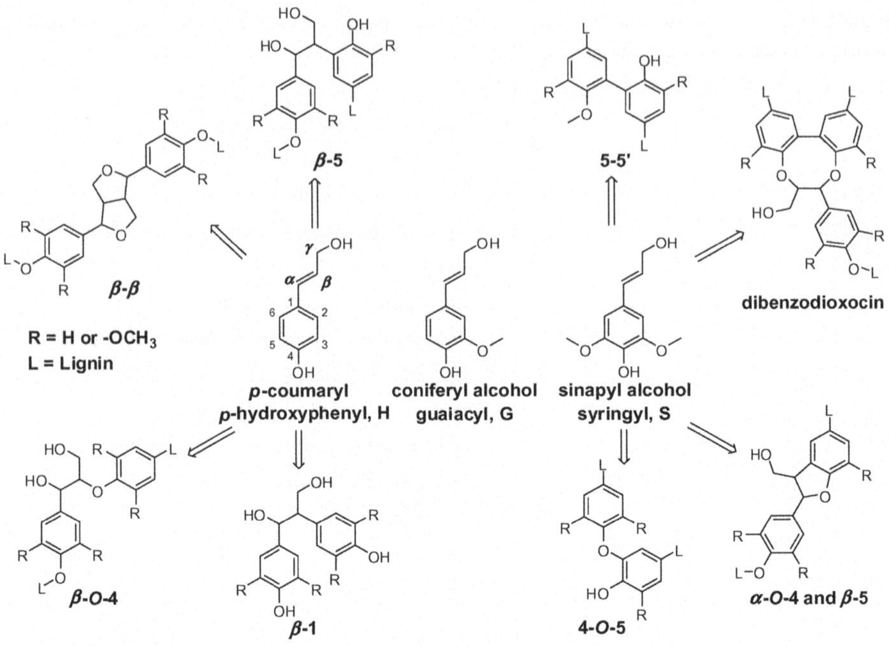

Fig. 4 Linkages in lignin

Table 3 Contents of different linkages in lignin

Linkages	Structures	Hardwood (%)	Softwood (%)
β-O-4	β-Aryl ether	45–65	42–50
β-5′	Phenylcoumaran	5–6	9–12
5-5′	Biphenyl	4–10	9.5–25
4-O-5′	Diaryl ether	6–7	3.5–8
α-O-4′	α-Aryl ether	4–8	6–12
β-1′	1,2-Diaryl propane	5–7	3–10
β-β′	Resinol	3–6	2–3
Dibenzodioxocin	Diary propane diether	1–2	3–7
Spirodienone		3–5	1–2

since it is the weakest among the inter-molecular linkages (Adler 1977). The β-O-4 ether together with two condensed linkages, namely phenylcoumarans (β-5′) and resinols (β-β′), generates linear lignin molecules. For these three linkages, the radical coupling reactions occur exclusively at the β-position, generating 4-O-phenolic function after rearomatization of the quinine methide intermediate (QMI) to facilitate the following coupling reactions. Subsequently, branching reactions take place at the positions of 4-O- and 5 of lignin oligomers via 4-O-5′ and 5-5′ linkages, respectively.

The diphenylethane dimers (β-1′) are relatively stable and account for about 5–10% of the total linkages in lignin (Ralph et al. 2004). The dibenzodioxocin structure was identified to be the main branching point (Karhunen et al. 1995, 1999) with abundance of 3.7% in softwood lignins (Argyropoulos et al. 2002). This octagonal moiety involves the formation of α- and β-ethers on the same 5-5′ biphenyl structures. Softwood lignins contain more resistant linkages than hardwood lignins, such as β-5′, 5-5′ and 4-O-5′ linkages, since more C_5 positions in G units available for coupling reactions to occur. Consequently, the condensation degree is higher for softwood than for hardwood lignins (Rodrigues Pinto et al. 2012). Less abundant linkages, such as biaryl ethers, non-cyclic benzyl alkyl ethers, and diphenyl ethers, totally account for about 6% of the linkages between phenylpropanoid units in softwood lignin (Freudenberg and Harkin 1964). Pinoresinol accounts for less than 5% of linkages in lignin (Adler and Lundquist 1963). It contains two ether groups and is produced by the β-β′ coupling of two coniferyl alcohol radicals followed by a double-ring closure (Marco et al. 2011).

2.3 *Lignin–Carbohydrate Complex (LCC)*

In 1954, Björkman first introduced LCC to describe the stable aggregation of lignin and polysaccharides and to represent the interaction between them in cell walls of plants. It mainly consists of polysaccharides chains and small lignin fragments (Yuan et al. 2011). The structural characteristics of LCC relate to the functional groups, free radicals, and the corresponding enzymes involved in construction of cell walls in plants (Cornu et al. 1994).

Linkages identified in LCCs include benzyl ethers, benzyl esters, phenyl glycoside bonds, and acetal bonds (Kondo et al. 1990; Tetsuo et al. 1984; Watanabe and Koshijima 1988; Xie et al. 2012). Balakshin et al. (Balakshin et al. 2011) have characterized the LCC linkages in milled wood lignin (MWL) extracted with acetic acid and cellulolytic enzyme lignin (CEL) from loblolly pine and white birch. The amounts of benzyl ether, benzyl ester, and phenyl glycoside bonds were determined, but acetal bonds have not been detected due to their hydrolysis under acidic conditions. The LCCs are believed to be the main hinder for entirely selective separation and isolation of lignin from biomass. It is extremely difficult to separate LCCs into their individual components by traditional treatments or chemical/enzymatic degradations. Therefore, LCCs could exist inevitably as impurities in isolated lignin or carbohydrates in paper industrial processes.

Carbohydrates contained in LCC are basically hemicellulose constituents. There are two hypothesizes describing the formation of linkages between lignin and hemicelluloses in secondary cell walls of plants, i.e., one is radical coupling, in which cross-linking reactions take place between xylan and lignin mediated by ferulate; the other one is rearomatization of the QMI by polysaccharide nucleophiles, which generates ether or ester bonds between hemicelluloses and lignin (Terrett and Dupree 2019).

2.4 Monolignol Radicals

Monolignol radicals are initially generated in the early stage of lignification, which subsequently produce various oligomers via different coupling reactions (Boerjan et al. 2003). The coupling of monomers/oligomers with monolignol radicals to form QMIs that act as new oligomeric radicals and further to generate higher oligomers (Grabber 2005). Seemingly, the formation of lignin undergoes a random way to produce irregular polymer involving a series of polymerization/termination reactions. Accordingly, the chemical structure of lignin is defined by the relative abundance of H/G/S units deposited at the lignification zone in the cell walls and distribution of inter-unit linkages formed. Stable radicals play a critical role in the lignin structural characteristics.

In 1960, stable radicals were first detected in lignin by using electron paramagnetic/spin resonance (EPR/ESR) when native lignin was suffering acidic and fungal attack (Rex 1960; Schonland 1959). The radicals were stable inside the lignin micelles for a very long time due to the protection of macroenvironment. Freudenberg (Freudenberg 1959) proposed a free radical mechanism for the polymerization of lignin, in which semiquinone radical intermediates were generated from the dehydrogenation or oxidative deprotonation of phenolic species. Steelink et al. (2002) suggested an improved process to stabilize and quantify semiquinone radicals and diamagnetic quinhydrone moieties in different lignins. Alkali and fungal lignins showed much greater spin concentrations than native, Bjorkman and Klason lignins (Levitin et al. 1955; Steelink 1964). Quinone species and corresponding radicals could be produced in both biological and chemical treatments of lignin (Marton 1964). EPR was proven sensitive to the detection of phenoxy radicals in lignin samples (Yang and Gaoni 2002).

The structures, properties, and contents of radicals in lignin are related closely to the botanical origin of lignin, the sample preparation method, the sample conditions, the extent of decay of lignin, and the nature of radical precursors and products (Colburn et al. 1963; Ebert and Law 1965; Košiková et al. 1993). For example, Kraft lignins from hardwoods had higher spin contents than softwoods. The spin content increased significantly after conversion of lignin to its metal salts. The generation of semiquinone radicals favored high pH, e.g., in Kraft treatment (Fitzpatrick and Steelink 1969; Steelink 1965). Generally, the radical content in lignins decreased within certain temperature range according to the sample conditions (Hatakeyama and Nakano 1970). For example, the radical content decreased when heating of MWL and organosolv lignin. However, inconsistent results were found at low-temperature region for oxidized organosolv lignin, dioxane lignin, and thiolignin (Hatakeyama and Nakano 1970). In these processes, thermal decomposition may overcome radical decay.

Various EPR spectroscopic techniques (Pillinger et al. 1996), ESR spin-trapping methods (Brai et al. 2009), and ^{31}P NMR spectroscopy (Zoia et al. 2011) are usually employed to determined stable radicals and related species or reactive radical intermediates in lignin.

2.5 *Biosynthesis of Lignin*

Generally, the lignification process of plant involves the biosynthesis of monolignols, the transport of monomeric precursors and monolignols to the position of polymerization in the cell walls, and dehydrogenation and polymerization of monolignols to form lignin macromolecules (Chen 2014). Lignin is primarily synthesized and deposited in the secondary cell wall of plants (Sarkanen and Ludwig 1971). The lignin contents vary in different layers in the cell wall, e.g., more than 50% for the middle lamella (ML) layer and 20% for the S layer. Nevertheless, most of lignin still accumulated in the S layer since its volume is far greater than that of ML layer.

There are two hypotheses referring to the lignin biosynthesis. One is the combinatorial random coupling between monolignol radicals (Neish 1968; Sarkanen and Ludwig 1971). The other is dirigent proteins-directed synthesis (Davin and Lewis 2005; Davin et al. 1997). Up to now, there is no plant species or tissue in which the biosynthetic pathway of lignin has been completely established (Sederoff et al. 1994). Generally, the lignin biosynthesis process contains four steps (CalvoFlores et al. 2015): the shikimate pathway producing phenylalanine and tyrosine, the phenylpropanoid common pathway generating *p*-hydroxycinnamoyl-CoA from phenylalanine and tyrosine, the synthesis of cinnamyl alcohols from *p*-hydroxycinnamoyl-CoA, and finally the oxidation of cinnamyl alcohols to produce 4-O-phenoxy radicals, which polymerize spontaneously to form lignin macromolecules.

2.5.1 The Shikimate Pathway

The shikimate pathway is highly transcriptionally controlled (Vered et al. 2010; Yanhui et al. 2006), as shown in Fig. 5. In this pathway, shikimic acid is first generated as an intermediate from glucose produced in the photosynthesis. Then, the shikimic acid is converted to phenylalanine and tyrosine through prephenic acid. Phenylalanine and tyrosine are the two amino acids as the starting materials for the cinnamic acid pathway. In the enzymatic steps, the glycolytic intermediate phosphoenolpyruvate and the pentose phosphate pathway intermediate erythrose-4-phosphate are metabolized into chorismate via 3-dehydroshikimate (Haslam and Kirby 1999). 3-Dehydroshikimate is a precursor for gallic acid involving in the gallotannin biosynthesis (Werner et al. 2004). Chorismate is the precursor for *p*-aminobenzoate, phenylalanine, tyrosine, and tryptophan (Basset et al. 2004; Knaggs 2003). Phenylalanine and tyrosine are produced from chorismate via arogenate after enzymatic reactions involving chorismate mutase, prephenate aminotransferase, and arogenate dehydrogenase or arogenate dehydratase.

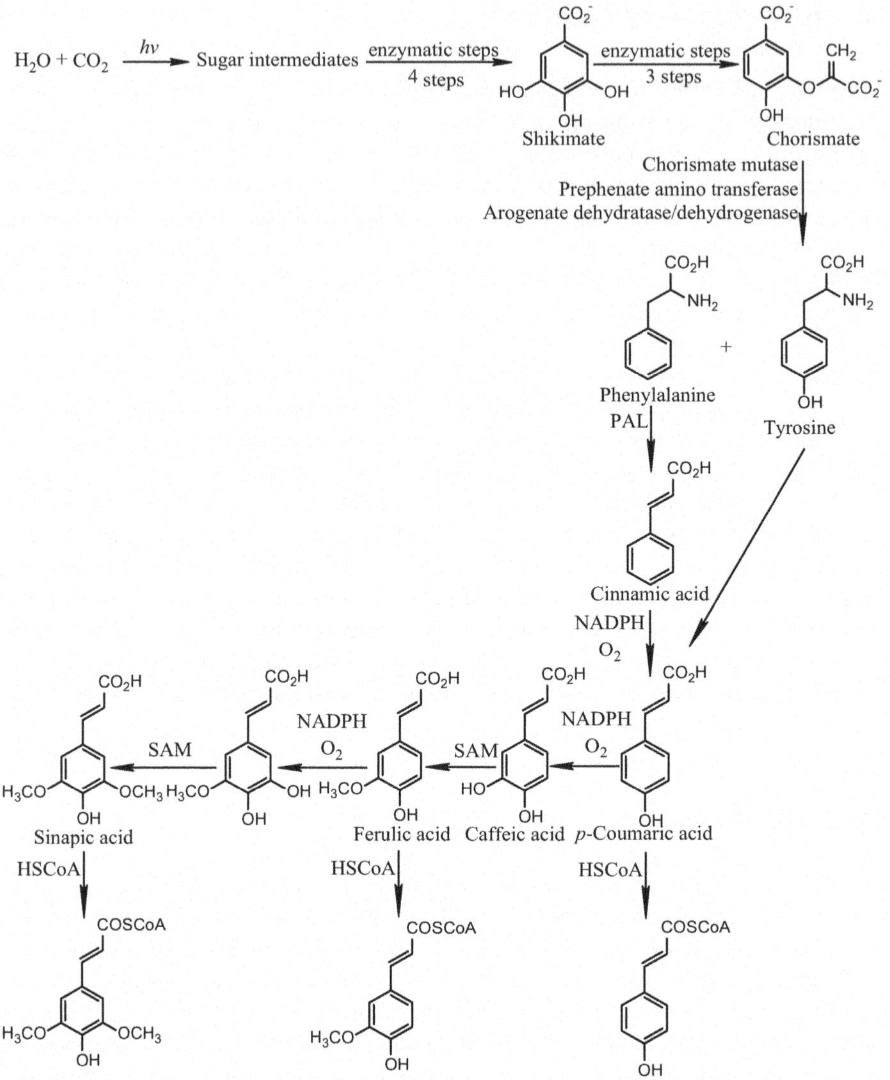

Fig. 5 Biosynthesis of lignin: the shikimate pathway and the phenylpropanoid pathway

2.5.2 The Phenylpropanoid Pathway

As main intermediates in the biosynthesis of lignin, phenylalanine and cinnamic acids are generated from carbohydrates through the shikimic acid pathway and cinnamic acid pathway, respectively (Hasegawa and Higuchi 2008; Higuchi 1990; Kratzl and Zauner 1962). Feruloyl-CoA is then produced from phenylalanine through a seven-step process in the phenylpropanoid pathway (CalvoFlores et al.

2015; Humphreys and Chapple 2002) (Fig. 5). Phenylalanine and tyrosine undergo deamination catalyzed by phenylalanine ammonialyase (PAL) and PAL isozyme to produce cinnamate and *p*-coumarate, respectively (Rosler et al. 1997). Cinnamate 4-hydroxylase (C4H) catalyzes cinnamate to produce *p*-coumarate. Tyrosine ammonialyase (TAL) can also catalyze tyrosine to *p*-coumaric acid (Higuchi et al. 1967). 4-Coumarate-CoA ligase (4CL) activates the *p*-coumaric acid to *p*-coumaroyl-CoA. This species is transesterified to its quinic or shikimic ester derivative catalyzed by shikimate/quinate hydroxycinnamoyltransferase (HCT). *p*-Coumaroyl shikimate or quinate is hydroxylated by *p*-coumarate 3-hydroxylase (C3H) and then transesterified again by HCT to caffeoyl-CoA. Cinnamic acid is hydroxylated to *p*-coumaric and caffeic acids. Then, the caffeic acid is methylated to ferulic acid by O-methyltransferase (OMT) (Higuchi et al. 1967; Hsi-Chuan et al. 2011). These pathways cover both softwood and hardwood lignins. The caffeic acid undergoes enzymatic reaction with 4CL and caffeoyl-CoA O-methyltransferase (CCoAOMT) generating feruloyl-CoA.

2.5.3 The Specific Pathway for Monolignols

The biosynthesis of monolignols is basing on the specific pathway. It involves two-step methylation reactions at the 3′ and 5′ positions catalyzed by CCoAOMT and caffeic acid O-methyltransferase (COMT), respectively. The rate-limiting step is the reduction of feruloyl-CoA to coniferaldehyde catalyzed by cinnamoyl-CoA reductase (CCR). The hydroxylation at the position 5 is catalyzed by ferulate 5-hydroxylase (F5H) producing 5-hydroxyconiferaldehyde (Osakabe et al. 1999). COMT can catalyze the transformation of 5-hydroxyconiferaldehyde, caffeic acid, 5-hydroxyl coniferylaldehyde, and 5-hydroxyconiferyl alcohol into sinapaldehyde, ferulic acid, sinapic acid, and sinapic alcohol, respectively.

The reduction of *p*-coumaraldehyde, coniferaldehyde, and sinapaldehyde to their corresponding alcohols, namely *p*-coumaryl, coniferyl, and sinapyl alcohols, respectively, is catalyzed by cinnamyl alcohol dehydrogenase (CAD). Specifically, CAD aggregates in cell layers where G lignin precipitates, while sinapyl alcohol dehydrogenase (SAD), F5H, and COMT locate in tissues with precipitation of S lignin. C3H and F5H are hydroxylases found in the biosynthesis of monolignols. C3H catalyzes coumaric acid to caffeic acid, while F5H catalyzes the hydroxylation of coniferaldehyde and coniferyl alcohol, which act as precursors of S lignin. Therefore, ferulates and diferulates are often present in grass lignins.

2.5.4 Lignification from Monolignols

Lignification is the last step in the biosynthesis of lignins from cinnamyl alcohol. After being synthesized, monolignols are transferred to the secondary cell wall of tissues in plants. Various enzymes, such as peroxidases, laccases, polyphenol oxidases, and coniferyl alcohol oxidases, are involved in the dehydrogenation of monolignols to produce the corresponding phenoxy radicals. After that, random chemically driven radical reactions take place without any enzymes resulting in amorphous and non-optical lignin. This may lead to the sense that lignin is fundamentally random.

However, the fact is that plants can control the amount, composition, and structure of lignin in the plant cell walls according to different species, tissues, locations, and growing stages (Müse et al. 1997), by reserving intrinsic characteristics. Thus, two major hypotheses emerged to explain lignifications. One is the random coupling model (Neish 1968; Sarkanen and Ludwig 1971), which suggests that monolignols connect to the lignin polymer gradually via random enzyme-catalyzed oxidative coupling, and the formation of lignin is regulated according to the types, amount, and chemical characteristics of monolignols. This hypothesis is the most widely accepted model for the lignifications process. The other hypothesis is the dirigent protein model (Davin and Lewis 2005; Davin et al. 1997), which proposes that the lignification is strictly under the control of proteins. This model can explain the high abundance of β-O-4 linkages and the highly ordered metabolism pathway of lignin. The formation of monolignols is associated with a series of ordered enzymatic reactions involving enzymes and proteins. The degree of random or ordered for lignification and the model of lignin polymerization are therefore questions of ongoing debate (Müse et al. 1997).

2.6 Computational Assays

In order to investigate structural characteristics and lignin decomposition-related processes, appropriate computational methods involving simulation approaches can be used in specific system by providing complementary information, such as accurate structures, energies, and molecular properties, with the combination of experimental techniques (Dellon et al. 2017; Sangha et al. 2012). Computational simulation has been applied in the lignin-related systems for various purposes, such as electron delocalization in monolignol radicals (Durbeej and Eriksson 2003b), the energetic behavior of β-O-4 linkage (Durbeej and Eriksson 2003a), the hydrogen bonding between monolignols (Rodrigo et al. 2011), the bond strengths of the benzodioxane linkage, and the degradation mechanism of lignin model compounds under pyrolytic conditions (Jarvis et al. 2011b).

Semi-empirical and molecular mechanic methods coupled with systematic or random conformer generation algorithms could be used in the study of conformations for higher multimers in lignification from energy point of view (Elder and Fort 2010). The most stable and reasonable conformer(s) with low energy could be refined with a more accurate quantum chemical method, such as density functional theory (DFT). By using DFT method, the RRSRS stereoisomer of spirodienone was found to be the most stable stereochemical configuration thermodynamically (Elder et al. 2017). The B3LYP method is the most popular method providing improved accuracy.

For long timescale processes without breaking or forming chemical bonds in condensed phase of mesoscopic systems, classical molecular dynamics (MD) simulations with standard software are competent. However, in systems where chemical reactions take place such as oxidative coupling of monolignols, quantum mechanical (QM) calculations are required. The level of accuracy is related to the computational configuration, the software settings, and the run time of computing. QM calculations have been used to study the mechanism of lignin biosynthesis, energies, and electronic properties of lignin model compounds (Jarvis et al. 2011b). MD methods can simulate complex molecular systems of 10^5 or more atoms for hundreds of nanoseconds. The CHARMM (Chemistry At Harvard Molecular Mechanics) force field is the most commonly used in MD simulation. For example, in the investigation of hydrogen bonds in aqueous solution containing guaiacyl dimer linked with β-O-4, MD simulation showed significant exchange of inter-molecular hydrogen bonds between dimmers and surrounding water molecules (Besombes and Mazeau 2004).

Semi-empirical QM method and combined QM/molecular mechanical (QM/MM) method (Hans Martin and Walter 2010) were treated as robust tools for describing the lignin structure and reactivity. In these investigations, MM method was used to study the majority of the enzyme-catalyzed lignin synthesis and QM method for the rest small portion (Senn and Thiel 2007).

2.7 Structural Models of Lignin

In the elucidation of chemical structure of lignins from different origins by various analytical strategies, isolation and depolymerization of lignins are inevitable. Because almost all the isolation and depolymerization processes lead to covalent bond cleavage in natural lignin according to specific mechanisms and conditions, the lignin structures would have great differences. The clear chemical formula of lignin could not be obtained since the extremely complex structure of lignin. Therefore, a hypothetical approximate structural model, deduced from the average abundance of monolignols and frequency of linkages, is necessary to represent the important structural features of lignin. It must be point out that the proposed lignin models are only statistical representations of the complicated lignin structures, and they could have great differences in categories of linkages and composition of functional groups, according to natural conditions of sources in plants, lignin isolation methods and conditions, and analytical techniques.

The first lignin model was proposed for spruce by Neish (1968) in 1968, which contained 18 units of monolignols. The most extensively used for softwood lignin structure model was suggested by Adler (1977) in 1977. It was a model for spruce lignin with 16 C_9 units. In 1995, Karhunen et al. (1995, 1999) quantified dibenzodioxocin by 2D NMR techniques and considered it to be the main branching point in softwood lignin. Brunow (Brunow 2005) recommended that dibenzodioxocin and 4-O-5′ linkage were the main branching structures in softwood lignin, and treated cinnamaldehyde and glycerol as the end units. More recently, Crestini et al. (2011) developed a softwood lignin model based on supramolecular aggregates of linear oligomers rather than a network polymer.

As mentioned above, the structures of lignin vary greatly according to S/G ratio. Higher relative content of S unit in lignin results in higher proportion of linear lignin, especially in hardwood lignin (Boerjan et al. 2003; Nimz 2010). For instance, typical abundance of G/S/H units in hardwood lignin is about 100:70:7.

The structures of lignins from grassy plants present quite significant variability. A remarkable characteristic is that these lignins are extensively acylated, especially at the γ position of S unit in side chain (up to 80%) with acetate or *p*-coumarate groups. Most of the acylated lignins preferred β-O-4 and α-O-4 linkages, leading to linear polymers (Dalimova and Abduazimov 1994; Río et al. 2008; Simionescu and Anton 1970). Due to the great diversity, no commonly accepted models for lignins from grassy plants accurately describe their structures. The detailed and accurate elucidation of structural characteristics of lignins from grassy plants relays on the comprehensive compositional analysis of the degraded fragments. In 2012, we proposed a lignin structure model for rice husk basing on the quantitative analysis of lignin-derived species in the pyrolytic oil by using gas chromatography/mass spectrometry (Lu et al. 2012) (Fig. 6). This model used dibenzodioxocin as the starting structure and combined 16 units of monolignols with various linkages according to the abundance of units and the frequencies of the inter-links.

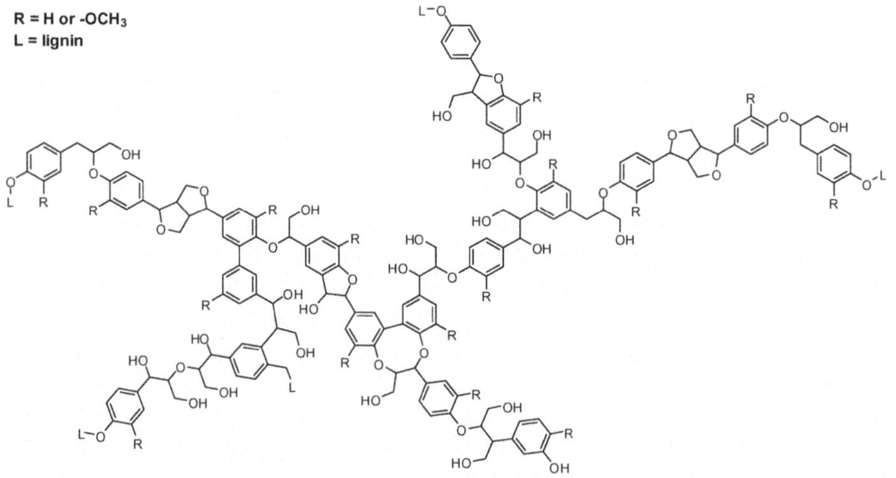

Fig. 6 A proposed lignin model

2.8 *Classification of Lignins*

Traditionally, basing on plant taxonomy, lignins can be classified into three different categories, namely angiosperm lignins (hardwood), gymnosperm lignins (softwood), and grass lignins. However, this classification has many exceptions. In addition, gymnosperm lignin contains more G unit, angiosperm lignin is generally a mixture of G and S units, and grass lignin is a mixture of all the three units. Hence, considering chemical composition, lignins can be divided into four types of groups: G type, G-S type, H-G-S type, and H-G type. Accordingly, lignins from softwood, hardwood, and grass are classified into G type, G-S type, and H-G-S type. Hardwood lignin contains more S unit than the other types; hence, the higher content of methoxyl groups makes this lignin less condensed and more amenable to chemical conversion.

3 Physicochemical Properties of Lignin

The physical and chemical properties of lignin vary significantly according to plant species and isolation processes. Compared to natural lignin, technical lignin has many advanced physicochemical features facilitating further analysis, processing, and conversion. For example, it has good solubility and distributability, multiple polydispersity, and good reactivity. Physicochemical characterization of lignin samples can be conducted using various analytical techniques, such as UV microscopy, FTIR spectroscopy, Raman spectroscopy, electron microscopy, thermogravimetry (TG), differential thermal analysis (DTA), differential scanning calorimetry (DSC), gas chromatography/mass spectrometry (GC/MS), and cross-polarization/magic angle spinning (CP/MAS) NMR spectroscopy.

3.1 *Heating Values*

Heating value of lignin is the detected net enthalpy of the reaction between the sample and oxygen under isothermal condition. Heating values of lignins are around 15–20 MJ kg^{-1}.

3.2 *Molecular Weight and Polydispersity*

Gel permeation chromatography (GPC), size exclusion chromatography (SEC), osmometric method, light-scattering method, super centrifugation, and high-performance liquid chromatography (HPLC) are usually the main techniques for the determination of the molecular weight of lignin. In fact, the accurate molecular weight of the original lignin is impossible to obtain since any isolation, separation, and identification method may induce degradation and changes of lignin structure to some extent, not to mention the complexity, variety, and polydispersity. By using mechanical, enzymatic, or chemical methods, the lignin can be degraded into different fragments with molecular weight distribution ranging from several hundreds to several millions.

The average molar mass of lignin, namely molecular weight (M_w), is important and common parameter in characterization of lignin structure. The number average molar mass, namely M_n, is also used. The M_w is more convenient (Chen 2014). However, the acceptance of using M_n or M_w to characterize the lignin is still under debate (CalvoFlores et al. 2015). This controversy is attributed to the complexity of lignin samples being analyzed, especially the degradation processes, isolation methods, and the analytical techniques used in the preparation of lignin. Inconsistent results are often obtained in some cases. This may due to the solubility of lignin sample in organic solvents in the measurement, since the M_w of soluble portion is smaller than that of the truly isolated lignin sample. In addition, extensive aggregation of lignin molecules is one of the main troubles for the determination of M_w. Generally, the M_w of isolated lignin is in the range of 10^3–10^5 Da.

The molecular weight of MWL from spruce varies greatly according to the grinding time and separation methods. The average M_w for softwood lignin is greater than 2×10^4 Da, which is higher than hardwood lignin. The average degree of polymerization of MWL was about 3–25 giving the molecular weight ranging from 500 to 5000 Da, basing on colligative properties. The measured M_w of lignosulfonates ranges from 10^3 to 10^5 Da, the maximum is in the order of magnitude of 10^6 Da. Kraft lignin has a much lower molecular weight. Vapor osmometry can be used to determine M_n. However, this technique suffers from supramolecular aggregation between non-hydrogen bonding solvent and lignin. Other methods, such as SEC and light-scattering techniques, show that the M_n is depended largely on lignin concentration, solvent system, pH, and lignin preparation methods.

Basing on the experimental determination, the soluble lignins and their derivatives present great dispersity. The dispersity index is defined as the ratio of M_w to M_n to express the width of molecular weight distributions. The smaller ratio values, the narrower mass diversity. Typical dispersity index of lignin varies from 2 to 11.

3.3 *Thermal Characterization*

Lignin is treated as an amorphous thermoplastic polymer. It has slight friability under high temperature and cannot form film in solution. It also has glassy transfer properties and can be described by using glass transition temperature (T_g). Similar to molecular weight of lignin, the determination of T_g depends on a number of factors (Chen 2014; Upton and Kasko 2015), such as thermal history, pretreatment methods, water content, cross-link density, molecular weight, and measurement methods, resulting in quite imprecise values. T_g curves for most lignin samples show slightly mass decreases between 0 and 150 °C due to vaporization of residual water. Generally, thermal decomposition of lignin starts from 200 °C. The extrapolated decomposition of technical lignins ranges from 209 °C to 330 °C, approximately. Under the T_g, lignin is in the solid glass phase; it begins to move when it is above the T_g and is softened to become sticky. The molecular motion of lignin (MML) is greatly influenced by hydrogen bonds between water molecules and the hydroxyl/phenol groups in the lignin chain. MML is significantly restricted when polar ionic and rigid groups are introduced into the molecular chain, leading to the increase of T_g.

Hatakeyma et al. (1969) measured the T_g of Björkman's lignin ranging from 97 to 117 °C, by using infrared (IR) spectrometry. However, the T_g was 92–142 °C when they investigated the same lignin with DSC method (Hatakeyama 1972). The large difference and variation of T_g indicated that not only the sample handling and preparation affected the measured results, but also the measurement method used. Taking all the factors into consideration, the T_g of lignin is reported to occur between 70 and 170 °C. This imprecise character of lignin indicates the complexity and variation of lignin samples (Hatakeyama et al. 1975).

Inconsistent results are also obtained for decomposition temperature (T_d) of lignin. The characteristic decomposition of lignin ranges from 360 to 480 °C, which is easily recognized from the other components in wood samples (Chauvette et al. 1985). TG/MS and pyrolysis-coupled GC/MS (Py-GC/MS) have been commonly used to investigate lignin decomposition behavior and products. Different processes for lignin preparation exhibit different T_d. For instance, the thermal decomposition of dimethyl sulfoxide lignin and alkali methylated lignin (International Conference on Thermal Analysis 1965) is started from 160 °C and 350 °C, respectively. Thermal behavior of lignin is influenced after various modification reactions. For example, the acetylated lignin shows lower T_g values because of the decline in the degree of inter-molecular association in the sample. The acetylation decreases the amount of hydroxyl groups and increases the rigidity of lignin (Lisperguer et al. 2009).

3.4 Solubility

Unmodified natural lignins from biomass are insoluble in most common solvents (Chen 2014). Therefore, in order to improve solubility, it is necessary to degrade lignin into fragments with lower M_w. A large number of polar functional groups exist in technical lignins, such as hydroxyls, generating strong inter- and intra-molecular hydrogen bonds between lignin and solvent, and leading to good solubility. The separated lignin can be divided into soluble and insoluble portions according to the efficiency of the separation methods, solvents, and other conditions (CalvoFlores et al. 2015; Chen 2014; Upton and Kasko 2015).

Under alkaline conditions, lignin is dissolved well due to the phenolic hydroxyl and carboxyl groups. However, acidic or enzymatic hydrolyzed lignin is not soluble effectively in any solvents. Separated Brauns' lignin and organosolv lignin can be dissolved in several organic solvents, such as methanol, ethanol, acetone, pyridine, dioxane, and DMSO. Alkali lignin and lignosulfonate can be dissolved in dilute alkali, water, and salt solution. Dioxane is a good solvent and can dissolve many lignins effectively. More than 50% of MWL from spruce can be extracted with dioxane. The best solvent system by far for most separated lignins is acetyl bromide and hexafluoroisopropanol in acetic acid.

3.5 Relative Density

The relative density of lignin is roughly ranging from 1.35 to 1.50. The determined values vary according to the method and liquid used in the measurement. For example, the relative density of sulfuric acid lignin isolated from pine is 1.451 and is 1.436 if measured by benzene; and the relative density of dioxane lignin is 1.33 when measured by water and is 1.391 by dioxane. The relative density of pine glycol lignin and pine hydrochloride lignin is 1.362 and 1.348, respectively.

3.6 Color

The color of native lignin is white or nearly colorless. The colors of technical lignins vary according to the separation and preparation process. For instance, the color of Brauns' lignin is light cream, and the colors of acid lignin, copper ammonia lignin, and periodate lignin vary from fawn to deep tan.

3.7 Functional Groups

In lignin, hydroxyl and carboxylic groups are the two important functional groups, which determine the hydrophobicity and significantly influence the hydrolysis behavior of lignin. Generally, lignin hydrophobicity increases with phenolic –OH (Ar–OH) and decreases with aliphatic –OH (al-OH) content. High Ar–OH and low al-OH contents introduce resistance to hydrolysis of lignin (Guo et al. 2014; Huang et al. 2016; Rahikainen et al. 2013; Yu et al. 2014). Carboxylic groups (–COOH) affect the physicochemical properties of lignin by providing high hydrophilicity, adjustable electrostatic charge, and hydrogen bonds (Moxley et al. 2012; Nakagame et al. 2011; Seiji et al. 2011). Carboxylation and sulfonation can increase the hydrophilicity of lignin (Yang and Pan 2016). Nevertheless, the correlations of Ar–OH and –COOH with enzymatic hydrolysis are complicated due to the various botanical origins and pretreatment methods. Proportions of main functional groups in hardwood and softwood lignins are listed in Table 4.

Table 4 Proportions of functional groups in lignins of hardwood and softwood

Functional groups	Hardwood	Softwood
Methoxyl	130–146	90–96
Phenolic hydroxyl	9–20	20–28
Benzyl hydroxyl	–	16
Aliphatic hydroxyl	–	120
Carbonyl	3–7	20
Carboxyl	11–13	–

3.8 Chemical Reactions

According to the functional groups contained in lignin, the chemical reactions cover oxidation, combination, halogenation, nitration, and chromogenic reaction, etc., taking place at various reaction positions, such as Ar-/al-OH, phenyl ring, benzyl alcohol, aryl ether bond, and alkyl ether bond. Generally, these chemical reactions can be classified into nucleophilic reactions and electrophilic reactions (CalvoFlores et al. 2015; Chen 2014; Kumar et al. 2009; Lupoi et al. 2015; Müse et al. 1997; Upton and Kasko 2015).

Most of the chemical reactions, which occur on the lignin side chains, are nucleophilic reactions, such as pulping process and lignin modification. For example, in the acidic sulfite pulping process, SO_2 aqueous solution is used to cleave aryl and alkoxy ether bonds resulting in the sulfonation of α-carbon. Under natural conditions, HSO_3^- and SO_3^{2-} break down the ether bond as nucleophiles. Nucleophilic reagents including OH^-, HS^-, and S^{2-} take part in cleavage of ether bonds under alkaline conditions. In this case, the phenolic –OH turns into phenolate anion Ar–O^- by releasing H^+, and the negative charged oxygen atom activates its *ortho*- and *para*- positions via inductive and conjugative effects, resulting in the cleavage of α-aryl ether bond. Consequently, a 1,2-diphenylethene structure is produced by the aromatization of methylene quinone.

Chemical reactions on the aromatic ring cover both electrophilic and nucleophilic reactions. The electrophilic reactions include substitution and oxidation reactions. Oxidative reagents, such as chlorine, chlorine dioxide, ozone, and nitro cation, oxidize the aromatic rings via electrophilic pathway into their corresponding *o*-quinone intermediates and finally into dicarboxylic acid derivatives. Nucleophilic reactions take place when using OH^-, ClO^-, and H_2O_2 as nucleophiles in the lignin degradation. An apparent character for oxidative processes is the formation of purple- or red-colored species, namely quinones. For example, the syringyl ring in lignin is oxidized into methoxy *o*-dihydroxybenzene and methoxy *o*-quinone when treated with acidic $KMnO_4$ and ammonia, sequentially. Coupling reactions also occur for technical lignin samples (Cui et al. 2013).

4 Depolymerization and Isolation Technologies

Natural lignin is a rigid, heterogeneous, cross-linking, and high-degree-condensed complex polymer, and it is extraordinarily difficult to study due to its insolubility in common solvents. Most techniques used for structural investigation are only applicable for fragments of degraded lignin in solution phase. Linkages or connecting bonds are cleaved according to their stabilities and the decomposition conditions used in the degradation processes.

Technical lignins are prepared from natural lignin by using various treatments. They have variable structures according to the botanical and geographical factors, the degradation and extraction methods, and conditions used in the treatment processes. For a given technical lignin sample, the structural variability refers to abundance and composition of monolignols, distribution or frequency of linkages and bonds, molecular weight, etc. Lignin can be qualitatively or quantitatively analyzed in situ, or in an isolated form in terms of with or without derivatization. The derivatization methods of lignin samples prior to analysis often include mechanical, chemical, physicochemical, and biological treatments, and/or their combinations. A gold standard procedure for lignin isolation does not exist. One should consider the chemical changes of lignin structure during the treatments of derivatization and isolation. Furthermore, due to the complex combination between lignin and polysaccharides with covalent bonds in plant cell walls, it is difficult to separate individual components into pure homogeneous fractions.

There are two ways to separate lignin from lignocellulosic biomass. One is to dissolve lignin into a solution by extraction or degradation; and the other is to hydrolyze cellulose and hemicelluloses under acidic condition leaving lignin as an insoluble residue. A fundamental contradiction arises when choosing separation/isolation methods to obtain lignin from biomass (Upton and Kasko 2015; Xu and Ferdosian 2017). On the one hand, the produced lignin should reserve original characteristics of natural lignin with smallest changes; on the other hand, it is desirable to obtain lignin as much as possible to facilitate further analysis or industrial utilization of biomass. These two aspects should be considered prudently when conducting the lignin isolation processes in practice. Many techniques with great progress are available currently for the isolation and purification of lignin; indeed, however, none of them can provide a 100% yield and structural authenticity (Xu and Ferdosian 2017).

4.1 Brauns' Lignin

In the isolation process, milled biomass material is first emerged in 96% of ethanol for several days, and then, the mixture is filtered to afford a light brown filtrate. After removing the solvent by evaporation and washing the residue filtrate with ether, finely divided Brauns' lignin with bright color will precipitate repeatedly from its solution in dioxane. This lignin is often treated as native lignin since very slight changes in chemical structures in this process. The drawbacks of this method include low yield (2–10%) and contamination by extractives (Dence 1992).

4.2 *Milled Wood Lignin (MWL)*

MWL, also known as Björkman's lignin, obtained by milling biomass material with a neutral organic solvent (water/dioxane), is considered to be the closest one to native lignin and the most representative lignin of the milled sample (CalvoFlores et al. 2015; Chen 2014; Rencoret et al. 2008). The MWL is mainly from the middle lamella of the plant cell walls where G unit accumulates. MWLs produced from both softwoods and hardwoods favor linear over network polymers (Claudia et al. 2011). Hardwood MWLs usually contain more carbohydrates than softwoods MWLs. It is generally rich in al-OH groups and poor in Ar–OH groups. In order to enhance the degradation and isolation of lignin, no matter which process employed, the first step will be the mechanical treatment, i.e., ball milling. Ball milling modifies the structure of lignin to a minimum extent. In the preparation of MWL, the linkages between lignin and polysaccharides are cleaved mechanically first with a ball-mill for tens of hours, and then the resulted lignin is dissolved in dilute NaOH solution followed by a water/dioxane mixture.

Generally, the yield of MWL is very low. In order to remove the polysaccharides and to increase the yield of MWL, enzymatic degradation was used before the extraction of lignin. However, the process took a very long time to accomplish (Dence 1992). Zhang et al. (2013) reported a three-step process to extract lignin from sugarcane bagasse by using 96% dioxane, 50% dioxane, and 80% dioxane containing 1% NaOH sequentially at their boiling temperatures. The total yield in lignin was up to 72%. Fang et al. (2018) carried out a mechanochemical approach using ring-and-puck milling with KOH in a solvent-free system to promote lignin degradation, in which considerable amount of β-O-4 linkages were cleaved generating significant amounts of aliphatic hydroxyls and phenols.

4.3 *Soda Lignin (Alkali Lignin)*

The soda process is usually applied for non-wood biomass materials at high temperature with NaOH solution. In this process, water-soluble phenolic and alkali stable species, such as quinine methides, stilbene, and styryl aryl moieties, are produced due to the cleavage of ether bonds in lignin. After acidification, lignin fragments can be isolated easily as precipitate. Soda lignin usually has higher purity but much lower molecular weight.

4.4 Kraft Lignin

The Kraft process generates most of the paper pulp worldwide. In this process, biomass material is treated with Na_2S/NaOH mixed solution (white liquor) at high temperature (155–175 °C) and high pressure for several hours (Gellerstedt and Zhang 2001; Gierer 1980). The lignin undergoes a large variety of reactions in the depolymerization generating phenolic fragments. Up to 90–95% of all lignin present in the biomass can be dissolved into the solution. The resulted fluid is rich in lignin, called black liquor. By using a strong acidic solution, Kraft lignin can be obtained as precipitate from black liquor. As a sulfonated lignin, more than two-thirds of the hydroxyl groups turn into their sulfated form. Therefore, the Kraft lignin can be dissolved in basic solution and highly polar organic solvents. The typical M_n of Kraft lignin is in the range of 1000–3000 Da, with a polydispersity between 2 and 4. Kraft lignin is often used as fuel to generate energy for pulping mills, but is rarely adopted in value-added utilizations, such as preparation of fine chemicals and advanced materials.

4.5 Sulfite Lignin (Lignosulfonates)

Sulfite lignin, known as lignosulfonates, is the sulfonated lignin recovered from sulfite pulping of biomass material. In the process, an aqueous solution containing SO_2 and a sulfate salt are used to cleave the linkages between lignin and polysaccharides at high temperature (140–170 °C). The resulted lignosulfonates are water-soluble anionic polyelectrolyte polymers. They present both hydrophobic and hydrophilic properties (Lora 2008) and are soluble in both acidic and basic aqueous solutions, as well as in high-polar organic solvents. The M_w of lignosulfonates varies from 1000 Da to 140,000 Da, with the majority lying between 5000 Da and 20,000 Da, which is higher than that of Kraft lignin. Due to harsh conditions used in the process, large structural modifications are introduced into lignin; hence, neither Kraft lignins nor lignosulfonates are suitable for studying the behaviors and the characteristics of natural lignin.

The production of Kraft, soda, and sulfite lignins are involved in the main types of industrial pulping processes. Soda process is mostly used on annual fiber feedstocks, while the other two are predominantly used on wood. Lignins obtained from annual fiber feedstocks have higher contents of carboxyl groups than lignins from wood. Sulfite pulping introduces sulfonate groups increasing solubility of lignin in water. Kraft and soda pulping produce soluble lignins as ionized aromatic hydroxyl salts with relatively higher purity, much lower molecular weight and hydrophilicity than lignosulfonates. Kraft lignins have higher sulfur content than soda lignins.

4.6 *Organosolv Lignin*

Organosolv lignin is extracted from biomass material directly with aqueous organic solvents and some additives or catalysts at the typical temperature ranging from 140 to 220 °C. Methanol, ethanol, acetone, glycols, water, acids and bases, as well as their mixtures, have been used in the process. Ethanol/water is an efficient extracting solvent system and widely used (Lawther et al. 1996). Alcell method is one of the very popular processes producing high-purity organosolv lignin, in which lignin is extracted by a mixture of water and ethanol (1:1, v/v) at 175–195 °C for 1 h with a liquid/solid biomass ratio about 4–7 by weight. The process is environmentally friendly because of the conditions used, such as sulfur free, low temperature, and low pressure. The yield of organosolv lignin from ethanol–water mixed solvent is higher than that of Kraft lignin. Most of the organosolv lignins are insoluble in acidic aqueous solutions, but dissolve in basic solutions or organic solvents.

Typical M_n of organosolv lignin is less than 1000 Da, and polydispersity ranges from 2.4 to 6.4, with a monomer molecular weight of 188 Da. Organosolv process offers high-purity lignin with similar structure to native lignin, which might facilitate value-added applications of lignin.

4.7 *Klason Lignin*

Klason method is treated as a standard method for determining the lignin content. In this process, most of the polysaccharides could be hydrolyzed and dissolved into 72% of H_2SO_4 as fermentable sugars, and the lignin is precipitated as solid residue. Many factors affect the process, such as pretreatment of the sample, temperature and time of hydrolysis, H_2SO_4 concentration, and dilution conditions (Hatfield et al. 2010).

4.8 *Hydrolytic Lignin*

Similar with Klason method, in this process, a dilute acid is used to hydrolyze and dissolve the polysaccharide linkages in biomass material. The hydrolytic lignin precipitates as insolvable residue. The carbohydrates in the complex mixture undergo enzymatic reaction to produce ethanol. Finally, the pure hydrolytic lignin is extracted from the system (Chatel and Rogers 2016). Theoretically, hydrolytic lignins are more reactive than lignosulfonates and Kraft lignins since more reactive functional groups contained such as phenolic/alcoholic hydroxyl (Chatel and Rogers 2016). However, in practice, its reactivity is significantly reduced by strong steric effect due to its large size and complex structure.

4.9 Acidolysis Lignin

Acidolysis lignin is released from biomass material after hydrolysis with 0.2 M HCl in aqueous 1,4-dioxane under room temperature. This lignin has only a very few carbohydrate impurities with minimal chemical changes compared to natural lignin. Thioacidolysis provides more effective degradation and increases the yield of lignin producing thioethylated oligomers and monomers, in which ethanethiol is used as the solvent instead of water. In the process, arylglycerol-β-aryl ethers are selectively cleaved by boron trifluoride etherate and ethanethiol. Thioacidolysis can be used to determine the average S:G:H ratio of lignin samples. For example, the S:G:H ratio was reported as 68:32:0.02 basing on the investigation of 104 poplar lignin samples (Robinson and Mansfield 2010).

4.10 Cellulolytic Enzyme Lignin (CEL)

In the production of CEL, finely milled wood is treated with hydrolytic enzymes to remove polysaccharides as much as possible prior to solvent extraction. The removal of the polysaccharides is increased up to 85–90%. CEL is relatively more representative of the total lignin than other processes (Dence 1992) due to its higher yields with less chemical structure changes. In order to enhance the enzymatic degradation of the polysaccharides and decrease the carbohydrates contaminants, a swelled enzymatic method was proposed (Chen et al. 1995). Typical M_n of CEL is about 2000 Da with the polydispersity around 6, and the average molecular weight of monomer is 187 Da.

4.11 Enzymatic Mild Acidolysis Lignin (EMAL)

EMAL is obtained from a refined CEL process combining an enzymatic treatment and a mild acidolysis of lignin (Wu 2004). After enzymatic hydrolysis, the insoluble portion is then washed with dilute acid to remove carbohydrates, reserving ether bonds within the lignin. Finally, the extraction is conducted in a dioxane/water mixed solvent containing 0.01 M hydrochloric acid in the presence of N_2. Generally, the yield of EMAL is about fourfold greater than that of the corresponding MWL and about twofold greater than CEL. Compared to CEL, EMAL has similar M_n and average monomer molecular weight with improved polydispersity (Anderson et al. 2006a, b).

4.12 Pyrolytic Lignin

Pyrolytic lignin is produced from thermal degradation of material occurring in the absence of oxygen with considerable high temperature (e.g., 450 °C) and very short residence time (less than 2 s) (Zhang et al. 2012; Fan et al. 2015; Lourenço et al. 2016; Marathe et al. 2019). Patwardhan et al. (2011) suggested a two-step mechanism of the pyrolysis of lignin, i.e., the decomposition of lignin macromolecules into monomeric phenolic compounds and the re-oligomerization of some monomeric compounds to higher oligomers. However, the detailed mechanism of pyrolysis of lignin is unclear since the complexity of the lignin structure and the violent conditions used in the process. Various kinds of reactions can occur simultaneously during pyrolysis. Quinones and specific radical species are formed during lignin and wood pyrolysis (Kibet et al. 2012).

The notable characteristic of pyrolytic lignin is the extraordinary low average molecular weight compared to natural lignin and other kinds of technical lignins. The M_n of pyrolytic lignin is ranging from 300 to 600 Da. Pyrolytic lignin can offer specific aromatic hydrocarbons that are not available from other processes.

4.13 Steam Explosion Lignin (SEL)

In the steam explosion process, biomass material is immersed in a steam (180–230 °C) under high pressures (1.38–3.00 MPa) within very short function times (1–20 min) followed by sudden pressure release in the presence of some chemicals, producing a water-insoluble lignin (Focher et al. 1991). This process is also a traditional method for preparing cellulose pulp. The yield of SEL from hardwood can reach up to 90% by using alkali washing or extraction with organic solvents. SEL presents low molecular weight and high solubility in organic solvents. It has attracted more and more attention as a promising product since its simple and cheap preparation process and is going to be commercially available in the near future (Garrote et al. 1999).

4.14 Ionic Liquid Extraction

Ionic liquids extract lignin from biomass selectively and environmentally (Joseph et al. 2013). They can disrupt certain linkages in biomass material by various specific interactions, such as hydrogen bond, dipole–dipole, and van der Waals forces. The advantages of ionic liquids include low melting points, good thermal stability, negligible vapor pressure, and tunable physicochemical properties. At the same time, the drawbacks are high price and the difficulty in separation of products after treatment.

Ethyl-methylimidazolium acetate ([EMIM][OAc]) provides outstanding performance on extracting lignin (Hyun et al. 2010). Selectivity is a key point that should be considered in the extraction of lignin by ionic liquids. Otherwise, further separation of extracted components is needed (Fort et al. 2007). 1-Ethyl-3-methylimidazolium cation and alkylbenzene sulfonate anions ([EMIM][BMS]) can extract lignin with very high yield, i.e., up to 93% (Tan et al. 2009). Pyridinium formate [PyFor] presents a high capacity for the dissolution of Kraft lignin (70%, w/w) at a relatively lower temperature (75 °C) (Rashid et al. 2016). CO_2-switched [DBU][MEASIL] can be used to generate carbohydrate-free lignin since it provides high selectivity in extracting the interlinked polysaccharide impurities from the technical lignin without interfering with the linkages and aromatic subunits in lignin during the process (Khokarale et al. 2016).

4.15 Derivatization Followed by Reductive Cleavage (DFRC)

The DFRC procedure is relatively a new analytical degradation method for characterizing lignin structure (Argyropoulos et al. 2002; Claudia et al. 2011; Lourenço et al. 2016; Lu and Ralph 1997). This method is an alternative way that can selectively cleave β-aryl ethers in lignin and release arylglycerol-β-ethers that can be quantified with GC conveniently. It contains successive steps including bromination of the benzylic positions and concomitant acetylation of free hydroxyl groups by acetyl bromide, reductive cleavage of the β-O-4 linkage in brominated intermediate with zinc and acetylation of newly formed free phenolic groups. An outstanding advantage of this method is that it depolymerizes lignin back to its monolignols (Lu and Ralph 1998; Sun 2010).

5 Instrumental Strategies

Lignins are complex polymers which vary in chemical structure according to their original conditions, preparation treatments, analytical methods, etc. In order to elucidate detailed and comprehensive chemical characteristics of lignin, various information for certain sample should be collected including the nature, proportion, and distribution of the building blocks (units, monolignols, or monomers) and inter-unit linkages, by using robust analytical techniques. Because of the selectivity and efficiency of degradation method, only a fraction of natural lignin in biomass sample is liberated and analyzed. Furthermore, the instrumental techniques only provide partially information of lignin sample due to their limited detection range. Due to the complexity and diversity of the components in lignin products, it is impossible to obtain complete characteristics of a sample by a single analytical method.

5.1 *Elemental Analysis*

Lignins obtained from various origins and treatments have similar elemental composition. Roughly, the contents of carbon, hydrogen, and oxygen are 60–65%, 5–7%, and 28–35%, respectively. Minor amounts of sulfur and nitrogen account less than 1%.

The main structure of lignin is composed of the phenylpropane units (H, G, and S units) connecting with various linkages. Thus, lignin can be expressed conveniently as a C_9-unit structure together with methoxyl group (Lenz 1968; Monteil-Rivera et al. 2013). By using the empirical expression, complex structural composition of lignins can be evaluated and compared. For example, the formulas for typical MWLs from softwood and hardwood are $C_9H_{8.3}O_{2.7}(OCH_3)_{0.97}$ and $C_9H_{8.7}O_{2.9}(OCH_3)_{1.58}$, respectively. Accordingly, these formulas highlight the fact that the monomeric units comprising lignins have a C_9 carbon skeleton and emphasize the significance of the methoxyl group both for identifying the lignin type and for indicating the contributions of componential monolignols (Dence 1992). The common contents of methoxyl in softwoods and hardwoods are in the ranges of 13–16% and 18–22%, respectively.

5.2 *Spectroscopic Methods*

In the characterization of chemical structure of lignin, both degradative and non-degradative methods are employed according to the goals of determination and the techniques used in the analysis. The advantages of spectroscopic methods are their ability to analyze whole lignin structure and to detect lignin moieties directly. Spectroscopic methods are the most widely used for the structural elucidation of lignin both in solid state and in solution, including UV spectroscopy, infrared (IR) spectroscopy, Raman spectroscopy, various NMR methods, etc.

5.2.1 UV Spectroscopy

The characteristic UV absorption spectra of lignin have been used for the qualitative and quantitative determination of native lignins and technical lignins and for tracing the delignification in pulping processes. Lignin has very strong UV absorption due to the π–π^* transition of the aromatic nuclei. UV spectroscopy can be used to estimate the amount of phenolic hydroxyl groups in lignin basing on the difference in absorption between 292 and 370 nm (Hansen et al. 2016; Sluiter et al. 2008; Xiong et al. 2017; Xu et al. 2010). Symmetrical syringyl unit leads to a blue shift for the maximum absorbance of lignin. An additional absorbance at approximately 370 nm is assigned to conjugated phenolic hydroxyl groups (Yang et al. 2016). In the UV spectra, softwood lignin presents maximum ranging 280–285 nm and hardwood

Table 5 UV absorptions of typical functional groups and structures in lignin

Absorptions/nm	Typical functional groups and structures in lignin
200–205	Conjugated bonds/aromatic rings
240	Free hydroxyl
275–285	Conjugated bonds/aromatic ring
320	Aromatic ring conjugated bond with C=C
	C=O groups conjugated to aromatic ring
325	Etherified ferulic acid
370	Conjugated phenolic hydroxyl groups

lignin at around 275 nm. Typical soluble lignin in hexafluoropropanol shows the maximum at about 205 nm. UV absorption maximums of typical structures in lignin are presented in Table 5.

5.2.2 IR Spectroscopy

IR spectroscopy is the most widely used technique in the analysis of functional groups. Its determination is based on the analyzed substances that containing chromophores. This technique is competent for both solid samples and solutions of organic solvents and requires only very small amount of sample. It is a non-destructive, non-invasive, high sensitivity, simple, and rapid technique. However, one of the challenges in the characterization of lignin structure by IR spectroscopy is that strong bands of residual non-lignin compounds overlapping the characteristic lignin bands. Table 6 lists the IR absorbance of typical structure in lignin.

The IR absorption at 1515 cm^{-1} is often used for the quantitative determination of lignin (Kolboe and Ellefsen 1962). FTIR together with attenuated total reflectance (ATR) can be used for the evaluation of structural features of lignin in different pretreatments and growth stages (Koivu et al. 2016; Santoni et al. 2015; Yang et al. 2012). Rapid and accurate qualitative and quantitative analyses by using IR spectroscopy are often applied for lignin derivatives, e.g., oxyethylated lignins (Passauer et al. 2016). FTIR spectroscopy is considered to be a fast and efficient way of analyzing lignin from biorefineries without sample pretreatment (Le et al. 2017). Signals from contaminants (cellulose, hemicelluloses, and silica) can be filtered out mathematically.

5.2.3 Raman Spectroscopy

Raman spectroscopy can be treated as sister spectroscopic method of IR spectroscopy providing complementary information. Generally, Raman and IR spectroscopy share similar spectra for assignments of the absorption bands. It is competent for analysis of the aqueous solutions and provides more fine information reflecting structural

Table 6 IR absorbance of typical lignin components

Wavenumbers/cm^{-1}	Assignments	Functional groups and structures in lignin
3400–3600	O–H stretching	Free hydroxyl group
3100–3400	O–H stretching	Associated hydroxyl group
2820–2960	C–H stretching	$-CH_2$, $-CH_3$
2920	O–H stretching	Carboxylic hydroxyl
2650–2890	C–H stretching	Methyl group in methoxyl
1771	C=O stretching	Aromatic carbonyl
1700–1750	C=O stretching	Unconjugated ketones, carbonyls, and ester groups
1722	C=O stretching	Aliphatic carbonyl
1650–1680	C=O stretching	Conjugated *p*-substituent carbonyl and carboxyl
1500–1600	Aromatic skeletal vibration	Benzene ring
1425, 1460	C–H deformation	$-CH_2$, $-CH_3$
1360–1370	C–H bending	$-CH_2$, $-CH_3$
1325–1330, 1230–1235	C–O stretching	Syringyl ring
1270–1275	C–O stretching and C=C vibration in aromatic ring	Guaiacyl ring
1215	C–O and C–C stretching	Ether
1140	C–H stretching	Guaiacyl
1130	C–H stretching	Syringyl
1088	C–O stretching	Secondary alcohol and aliphatic ether
1030	C–O and C–H tretching	Aromatic ring and primary alcohol
750–860	C–H stretching	Aromatic ring

features of samples (Agarwal et al. 2005). Raman spectroscopy is competent to non-invasive microanalysis of native lignin structure even with no sample preparation (Perera et al. 2012). When analyzing a lignin sample in different solutions, the environmental effects of the solvents should be taken into consideration (Larsen and Søren 2011).

5.2.4 NMR Spectroscopy

Structural characterization of lignins by using NMR spectroscopy has attracted more and more attentions in the recent three decades (Glasser 2000). The development of NMR techniques has dramatically expanded the landscape of lignin structure characterization. It can provide more detailed, precise, and comprehensive information on qualitative and quantitative results with high resolution for the composition of H/G/S units, other monolignols, and the frequencies of linkages between them in the lignin. It can also recognize lignin molecular characteristics that cannot be obtained by other techniques, for example, in the discovery of new moieties, such as dibenzodioxocine and spirodienone structures in lignin (Karhunen et al. 1995, 1999; Marco et al. 2011; Ralph et al. 2001; Zhang and Gellerstedt 2001; Zhang et al. 2006).

^{1}H NMR, ^{13}C NMR, ^{19}F NMR, ^{31}P NMR, and various modes of 2D NMR techniques have been used in the structural and compositional analyses of various lignin samples. Among them, ^{1}H NMR and ^{13}C NMR are treated to be common tools for the analysis of lignin; and ^{13}C NMR and 2D heteronuclear single-quantum coherence (HSQC) NMR can provide accurate quantitative results on the functional groups and side-chain moieties for both solid and liquid lignin samples.

^{1}H NMR

^{1}H NMR is a dominant choice in the structural investigation of lignin, especially in the early days, due to its instrumental available, high sensitivity, simple preparation of samples, and fast scanning speed. ^{1}H NMR determines the chemical environment of proton in various H-containing moieties, such as benzylic hydroxyl, aliphatic and aromatic hydroxyl, and aliphatic hydrogen atoms. Typical functional groups' assignments in the ^{1}H NMR spectrum are shown in Table 7. Accordingly, the signals around 3.10 ppm are attributed to the protons in anhydroxylose units (An et al. 2017). The chemical shifts in the range of 3.3–4.0 ppm are attributed to protons in methoxyl groups. The peaks in the range of 4.0–6.3 ppm belong to aliphatic protons in β-O-4, β-β′, and β-5′ linkages. The chemical shifts around 7.5, 7.0, and 6.5 ppm can be assigned to aromatic protons in H, G, and S units, respectively (Fernández-Costas et al. 2014; Wang and Chen 2013).

However, limitations exist as well to this technique for applying to structural characterization of lignin, including the narrow range of chemical shifts, extensive signal overlapping, and proton-coupling effects. Derivatization methods for quantitative determination of the H-containing functional groups in lignin are usually carried out (Li and Lundquist 1994).

^{13}C NMR

^{13}C NMR overcomes the overlapping resonances significantly compared to ^{1}H NMR spectra by providing qualitative and quantitative results for both solid and liquid samples. Other advantages for ^{13}C NMR method include wide range of chemical shift, no coupling effects, high resolution, and better quantitative efficiency. Table 8 lists the chemical shift assignments to chemical structures in lignin in ^{13}C NMR spectra. Basing on the chemical environment of carbon atoms in different moieties, the chemical shifts can be divided roughly into three regions: The range of 20–90 ppm

Table 7 Typical functional groups in lignin assigned chemical shifts in ^{1}H NMR spectrum

Chemical shifts/ppm	Structures and functional groups in lignin
9.7–9.9	Cinamaldehydes and benzaldehydes
7.5–7.6	Aromatics *ortho* to C=O
7.2–7.5	Aromatic-H in coumaryl
6.9–7.0	Aromatics and vinyl
6.7–7.1	Aromatic-H in guaiacyl
6.2–6.7	Aromatic-H in syringyl
5.8–6.2	Benzylic hydroxyl in β-O-4 and β-1′
5.4–5.5	H_α in β-5′ and noncyclic benzyl aryl ethers and H_β in aryloxypropiophenones
4.9–5.1	Carbohydrates
4.6–4.7	H_α and H_β in β-O-4 and β-β′
4.4	H_γ in β-O-4 and β-5′
3.7–3.8	OCH_3 in guaiacyl and syringyl
3.3–4.0	Methoxyl
3.0–3.1	H_β in β-1′
2.2–2.4	Phenolic hydroxyl
1.6–2.2	Aliphatic hydroxyl

is assigned to carbons in saturated side chains, the range of 100–160 ppm is attributed to aromatic carbons, and the range of 160–200 ppm is for unsaturated bonds. It is worth noting that the characteristic chemical shift for carbon in methoxyl group is around 56.3 ppm.

The first use of ^{13}C NMR for the lignin structure characterization was in 1970s (Lüdemann and Nimz 1973). By using quantitative ^{13}C NMR, accurate results could be obtained including the ratio of H/G/S unit, content of various linkages, degree of condensation, and distribution of oligomers. The drawbacks of ^{13}C NMR are the low sensitivity, large sample amount, and long acquisition time. In addition, it usually shows weak and broad signal peaks for solid or insoluble samples. Therefore, lignin is often subjected to acetylation by anhydride/pyridine solution before analysis (Lu and Ralph 2003, 2010). Cross-polarization/magic angle spinning (CP/MAS) method (Schaefer and Stejskal 1976) is extensively used for rapid characterization of lignin structure with improved resolution but insufficient quantitative results (Ben and Ragauskas 2012). In addition, by using quantitative ^{13}C NMR, it is difficult to quantify minor components in lignin due to the overlapping of signals. Sodium 3-(trimethylsilyl)propionate is often used as internal standard to determine the content of lignin in samples by using solid-state ^{13}C CP/MAS NMR technique with reduced analysis time (Xin et al. 2015). By combining high-field dynamic nuclear polarization (DNP) and MAS, the sensitivity and resolution of the solid-state NMR method could be improved significantly (Lee et al. 2015; Mollica et al. 2015; Perras et al. 2016).

Table 8 Assignments of chemical shifts in ^{13}C NMR spectra to functional groups and structures in lignin

Chemical shifts/ppm	Assignments
167–178	Unconjugated –COOH
166.5	C_9 in *p*-hydroxyphenyl
162–168	Conjugated –COOH
160	C_4 in *p*-hydroxyphenyl
148	C_3 in guaiacyl
140–155	C_3, C_4 aromatic ether or hydroxyl
127–140	C_1 in aromatic C–C
128–129	C_2 and C_6 in *p*-coumarate
123–127	C_5 in aromatic C–C
118.4–119.4	C_6 in guaiacyl
114.7–115.1	C_5 in guaiacyl
110.4–111.1	C_2 in guaiacyl
106–114	C_2, aromatic C–H
83.8–84.6	C_β in β–O–4
78–90	Aliphatic C_β–O
67–78	Aliphatic C_α–O
60.2	C_γ in β–O–4
54–57.5	Methoxyl
53.4–54.0	C_β in β-β′ and β-5′

^{19}F NMR

Before the analysis with ^{19}F NMR technique, specific functional groups in lignin should be derived into corresponding F-containing moieties by using fluorinating reagents (Ahvazi and Argyropoulos 1996a, b; Barrelle 1993; Barrelle et al. 1992; Constant et al. 2017; Sevillano et al. 2001). For example, in the quantification of carbonyls in industrial humans and lignins, the carbonyl groups were transformed to corresponding hydrazone with 4-(trifluoromethyl)phenylhydrazine before ^{19}F NMR analysis.

^{31}P NMR

As a relative new and powerful method developed by Argyropoulos et al. ^{31}P NMR spectroscopy (Ahvazi et al. 2016; Argyropoulos 1993, 1994; Pu et al. 2011; Springer et al. 2016) has been used to investigate soluble lignins and carbohydrate samples after phosphitylation of hydroxyl groups with phosphorous reagents. By using this technique, the quantification of aliphatic, phenolic, and carboxylic hydroxyl groups, as well as the contents of H, G, and S units in lignin can be obtained. Quantitative ^{31}P NMR combining with DFRC is often used to determine arylglycerol-α-aryl ether

Table 9 Assignments of chemical shifts in ^{31}P NMR spectra to hydroxyl groups in lignin (after phosphitylation) (Ahvazi et al. 2016; Argyropoulos 1994; Pu et al. 2011; Springer et al. 2016)

Chemical shifts/ppm	Assignments of hydroxyl groups
145.5–150.0	Aliphatic –OH
136.5–144.7	Phenols
140.0–144.5	C_5 substituted
143.5	β–5′
142.7	Syringyl
142.3	4–O–5′
141.2	5–5′
139.0–140.0	Guaiacyl
138.2–139.0	Catechol
137.3–138.2	*p*-Hydroxyphenyl
133.6–136.6	Carboxylic acid –OH

and other linkages (Argyropoulos et al. 2002; Claudia et al. 2011). High content of phenolic OH indicates the presence of condensed aromatic units (Mbotchak et al. 2015). Typical assignments of the chemical shifts in ^{31}P NMR spectra are listed in Table 9.

2D NMR

2D NMR methods can overcome the overlapping of resonances and provide higher resolution and semi-quantitative results in the characterization of lignin (Ando et al. 2015, 2016; Bozell et al. 2011; Brech et al. 2016; Crestini et al. 2017; Crouvisier-Urion et al. 2016; Zeng et al. 2013; Jorge et al. 2015; Khokarale et al. 2016; Lourenço et al. 2016; Marco et al. 2011; Mattsson et al. 2016; Saito et al. 2016; Yann et al. 2015; Zhang and Gellerstedt 2011). Heteronuclear single-quantum coherence (HSQC) spectroscopy, heteronuclear multiple-quantum coherence (HMQC) spectroscopy, total correlation spectroscopy (TOCSY), rotating-frame Overhauser experiment spectroscopy (ROESY), heteronuclear correlation (HETCOR) spectroscopy, homonuclear Hartmann–Hahn (HOHAHA) spectroscopy, and heteronuclear multiple-bond coherence (HMBC) spectroscopy have been carried out for the structure characterization of lignin. ^{13}C–^{1}H correlated HSQC-NMR, by combining sensitivity of ^{1}H NMR and high resolution of ^{13}C NMR, is able to identify the structures of lignin and polysaccharides clearly in cell walls without isolating them. It is by far the most extensively used 2D NMR mode revealing the frequencies of different monolignols and the inter-unit bonding patterns in lignin (Ando et al. 2015, 2016; Karhunen et al. 1995, 1999; Khokarale et al. 2016; Marco et al. 2011; Zhang and Gellerstedt 2011).

By using 2D HSQC-NMR, structural changes of lignin and the other components in biomass during chemical treatments can be easily monitored (Ando et al. 2016; Jorge et al. 2015; Mattsson et al. 2016; Saito et al. 2016). For example, the cleavages of aliphatic C–O and C–C bonds and the repolymerization of monomers and oligomers were observed simultaneously by using high-resolution 2D HSQC-NMR in the depolymerization of Kraft lignin in subcritical water (Mattsson et al. 2016). The combination of 2D HSQC-NMR and quantitative ^{13}C NMR overcomes the overlap of signals and reduces error in quantification detection, and has been considered as the most powerful approach for describing the structural characteristics of various lignins (Crestini et al. 2017; Zeng et al. 2013; Zhang and Gellerstedt 2011). Solid-state ^{1}H–^{13}C CP/MAS NMR is able to detect the low-energy dipole–dipole interactions (Crouvisier-Urion et al. 2016). DNP-enhanced 2D homonuclear ^{13}C–^{13}C correlation solid-state NMR spectra can provide bond activity in both raw and treated lignin samples at an atomic level (Perras et al. 2016).

5.2.5 Chromatographic Techniques

Chromatography is an effective technique for analysis, separation, and isolation of complex samples and has been widely used in the characterization of lignin structures for various physicochemical properties, such as molecular weight distribution and polydispersity, composition of monolignols and oligomers, and linkage types. Chromatographic techniques are employed in lignin structural investigations including GPC, SEC, GC, LC, and two-dimensional (2D) chromatography. These techniques equipped with various advanced detectors combining appropriate sample preparation procedures provide powerful analytical protocols for elucidation of structural features of lignins with high resolution and separation capacity.

Gel Permeation Chromatography (GPC)

GPC is commonly used to evaluate the size and molecular weight of lignin products (Forss et al. 2009). In the analysis, lignin sample is dissolved into appropriate solvents, and the separation of molecules with different molecular weight is realized according to their distribution coefficients between column stationary phase (e.g., cross-linked polydextran gels) and mobile phase. Molecules with small molecular size penetrate the gel particle pores, but large molecules are excluded and pass through the column directly (elute first). Various detectors are used for collecting information of the elutes. An UV detector with wavelength range of 240–300 nm is often used to monitor the effluent for detecting the lignin-related molecules (Chen et al. 2018). In the molecular weight determination of lignin, GPC coupled with UV detector is by far the most widespread analytical method, although with many limitations, e.g., the reproducibility is poor. In addition, it cannot identify the aggregation. In order to improve the solubility and enhance the separation of technical lignins, acetylation or methylation pretreatment is often applied. The light-scattering method has also been considered to be one of the most popular techniques for determining the M_w and the actual shape of lignin molecules in solution.

Size Exclusion Chromatography (SEC)

SEC is another chromatographic method used for lignin molecular weight determination. However, the results obtained usually with poor accuracy and reproducibility (Janne et al. 2012; Savy and Piccolo 2014). In order to improve status of lignin fragments during analysis, derivatization methods such as acetylation and acetobromination are often adopted (Janne et al. 2012). Compared with UV detectors, multi-angle laser light-scattering detector provides more detailed and reliable results on molecular weight distribution and dynamic behaviors (Gidh et al. 2006a, b; Sofía et al. 2008).

Gas Chromatography (GC)

GC is a powerful technology traditionally used in separation of volatile compounds according to their different boiling points and structural features. GC concentrates on detecting the volatile organic species with lower polarity and lower boiling point (<350 °C) with molecular weight range of 50–500 Da. GC coupled with various detectors has long been extensively used in the characterization of lignin-related complex samples (Fan et al. 2015; Galkin and Samec 2015; Karl and Ronald 2012; Lu et al. 2012; Lupoi et al. 2015; Río et al. 2005; Wang et al. 2014; Xu et al. 2012). Different detectors have different sensitivities and resolutions to specific species or functional groups with inherent limitations. Hundreds of species can be identified simultaneously in a single run for a lignin sample. One of the challenges encountered in GC analysis is assignment of species in chromatogram. Species in sample could be partially characterized. For example, the resulting total ion chromatogram from GC/MS analysis is complicated due to the overlapping of peaks and the identification of components relied heavily on the basis of the database.

In recent years, Py-GC/MS is prevalently used and proven to be a powerful technique for lignin analysis, such as content, molecular weight distribution, composition of monomers and oligomers, frequency of linkages, and thermodynamic behaviors (Zhang et al. 2012; Fan et al. 2015; Fortin et al. 2015; Lv et al. 2018; Marathe et al. 2019; Nsimba et al. 2013; Patwardhan et al. 2011; Río et al. 2012; Van et al. 2017). Given the qualitative and quantitative results from analysis, combining the characters of reactions during pyrolysis, one can propose structure model for lignin (Lu et al. 2012). Furthermore, pyrolysis is considered as an effective way to produce benzene-containing species, which are mainly derived from lignin.

Liquid Chromatography (LC)

LC offers analysis of non-volatile, low molecular weight, and low thermal stability compounds according to the difference in polarity of species. Traditional detectors such as UV detector and refractive index detector have been used to identify various species of interests. LC coupled with mass spectrometry (LC/MS) is one of the most popular techniques for the determination of lignin-derived compounds with high sensitivity and resolution. Basing on the compositional characters of lignin samples, negative-ion-mode electrospray ionization with NaOH dopant is appropriate for the efficient ionization of the compounds (Haupert et al. 2012). In order to improve resolution and accuracy in the characterization of components in sample, tandem mass spectrometry (MS^n, up to MS^7), and ultra-high-performance liquid chromatography (UHPLC) are employed in recent years (Eduardo et al. 2012; Jarrell et al. 2014; Ouyang et al. 2014; Owen et al. 2012; Sun et al. 2016).

Two-Dimensional (2D) Chromatography

Two-dimensional chromatographic methods (GC × GC and LC × LC) are powerful techniques in the structural investigation of lignin samples by providing qualitative and quantitative results with higher separation power, higher resolution, and peak capacity. However, a great challenge is the identification of individual species. In the analysis, a coordinate with *X*- and *Y*-axis is used to present retention behavior of species with respect to the two columns. The quantitative result of species is calculated according to the peak density and intensity.

In order to improve resolution and efficiency of 2D chromatographic techniques, mass spectrometric detectors with different ionization and analyzing modes are used universally to identify compounds, e.g., GC × GC/MS (Kalogiannis et al. 2015; Michailof et al. 2014). In the compositional characterization of species in lignin-derived sample, GC × GC coupled with TOFMS (Onorevoli et al. 2014; Tessarolo et al. 2016) and Q-MS (Zhang et al. 2017) is frequently used. GC × GC/TOFMS provides a higher number of identified species compared to GC × GC/Q-MS (Onorevoli et al. 2014). GC × GC system coupled with simultaneous dual detectors, namely FID and MS, provides qualitative and quantitative results for the decomposition products of lignin-related samples in a single run (Purcaro et al. 2016; Windt et al. 2009), especially for the compound with minor content.

The main challenges of LC × LC are to obtain a high degree of orthogonality and to maximize peak capacity. The commonly used modes for LC separation include reversed-phase liquid chromatography (RPLC), normal-phase liquid chromatography (NPLC), SEC, supercritical fluid chromatography (SFC), hydrophilic interaction liquid chromatography (HILC), and ion-exchange chromatography (IEC). In LC × LC analysis, the retention space is being delimited by the retention times of the least and the most retained compounds in both dimensions (Débora et al. 2014; François and Sandra 2009; Masle et al. 2014). Although RPLC × RPLC has relatively low degree of orthogonality issue, it is still preferred (François et al. 2009).

Brudin et al. (2008) used RPLC × SEC method to determine formulation characteristics of sulfonated lignins. They found that retention behavior of molecules in RPLC was not determined by the degree of sulfonation of similar-size molecules. Setvenson et al. (2012) constructed HPLC × SFC system and obtained a peak capacity of 2400 in 280 min. Recently, in order to address the relatively low degree of orthogonality issue with RPLC × RPLC, Sun et al. (2018) built an online system by the combination of SFC and RPLC × SFC with trapping column-assisted modulation for depolymerised lignin analysis. A high degree of orthogonality with coverage factors 0.79 was obtained for the standard mixture of 40 compounds with the diol SFC column.

The identification of species was significantly improved by coupling several sensitive detectors or using advanced data handling methods. Meanwhile, the combined analytical system with higher resolution needs to be set up, and the quantitative analysis methods need to be established.

5.2.6 Mass Spectroscopy (MS)

MS is frequently used as detector which follows chromatographic technique to assign components in lignin-related samples. Ionization methods of MS include electrospray ionization (ESI), atmospheric pressure photoionization (APPI), atmospheric pressure chemical ionization (APCI), electron ionization (EI), and matrix assisted laser desorption ionization (MALDI). Among them, ESI with negative mode has been used most frequently to detect a relative wider range of compounds in lignin samples. In recent years, the most widely used mass analyzers are quadrupole (Q-MS), ion trap (ITMS), and Fourier transform-ion cyclotron resonance (FT-ICR) mass spectrometers.

Py-GC/MS is conventional and dominating analytical technique for the investigation of lignin thermochemical degradation and for the identification of the monomers, oligomers, and linkages (Lourenço et al. 2015a; Ohra-Aho and Linnekoski 2015; Windt et al. 2009). Tandem mass spectroscopy (MS^n) and high-resolution mass spectroscopy (HRMS) have been successfully introduced into chemical structure characterization of lignins. For example, linear quadrupole ion trap-Fourier transform-ion cyclotron resonance hybrid mass spectrometer (LQIT-FT-ICR MS) and triple-quadrupole mass spectrometer (TQMS) were combined with LC to determine lignin-derived compounds (Gierer 1980; Morreel et al. 2010). Monomers, dimmers, trimers, tetramers, and pentamers were identified with ESI source run in negative mode.

Since lignin samples are complex mixtures of large number of similar phenolic compounds, the selection of tentative lignin oligomers in the MS data is challenging. Tandem MS at all stages improves the structure identification of lignin oligomers without the need of chemical standards (Prothmann et al. 2018). By using this method combined with a principal component analysis-quadratic discriminant analysis classification model (Prothmann et al. 2018), 36 tentative lignin oligomers including dimers, trimers, and tetramers were identified out of 587 detected peaks in the Kraft lignin sample. In a systematic optimization process, ionization efficiency was significantly improved for 25 lignin oligomers allowing the MS^7 to be conducted.

HRMS techniques are used to identify individual component in complex mixtures by providing accurate mass measurement for the compounds with sufficiently high mass resolving power according to the monoisotopic mass. Orbitrap MS, Q-TOFMS, and FT-ICR MS are extensively employed in the analysis of lignin samples.

In Orbitrap MS analysis, the produced ions from the ionization source are passing through a series of quadrupoles and deflection lenses before introduced to analyzer. Q-TOFMS is fast and sensitive technique with simple operating principle. FT-ICR MS provides non-specific identification of molecules within a wide range of molecular weight (200–1000 Da) with considerable high mass resolution (better than 0.003 Da). An outstanding advantage of FT-ICR MS is that high-molecular-weight species and heteroatom-containing compounds can be quantified effectively. The main factor limit of its application is the high cost of the instrument.

Orbitrap MS and Q-TOFMS can provide satisfactory results for conventional use, while FT-ICR MS has been proven to be an ideal technique and is a final choice when detailed and comprehensive results are required (Abdelnur et al. 2013; Lupoi et al. 2015; Tessarolo et al. 2016). The former two methods favor volatile compounds with fewer carbon and oxygen atoms, while the last is more efficient in distinguishing ions with higher m/z and a much broader range of polarity (Jarvis et al. 2012). Consider the mass resolution and accuracy, the HRMSs follow the order FT-ICR MS > Orbitrap MS > Q-TOFMS.

More detailed and comparative usages of MS techniques related to the investigation of lignin samples were reviewed by Lupoi et al. (2015).

5.2.7 XPS

X-ray photoelectron spectroscopy (XPS) can be used to characterize the surface chemistry and degree of disorder of the poly-aromatic stacks, bonding states of atoms as well as their relative contents, etc. The distribution of lignin content on the surface of biomass (5–10 nm deep) can be estimated semi-quantitatively based on oxygen–carbon ratio (O/C) and aliphatic carbon component as well as elemental compositions (Ma et al. 2015; Zhang et al. 2014). Lignin is mainly responsible for the C_1 (C–C or C–H) and O_1 (C=O) chemical shifts. Higher O/C ratio indicates higher cellulose and/or hemicellulose content, while lower O/C suggests the presence of more lignin. The lignification and polymerization of lignin cause a drastic increase for chemical shift of C_1 and a decrease for C_2, C_3, and O_2 that leads to a lower O/C ratio (Ma et al. 2015).

5.2.8 SEM, TEM, and AFM

Scanning electron microscopy (SEM) has been widely used directly to image the microstructure, morphological changes, and surface dispersion patterns of lignin in various processes (Prasad and Claus 2010; Selig et al. 2007). Lignin droplets and patches which precipitate on the surface of biomass with a size range from about 0.05 to 2 mm can be observed. Field-emission SEM is usually used to provide high resolution of the fractures (Lei et al. 2012). The microscope could be equipped with energy-dispersive X-ray analyzer, which provides elemental and chemical composition information (Mansur et al. 2005).

Transmission electron microscopy (TEM) can be applied to determine lignification process and the ultra-structure and distribution of lignin in cell walls of plants (Ma et al. 2013) by providing relative content of lignin. For example, lignin content in the cell corner middle lamella is higher than other layers.

Atomic force microscopy (AFM) is an efficient technique for visualizing the molecular conformation and characterizing the topographical ultra-structure in lignin-related samples. It can provide detailed lignin distribution patterns (Gidh et al. 2006a; Hollertz et al. 2013). Lambert et al. (2018) introduced dynamic in situ real-time AFM to investigate lignocellulosic biomass recalcitrance mechanisms affected by lignin at nanoscale. It was found evidently that lignin content and lignin distribution affected hydrolysis efficiency.

6 Summary and Prospects

Great progress has been made for the understanding of structural characteristics of lignin during last several decades. Various technologies have been introduced into the degradation or depolymerization, modification, isolation, and detailed characterization of lignin-related samples. This chapter presents a comprehensive overview of structural features of lignin and related developments in the studies including experimental and computational assays, physicochemical properties investigation, isolation and separation methods, and the instrumental technologies. The biggest challenge in describing lignin structure is its intrinsic complexity and diversity due to the random cross-linked structure and different sources. In addition, the isolation methods from other components in biomass and the in situ or ex situ characterization of lignin also affect the results. So many factors influence the cognition of a certain lignin sample.

According to different aims, i.e., structural investigation, fuel production, paper pulping, or fine chemical production, different treatments are used to isolate lignin. Mild conditions with minimizing cleavage of chemical bonds are conducive to structural investigation, while production of low-molecular-weight chemicals favors harsh conditions. Appropriate analytical procedures and instruments should be chosen basing on the characteristic of lignin sample to be analyzed. Although no single method or technique may stand alone as the superlative one, the investigations have facilitated the construction of a more expansive protocol to comprehensive and profound elucidation of the lignin structure from various aspects. For example, the progress in advanced 2D NMR has brought about vast improvement in the understanding structural features of lignin such as simpler sample preparation before analysis, discovery of new monolignols, and qualitative and quantitative analysis of classes of species in lignin sample to set up fingerprint plot for plants. Various HRMS and MS^n techniques strengthen analytical efficiency significantly by providing considerable high resolution and high throughput, especially for the completely componential description and deeply revealing of correlation of species in lignin sample. Furthermore, in situ non-destructive analysis of lignin could be conducted by morphological techniques. For a real lignin sample investigation, combination and coupling of the techniques are necessary to meet accurate and comprehensive detection. Very recently, new structure is still being discovered, i.e., diferuloylputrescine, by using combination of DFRC method and 2D NMR technique (Del Río et al. 2018). This structure participates in radical coupling reactions of lignifications.

Because lignin is combinatorial, sequencing of lignin is meaningless, except for individual molecules. However, understanding how the monolignols coupling in the lignification provides insight into the factors that determine lignin structure. According to Freudenberg's concept, coupling of monolignols is dependent on their speed of delivery to the cell wall, their chemical characteristics, and the oxidative environment in the wall, among other factors. There is still a violent debate on the scheme of polymerization of monolignols to produce lignin (Müse et al. 1997; Tobimatsu and Schuetz 2019), namely whether the formation of lignin undergoes a random coupling process or a dirigent-like protein controlled pathway. Recent view on the polymerization of monolignols is that the combinatorial chemistry of the process under tight biological control. Specific laccase and/or peroxidase proteins are secreted to facilitate the spatial control of lignin deposition in distinct cell wall domains. However, to what degree their substrate specificities influence the incorporation of different monolignol types into lignin polymers remains unknown (Tobimatsu and Schuetz 2019).

The investigation of chemical structural characteristics of lignins is a long way to go. By combining of conventional chemical degradation methods and instrumental techniques, a great improvement on lignin structure study has been emerged. One new trend in lignin structure characterization and application research is to steer biosynthetic pathways by incorporation of species into the lignin polymer and further contribute to analysis or degradation processes. For example, a designed monolignol, 3-O-propargylcaffeyl alcohol was incorporated into the process of lignification. It was compatible with lignification and did not alter the fidelity of lignifications (Natalie et al. 2014). This investigation sets up a general platform for using different spectroscopic and imaging approaches to interrogate plant cell wall lignification. The "negligible" auxiliary for both in vitro and in vivo studies of lignification will shed new light on this complex polymer. Another example is the use of an artificial lignin-like material. An all-G β-O-4 polymer was designed to model softwood lignin (Montgomery et al. 2017), which provided a stepping stone between the simplicity of model compounds and the complexity of lignin. It was assumed that lighter fractions of lignin with simpler linear structures that could be more suitable materials for exploring lignin depolymerization reactions than the crude unfractionated lignin.

References

Abdelaziz OY, Brink DP, Prothmann J et al (2016) Biological valorization of low molecular weight lignin. Biotechnol Adv 34:1318–1346
Abdelnur PV, Vaz BG, Rocha JD et al (2013) Characterization of bio-oils from different pyrolysis process steps and biomass using high-resolution mass spectrometry. Energy Fuels 27:6646–6654
Adler E (1977) Lignin chemistry—past, present and future. Wood Sci Technol 11:169–218
Adler E, Lundquist K (1963) Spectrochemical estimation of phenylcoumaran elements in lignin. Acta Chem Scand 17:13–26
Agarwal UP, Atalla RH (1986) In-situ Raman microprobe studies of plant cell walls: macromolecular organization and compositional variability in the secondary wall of Picea mariana (Mill.) B.S.P. Planta 169:325–332
Agarwal UP, Reiner RS, Pandey AK et al (2005) Raman spectra of lignin model compounds. In: 59th Appita annual conference and exhibition: incorporating the 13th ISWFPC (international symposium on wood, fibre and pulping chemistry), Auckland, New Zealand, 16–19 May 2005: proceedings, 2005. Appita Inc., p 1
Ahvazi B, Argyropoulos D (1996a) Quantitative trifluoromethylation of carbonyl-containing lignin model compounds. J Fluor Chem 78:195–198
Ahvazi BC, Argyropoulos DS (1996b) ^{19}F nuclear magnetic resonance spectroscopy for the elucidation of carbonyl groups in lignins. 1. Model compounds. J Agric Food Chem 44:2167–2175
Ahvazi B, Cloutier E, Wojciechowicz O et al (2016) Lignin profiling: a guide for selecting appropriate lignin as precursors in biomaterials development. ACS Sustain Chem Eng.4: 5090-5105. https://doi.org/10.1021/acssuschemeng.6b00873
Akiyama T, Sugimoto T, Matsumoto Y et al (2002) Erythro/threo ratio of β-O-4 structures as an important structural characteristic of lignin. I: improvement of ozonation method for the quantitative analysis of lignin side-chain structure. J Wood Sci 48:210–215
An L, Wang G, Jia H et al (2017) Fractionation of enzymatic hydrolysis lignin by sequential extraction for enhancing antioxidant performance. Int J Biol Macromol 99:674–681
Azarpira A, Lu FC, Ralph J (2011) Reactions of dehydrodiferulates with ammonia. Org Biomol Chem 9:6779–6787
Anderson G, Ilari F, Lucia LA et al (2006a) Comparative evaluation of three lignin isolation protocols for various wood species. J Agric Food Chem 54:9696–9705
Anderson G, Ilari F, Lucia LA et al (2006b) Toward a better understanding of the lignin isolation process from wood. J Agric Food Chem 54:5939–5947
Ando D, Nakatsubo F, Takano T et al (2015) Elucidation of LCC bonding sites via γ-TTSA lignin degradation: Crude milled wood lignin (MWL) from Eucalyptus globulus for enrichment of lignin xylan linkages and their HSQC-NMR characterization. Holzforschung 70:489. https://doi.org/10.1515/hf-2015-0112
Ando D, Nakatsubo F, Yano H (2016) Acetylation of ground pulp: monitoring acetylation via HSQC-NMR spectroscopy. ACS Sustain Chem Eng 5:1755–1762
Angela Z, Kristen G, Rui K et al (2010) Increase in 4-coumaryl alcohol units during lignification in alfalfa (Medicago sativa) alters the extractability and molecular weight of lignin. J Biol Chem 285:38961–38968
Argyropoulos D (1993) Quantitative phosphorus-31 NMR analysis of six soluble lignins. J Wood Chem Technol 14:65–82
Argyropoulos D (1994) Quantitative phosphorus-31 NMR analysis of lignins, a new tool for the lignin chemist. J Wood Chem Technol 14:45–63
Argyropoulos DS, Lubo J, Lívia K et al (2002) Abundance and reactivity of dibenzodioxocins in softwood lignin. J Agric Food Chem 50:658–666
Austin AT, Ballaré CL (2010) Dual role of lignin in plant litter decomposition in terrestrial ecosystems. Proc Natl Acad Sci USA 107:4618–4622
Balakshin M, Gracz H, Chang HM et al (2011) Quantification of lignin–carbohydrate linkages with high-resolution NMR spectroscopy. Planta 233:1097–1110

Barrelle M (1993) A new method for the quantitative 19F NMR spectroscopic analysis of hydroxyl groups in lignins. Holzforschung-Int J Biol Chem Phys Technol Wood 47:261–267
Barrelle M, Fernandes JC, Froment P et al (1992) An approach to the determination of functional groups in oxidized lignins by 19F NMR
Barros J, Serk H, Granlund I et al (2015) The cell biology of lignification in higher plants. Ann Bot 115:1053–1074
Basset GJC, Quinlivan EP, Ravanel S et al (2004) Folate synthesis in plants: the p-aminobenzoate branch is initiated by a bifunctional PabA-PabB protein that is targeted to plastids. Proc Natl Acad Sci USA 101:1496–1501. https://doi.org/10.1073/pnas.0308331100
Ben H, Ragauskas AJ (2012) Torrefaction of loblolly pine. Green Chem 14:72–76
Benedikt R, Bamberger M (1890) Über eine quantitative Reaction des Lignins. Monatshefte für Chemie und verwandte Teile anderer Wissenschaften 11:260–267. https://doi.org/10.1007/bf01540738
Bente F (1875) Ueber die Constitution des Tannen- und Pappelholzes. Ber Dtsch Chem Ges 8:476–479. https://doi.org/10.1002/cber.187500801147
Besombes S, Mazeau K (2004) Molecular dynamics simulations of a guaiacyl β-o-4 lignin model compound: examination of intramolecular hydrogen bonding and conformational flexibility. Biopolymers 73:301–315
Besombes S, Mazeau K (2005a) The cellulose/lignin assembly assessed by molecular modeling. Part 1: adsorption of a threo guaiacyl beta-O-4 dimer onto a I beta cellulose whisker. Plant Physiol Biochem 43:299–308
Besombes S, Mazeau K (2005b) The cellulose/lignin assembly assessed by molecular modeling. Part 2: seeking for evidence of organization of lignin molecules at the interface with cellulose. Plant Physiol Biochem 43:277–286
Blanch HW, Simmons BA, Kleinmarcuschamer D (2011) Biomass deconstruction to sugars. Biotechnol J 6:1086–1102
Bo J, Wang W, Feng G et al (2016) Comparison of the substrate enzymatic digestibility and lignin structure of wheat straw stems and leaves pretreated by green liquor. Biores Technol 199:181–187
Boerjan W, Ralph J, Baucher M (2003) Lignin biosynthesis. Annu Rev Plant Biol 54:519–546
Bozell JJ, O'Lenick CJ, Stacy W (2011) Biomass fractionation for the biorefinery: heteronuclear multiple quantum coherence-nuclear magnetic resonance investigation of lignin isolated from solvent fractionation of switchgrass. J Agric Food Chem 59:9232
Brai M, Longo A, Maccotta A et al (2009) Electronic paramagnetic resonance power saturation of wooden samples. J Appl Phys 105:24–47
Brech YL, Raya J, Delmotte L et al (2016) Characterization of biomass char formation investigated by advanced solid state NMR. Carbon 108:165–177
Breznak JA, Brune A (1994) Role of microorganisms in the digestion of lignocellulose by termites. Annu Rev Entomol 39:453–487. https://doi.org/10.1146/annurev.en.39.010194.002321
Brudin S, Berwick J, Duffin M et al (2008) One-dimensional and two-dimensional liquid chromatography of sulphonated lignins. J Chromatogr A 1201:196–201
Brunow G (2005) Methods to reveal the structure of lignin. Part 1. Lignin, humic substances and coal. Wiley-VCH Verlag GmbH & Co. KgaA., Weinheim, Germany
Buanafina MM, De O (2009) Feruloylation in grasses: current and future perspectives. Mol Plant 2:861–872
Bule MV, Gao AH, Bill H et al (2013) Structural modification of lignin and characterization of pretreated wheat straw by ozonation. J Agric Food Chem 61:3916–3925
Buranov AU, Mazza G (2008) Lignin in straw of herbaceous crops. Ind Crops Prod 28:237–259
Calvo-Flores FG, Dobado JA, Isac-García J et al (2015) Lignin and lignans as renewable raw materials: chemistry, technology and applications. Wiley, Chichester, UK
Campbell MM, Sederoff RR (1996) Variation in lignin content and composition (mechanisms of control and implications for the genetic improvement of plants). Plant Physiol 110:3

Chang GY, Kim H, Lu F et al (2016) Understanding the physicochemical characteristics and the improved enzymatic saccharification of corn stover pretreated with aqueous and gaseous ammonia. Bioenergy Res 9:67–76
Chatel G, Rogers RD (2016) Review: oxidation of lignin using ionic liquids—an innovative strategy to produce renewable chemicals. ACS Sustain Chem Eng 2:322–339
Chauvette G, Heitz M, Rubio M et al (1985) TG/DTG as a rapid method for the characterization of solid residues derived from liquefaction of lignocellulosics. Thermochim Acta 84:1–5
Chen H (2014) Chemical composition and structure of natural lignocellulose. In: Biotechnology of Lignocellulose. Springer, Dordrecht. 25–71
Chen Y, Zhang X et al (2006) Overexpression of the wounding-responsive gene AtMYB15 activates the shikimate pathway in Arabidopsis. J Integr Plant Biol 48:1084–1095
Chen JY, Shimizu Y, Takai M et al (1995) A method for isolation of milled-wood lignin involving solvent swelling prior to enzyme treatment. Wood Sci Technol 29:295–306
Chen WJ, Zhao BC, Wang YY et al (2018) Revealing the topochemistry and structural features of lignin during the growth of Eucalyptus grandis × Eucalyptus urophylla. ACS Sustain Chem Eng 6:9198–9207
Chiaramonti D, Prussi M, Ferrero S et al (2012) Review of pretreatment processes for lignocellulosic ethanol production, and development of an innovative method. Biomass Bioenerg 46:25–35
Claudia C, Federica M, Marco S et al (2011) Milled wood lignin: a linear oligomer. Biomacromol 12:3928–3935
Colburn CB, Ettinger R, Johnson FA (1963) Isolation and storage of free radicals on molecular sieves. I. The electron paramagnetic resonance spectrum of nitrogen dioxide. Inorg Chem 2:1305–1306
Constant S, Lancefield CS, Weckhuysen BM et al (2017) Quantification and classification of carbonyls in industrial humins and lignins by 19 F NMR. ACS Sustain Chem Eng 5:965–972. https://doi.org/10.1021/acssuschemeng.6b02292
Cornu A, Besle JM, Mosoni P et al (1994) Lignin-carbohydrate complexes in forages: structure and consequences in the ruminal degradation of cell-wall carbohydrates. Reprod Nutr Dev 34:385–398
Crestini C, Lange H, Sette M et al (2017) On the structure of softwood kraft lignin. Green Chem 19:4104–4121
Crouvisier-Urion K, Bodart PR, Winckler P et al (2016) Biobased composite films from chitosan and lignin: antioxidant activity related to structure and moisture. ACS Sustain Chem Eng 4:6371–6381. https://doi.org/10.1021/acssuschemeng.6b00956
Cui C, Sadeghifar H, Sen S et al (2013) Toward thermoplastic lignin polymers; Part II: Thermal & polymer characteristics of kraft lignin & derivatives. 8:23. https://ojs.cnr.ncsu.edu/index.php/BioRes/article/view/3451
Dalimova GN, Abduazimov KA (1994) Lignins of herbaceous plants. Chem Nat Compd 30:146–159
Davin LB, Lewis NG (2005) Lignin primary structures and dirigent sites. Curr Opin Biotechnol 16:407–415
Davin LB, Wang HB, Crowell AL et al (1997) Stereoselective bimolecular phenoxy radical coupling by an auxiliary (dirigent) protein without an active center. Science 275:362–366
Débora T, Francesco C, Francesca R et al (2014) Complementary analytical liquid chromatography methods for the characterization of aqueous phase from pyrolysis of lignocellulosic biomasses. Anal Chem 86:11255–11262
Del Río JC, Prinsen P, Rencoret J et al (2012) Structural characterization of the lignin in the cortex and pith of elephant grass (Pennisetum purpureum) stems. J Agric Food Chem 60:3619–3634
Del Río JC, Rencoret J, Gutiérrez A et al (2018) Structural characterization of lignin from maize (Zea mays L.) fibers: evidence for diferuloylputrescine incorporated into the lignin polymer in maize kernels. J Agric Food Chem 66:4402–4413
Dellon L, Yanez-Mckay AJ, Li W et al (2017) Computational generation of lignin libraries from diverse biomass sources. Energy Fuels 31:8263–8274
Dence CW (1992) The determination of lignin. In: Methods in lignin chemistry. Springer, pp 33–61

Donaldson LA (2001) Lignification and lignin topochemistry—an ultrastructural view. Phytochemistry 57:859–873

Durbeej B, Eriksson LA (2003a) A density functional theory study of coniferyl alcohol intermonomeric cross linkages in lignin—three-dimensional structures, stabilities and the thermodynamic control hypothesis. Holzforschung 57:150–164

Durbeej B, Eriksson LA (2003b) Spin distribution in dehydrogenated coniferyl alcohol and associated dilignol radicals. Holzforschung 57:59–61. https://doi.org/10.1515/hf.2003.009

Ebert M, Law J (1965) Electron spin resonance spectra of polymethyl methacrylate after exposure to nitric oxide and nitrogen dioxide. Nature 205:1193–1196

Eduardo K, Paulo M, Sawaya ACHF (2012) Analysis of soluble lignin in sugarcane by ultrahigh performance liquid chromatography-tandem mass spectrometry with a do-it-yourself oligomer database. Anal Chem 84:7015–7020

Elder T, Fort RCJ (2010) Reactivity of lignin-correlation with molecular orbital calculations.In Heitner C, Dimmel DR, Schmidt JA eds. Lignin and Lignans: Advances in Chemistry. Boca Raton, FL: CRC Press. 321–347

Elder T, Berstis L, Beckham GT et al (2017) Density functional theory study of spirodienone stereoisomers in lignin. ACS Sustain Chem Eng 5: 7188–7194. https://doi.org/10.1021/acssuschemeng.7b01373

Fan X, Zhu JL, Zheng AL et al (2015) Rapid characterization of heteroatomic molecules in a bio-oil from pyrolysis of rice husk using atmospheric solids analysis probe mass spectrometry. J Anal Appl Pyrol 115:16–23

Fang Z, Mobley JK, Meier MS (2018) Dramatic simplification of lignin heteronuclear single quantum coherence spectra from ring-and-puck milling followed by oxidation. Energy Fuels 32:11632–11638

Faulon JL, Hatcher PG (1994) Is there any order in the structure of lignin? Energy Fuels 8:402–407

Fengel D, Wegener G, Fengel D et al (1984) Wood: chemistry, ultrastructure, reactions. Holz als Roh- und Werkstoff 42:314

Fernández-Costas C, Gouveia S, Sanromán MA et al (2014) Structural characterization of Kraft lignins from different spent cooking liquors by 1D and 2D nuclear magnetic resonance spectroscopy. Biomass Bioenerg 63:156–166

Fitzpatrick JD, Steelink C (1969) Benzosemiquinone radicals in alkaline solutions of hardwood lignins. Tetrahedron Lett 10:5041–5044

Focher B, Marzetti A, Crescenzi V (1991) Steam explosion techniques: fundamentals and industrial applications. In: Proceedings of the international workshop on steam explosion techniques: fundamentals and industrial applications, Milan, Italy, 20–21 Oct 1988. CRC Press

Forss K, Kokkonen R, Sagfors PE (2009) Determination of the molar mass distribution of lignins by gel permeation chromatography. In Lignin, ACS Symposium Series. 397, 124–133

Fort DA, Remsing RC, Swatloski RP et al (2007) Can ionic liquids dissolve wood? Processing and analysis of lignocellulosic materials with 1-*n*-butyl-3-methylimidazolium chloride. Green Chem 9:63–69

Fortin M, Beromi MM, Lai A et al (2015) Structural analysis of pyrolytic lignins isolated from switchgrass fast-pyrolysis oil. Energy Fuels 29:8017–8026

François I, Sandra P (2009) Comprehensive supercritical fluid chromatography × reversed phase liquid chromatography for the analysis of the fatty acids in fish oil. J Chromatogr A 1216:4005–4012

François I, Sandra K, Sandra P (2009) Comprehensive liquid chromatography: fundamental aspects and practical considerations—a review. Anal Chim Acta 641:14–31

Freudenberg K (1959) Biosynthesis and constitution of lignin. Nature 183:1152–1155

Freudenberg K, Harkin JM (1964) Ergänzung des Konstitutionsschemas für das Lignin der Fichte. Holzforschung-Int J Biol Chem Phys Technol Wood 18:166–168

Freudenberg K, Neish AC (1968) Constitution and biosynthesis of lignin. Constitution and biosynthesis of lignin

Galkin MV, Samec JS (2015) Selective route to 2-propenyl aryls directly from wood by a tandem organosolv and palladium-catalysed transfer hydrogenolysis. Chemsuschem 7:2154–2158
Gandini A, Belgacem MN (2008) Lignins as components of macromolecular materials. In: Monomers, polymers and composites from renewable resources. Elsevier, pp 243–71
Garrote G, Domínguez H, Parajó JC (1999) Hydrothermal processing of lignocellulosic materials. Holz als Roh- und Werkstoff 57:191–202
Gellerstedt G, Henriksson G (2008b) Lignins: major sources, structure and properties. In: Monomers, polymers and composites from renewable resources. Elsevier, pp 201–24
Gellerstedt G, Zhang L (2001) Chemistry of TCF-bleaching with oxygen and hydrogen peroxide. ACS Symp 785:61–72
Gidh AV, Decker SR, See CH et al (2006a) Characterization of lignin using multi-angle laser light scattering and atomic force microscopy. Anal Chim Acta 555:250–258
Gidh AV, Decker SR, Vinzant TB et al (2006b) Determination of lignin by size exclusion chromatography using multi angle laser light scattering. J Chromatogr A 1114:102–110
Gierer J (1980) Chemical aspects of kraft pulping. Wood Sci Technol 14:241–266
Glasser WG (2000) Classification of lignin according to chemical and molecular structure. ACS Symp 742:216–238
Glasser WG, Hsu OHH, Reed DL et al (1981) Lignin-derived polyols, polyisocyanates, and polyurethanes
Grabber JH (2005) How do lignin composition, structure, and cross-linking affect degradability? A review of cell wall model studies this paper was originally presented at the lignin and forage digestibility symposium, 2003 CSSA annual meeting, Denver, CO. Crop Sci 45:820–831. https://doi.org/10.2135/cropsci2004.0191
Grabber JH, Hatfield RD, Ralph J et al (1995) Ferulate cross-linking in cell walls isolated from maize cell suspensions. Phytochemistry 40:1077–1082
Grabber JH, Schatz PF, Kim H et al (2010) Identifying new lignin bioengineering targets: 1. Monolignol-substitute impacts on lignin formation and cell wall fermentability. BMC Plant Biol 10:114
Guo F, Shi W, Sun W et al (2014) Differences in the adsorption of enzymes onto lignins from diverse types of lignocellulosic biomass and the underlying mechanism. Biotechnol Biofuels 7:38
Hans Martin S, Walter T (2010) QM/MM methods for biomolecular systems. Angew Chem Int Ed 48:1198–1229
Hansen B, Kusch P, Schulze M et al (2016) Qualitative and quantitative analysis of lignin produced from beech wood by different conditions of the organosolv process. J Polym Environ 24:85–97
Harmatha J (2005) Structural abundance and biological significance of lignans and related plant phenylpropanoids. Chemicke Listy 99:622–32 doi: <Go to ISI>://WOS:000232095100002
Harmatha J, Zidek Z, Kmonickova E et al (2011) Immunobiological properties of selected natural and chemically modified phenylpropanoids. Interdiscip Toxicol 4:5–10. https://doi.org/10.2478/v10102-011-0002-1
Hasegawa M, Higuchi T (2008) Formation of lignin from glucose in Eucalyptus tree. Nihon Ringaku Kai Shi = Journal of the Japanese Forestry Society 42:305–08
Haslam E, Kirby G (1999) The Shikimate Pathway. Annu Rev Plant Physiol Plant Mol Biol 50:473–503
Hatakeyama H (1972) Thermal analysis of lignin by differential scanning calorimetry. Cellulose Chem Technol 6:521–529
Hatakeyama H, Nakano J (1970) Electron spin resonance studies on lignin and lignin model compounds. Cellulose chemistry and technology. 4:281–291
Hatakeyama H, Iwashita K, Meshitsuka G et al (1975) Effect of molecular weight on glass transition temperature of lignin. J Jpn Wood Res Soc. 21:618–623
Hatakeyma H, Nakano J, Hatano A et al (1969) Variation of infrared spectra with temperature for lignin and lignin model compounds. Berliner Und Münchener Tierärztliche Wochenschrift.52:1724–1728

Hatfield R, Vermerris W (2001) Lignin formation in plants. The dilemma of linkage specificity. Plant Physiol 126:1351–1357

Hatfield RD, Jung HJG, Ralph J et al (2010) A comparison of the insoluble residues produced by the Klason lignin and acid detergent lignin procedures. J Sci Food Agric 65:51–58

Haupert LJ, Owen BC, Marcum CL et al (2012) Characterization of model compounds of processed lignin and the lignome by using atmospheric pressure ionization tandem mass spectrometry. Fuel 95:634–641

Higuchi T (1990) Lignin biochemistry: Biosynthesis and biodegradation. Wood Sci Technol 24:23–63

Higuchi T, Ito Y, Kawamura I (1967) p -hydroxyphenylpropane component of grass lignin and role of tyrosine-ammonia lyase in its formation. Phytochemistry 6:875–881

Hollertz, Rebecca, Arwin et al. (2013) Dielectric properties of lignin and glucomannan as determined by;spectroscopic ellipsometry and Lifshitz estimates of non-retarded;Hamaker constants. Cellulose 20:1639–1648

Hsi-Chuan C, Quanzi L, Shuford CM et al (2011) Membrane protein complexes catalyze both 4- and 3-hydroxylation of cinnamic acid derivatives in monolignol biosynthesis. Proc Natl Acad Sci USA 108:21253–21258

Huang C, He J, Min D et al (2016) Understanding the Nonproductive Enzyme Adsorption and Physicochemical Properties of Residual Lignins in Moso Bamboo Pretreated with Sulfuric Acid and Kraft Pulping. Appl Biochem Biotechnol 180:1508–1523. https://doi.org/10.1007/s12010-016-2183-8

Humphreys JM, Chapple C (2002) Rewriting the lignin roadmap. Curr Opin Plant Biol 5:224–229

Hüttermann A, Mai C, Kharazipour A (2001) Modification of lignin for the production of new compounded materials. Appl Microbiol Biotechnol 55:387–394

Hyun LS, Doherty TV, Linhardt RJ et al (2010) Ionic liquid-mediated selective extraction of lignin from wood leading to enhanced enzymatic cellulose hydrolysis. Biotechnol Bioeng 102:1368–1376

International Conference on Thermal Analysis (1965) Nature 205:26–26

Janne A, Tarja T, Argyropoulos DS (2012) Accurate and reproducible determination of lignin molar mass by acetobromination. J Agric Food Chem 60:8968–8973

Jarrell TM, Marcum CL, Sheng H et al (2014) Characterization of organosolv switchgrass lignin by using high performance liquid chromatography/high resolution tandem mass spectrometry using hydroxide-doped negative-ion mode electrospray ionization. Green Chem 16:2713–2727

Jarvis MW, Daily JW, Hans-Heinrich C et al (2011) Direct detection of products from the pyrolysis of 2-phenethyl phenyl ether. J Phys Chem A 115:428–438

Jarvis J, Mckenna A, Hilten R et al (2012) Characterization of pine pellet and peanut hull pyrolysis bio-oils by negative-ion electrospray ionization Fourier transform ion cyclotron resonance mass spectrometry. Energy Fuels 26:3810–3815

Jorge R, Pepijn P, Ana G et al (2015) Isolation and structural characterization of the milled wood lignin, dioxane lignin, and cellulolytic lignin preparations from brewer's spent grain. J Agric Food Chem 63:603–613

Joseph Z, Bruijnincx PCA, Jongerius AL et al (2013) The catalytic valorization of lignin for the production of renewable chemicals. Chem Rev 110:3552–3599

Jung HG, Phillips RL (2010) Putative Seedling Ferulate Ester (sfe) Maize Mutant: Morphology, Biomass Yield, and Stover Cell Wall Composition and Rumen Degradability. Crop Sci 50:403–418. https://doi.org/10.2135/cropsci2009.04.0191

Jung HG, Buxton DR, Hatfield RD et al (1993) Application of new methods for the investigation of lignin structure

Jung HG, Mertens DR, Phillips RL (2011) Effect of reduced ferulate-mediated lignin/arabinoxylan cross-linking in corn silage on feed intake, digestibility, and milk production. J Dairy Sci 94:5124–5137

Kalogiannis KG, Stefanidis SD, Michailof CM et al (2015) Pyrolysis of lignin with 2DGC quantification of lignin oil: Effect of lignin type, process temperature and ZSM-5 in situ upgrading. J Anal Appl Pyrol 115:410–418
Karhunen P, Rummakko P, Sipilä J et al (1995) The formation of dibenzodioxocin structures by oxidative coupling. A model reaction for lignin biosynthesis. Tetrahedron Lett 36:4501–4504
Karhunen P, Mikkola J, Pajunen A et al (1999) The behaviour of dibenzodioxocin structures in lignin during alkaline pulping processes. Nord Pulp Pap Res J 14:123–128
Karl K, Ronald B (2012) Characterization of lignin by gas chromatography and mass spectrometry using a simplified CuO oxidation method. Anal Chem 84:459–464
Kerr RG (1995) Natural products: their chemistry and biological significance. Econ Bot 49:308
Khokarale SG, Le-That T, Mikkola JP (2016) Carbohydrate free lignin: a dissolution-recovery cycle of sodium lignosulfonate in a switchable ionic liquid system. Acs Sustain Chem Eng 4:2337–2344
Kibet J, Khachatryan L, Dellinger B (2012) Molecular products and radicals from pyrolysis of lignin. Environ Sci Technol 46:12994–13001
Klason P (1920) Beitrag zur Kenntnis der Konstitution des Fichtenholz-Lignins. Berichte der deutschen chemischen Gesellschaft (A and B Series) 53:1864–1873. https://doi.org/10.1002/cber.19200530935
Knaggs AR (2003) The biosynthesis of shikimate metabolites. Cheminform 34:119–136
Koivu KAY, Sadeghifar H, Nousiainen P et al (2016) The effect of fatty acid esterification on the thermal properties of softwood kraft lignin. Acs Sustain Chem & Eng 4:5238–5247. acssuschemeng.6b01048
Kolboe S, Ellefsen O (1962) Infrared investigations of lignin. A discussion of some recent results. Tappi 45:163–166
Kondo R, Sako T, Iimori T et al (1990) Formation of glycosidic lignin-carbohydrate complex in the enzymatic dehydrogenative polymerization of coniferyl alcohol. Mokuzai Gakkaishi = Journal Jpn Wood Res Soc 36:332–338
Košiková B, Miklešová K, Demianová V (1993) Characteristics of free radicals in composite lignin/polypropylene films studied by the EPR method. Eur Polymer J 29:1495–1497
Kratzl K, Zauner J (1962) Incorporation of radioactivity in wood studied by means of xylitol-C14. Holzforsch Holzverwert 14:108–111
Kumar S, Mohanty AK, Erickson L et al (2009) Lignin and its applications with polymers. J Biobased Mater Bioenergy 3:1–24
Lambert E, Aguié-Béghin V, Dessaint D et al (2020) Real time and quantitative imaging of lignocellulosic films hydrolysis by atomic force microscopy reveals lignin recalcitrance at nanoscale. Biomacromolecules. 20:515–527
Lange H, Decina S, Crestini C (2013) Oxidative upgrade of lignin—recent routes reviewed. Eur Polymer J 49:1151–1173. https://doi.org/10.1016/j.eurpolymj.2013.03.002
Lapierre C (1993) Application of new methods for the investigation of lignin structure. In: Jung HG et al. 1705 ed. Forage cell wall structure and digestibility. Madison, WI: ASA, CSCA, SSSA 133–166
Larsen KL, Søren B (2011) Environmental effects on the lignin model monomer, vanillyl alcohol, studied by Raman spectroscopy. J Phys Chem B 115:11470–11480
Lawther J, Sun R-C, Banks W (1996) Extraction and comparative characterization of ball-milled lignin (LM), enzyme lignin (LE) and alkali lignin (LA) from wheat straw. Cellul Chem Technol 30:395–410
Le DM, Nielsen AD, Sørensen HR et al (2017) Characterisation of authentic lignin biorefinery samples by fourier transform infrared spectroscopy and determination of the chemical formula for lignin. Bioenergy Res 1–11
Lee D, Hediger S, Paëpe GD (2015) Is solid-state NMR enhanced by dynamic nuclear polarization? Solid State Nucl Magn Reson 66–67:6–20
Lei X, Zhao Y, Li K (2012) Improved surface properties of CTMP fibers with enzymatic pretreatment of wood chips prior to refining. Cellulose 19:2205–2215

Leisola M, Pastinen O, Axe DD (2012) Lignin–designed randomness. bio-complexity 2012. https://doi.org/10.5048/bio-c.2012.3
Lenz B (1968) Application of nuclear magnetic resonance spcetroscopy to characterization of lignin. Tappi 51:511–519
Levitin N, Thompson N, Purves C (1955) The oxidation of spruce periodate lignin with sodium chlorite and with chlorine dioxide. Pulp and Paper Mag Can 56:117
Li S, Lundquist K (1994) A new method for the analysis of phenolic groups in lignins by ′H NMR spectrometry. 9:191–195. https://doi.org/10.3183/npprj-1994-09-03-p191-195
Li X, Ximenes E, Kim Y et al (2010) Lignin monomer composition affects Arabidopsis cell-wall degradability after liquid hot water pretreatment. Biotechnol Biofuels 3:27
Li M, Foster C, Kelkar S et al (2012) Structural characterization of alkaline hydrogen peroxide pretreated grasses exhibiting diverse lignin phenotypes. Biotechnol Biofuels 5:38
Li M, Si S, Hao B et al (2014) Mild alkali-pretreatment effectively extracts guaiacyl-rich lignin for high lignocellulose digestibility coupled with largely diminishing yeast fermentation inhibitors in Miscanthus. Biores Technol 169:447–454
Li M, Pu Y, Ragauskas AJ (2016) Current understanding of the correlation of lignin structure with biomass recalcitrance. Front Chem 4:45
Lisperguer J, Perez P, Urizar S (2009) Structure and thermal properties of lignins: characterization by infrared spectroscopy and differential scanning calorimetry. J Chil Chem Soc 54:460–63. https://scielo.conicyt.cl/scielo.php?script=sci_arttext&pid=S0717-97072009000400030&nrm=iso
Lora J (2008) Chapter 10—Industrial commercial lignins: sources, properties and applications. monomers polymers & composites from renewable resources, pp 225–241
Lourenço A, Gominho J, Marques AV et al (2013) Variation of lignin monomeric composition during kraft pulping of eucalyptus globulus heartwood and sapwood. J Wood Chem Technol 33:1–18
Lourenço A, Neiva DM, Gominho J et al (2015a) Biomass production of four Cynara cardunculus clones and lignin composition analysis. Biomass Bioenerg 76:86–95
Lourenço A, Neiva DM, Gominho J et al (2015b) Characterization of lignin in heartwood, sapwood and bark from Tectona grandis using Py–GC–MS/FID. Wood Sci Technol 49:159–175
Lourenço A, Rencoret J, Chemetova C et al (2016) Lignin composition and structure differs between Xylem, Phloem and Phellem inQuercus suberL. Frontiers in Plant Science 7:1612–1626
Lu F, Ralph J (1997) Derivatization followed by reductive cleavage (DFRC Method), a new method for lignin analysis: protocol for analysis of DFRC monomers. J Agric Food Chem 45:2590–2592. https://doi.org/10.1021/jf970258h
Lu FC, Ralph J (1998) The DFRC method for lignin analysis. 2. Monomers from isolated lignins. J Agric Food Chem 46:547
Lu FC, Ralph J (2003) Non-degradative dissolution and acetylation of ball-milled plant cell walls: high-resolution solution-state NMR. Plant J Cell Mol Biology 35:535–544
Lu Y, Wei XY, Cao JP et al (2012) Characterization of a bio-oil from pyrolysis of rice husk by detailed compositional analysis and structural investigation of lignin. Biores Technol 116:114–119
Lüdemann HD, Nimz H (1973) Carbon-13 nuclear magnetic resonance spectra of lignins. Biochem Biophys Res Commun 52:1162–1169
Lupoi JS, Singh S, Parthasarathi R et al (2015) Recent innovations in analytical methods for the qualitative and quantitative assessment of lignin. Renew Sustain Energy Rev 49:871–906
Luterbacher JS, Alonso DM, Dumesic JA (2014) Targeted chemical upgrading of lignocellulosic biomass to platform molecules. Green Chem 16:4816–4838
Lv X, Li Q, Jiang Z et al (2018) Structure characterization and pyrolysis behavior of organosolv lignin isolated from corncob residue. J Anal Appl Pyrolysis. 136:115–124
Ma J, Ji Z, Zhou X et al (2013) Transmission electron microscopy, fluorescence microscopy, and confocal raman microscopic analysis of ultrastructural and compositional heterogeneity of Cornus alba L. wood cell wall. Microsc Microanal 19:243–253
Ma X, Yang X, Zheng X et al (2015) Toward a further understanding of hydrothermally pretreated holocellulose and isolated pseudo lignin. Cellulose 22:1687–1696

Mann DGJ, Labbé N, Sykes RW et al (2009) Rapid assessment of lignin content and structure in switchgrass (*Panicum virgatum* L.) grown under different environmental conditions. Bioenergy Res 2:246–256
Mansfield SD, Kim H, Lu FC et al (2012a) Whole plant cell wall characterization using solution-state 2D NMR. Nat Protoc 7:1579–1589
Mansfield SD, Kang KY, Chapple C (2012b) Designed for deconstruction–poplar trees altered in cell wall lignification improve the efficacy of bioethanol production. New Phytol 194:91–101
Mansur HS, Mansur AAP, Bicallho SMCM (2005) Lignin-Hydroxyapatite/tricalcium phosphate biocomposites: SEM/EDX and FTIR characterization. Key Eng Mater 284–286:745–748
Marathe PS, Westerhof RJM, Kersten SRA (2019) Fast pyrolysis of lignins with different molecular weight: Experiments and modelling. Appl Energy. 236:1125–1137
Marco S, Rainer W, Claudia C (2011) Elucidation of lignin structure by quantitative 2D NMR. Chemistry 17:9529–9535
Marques G, Rencoret J, Gutiérrez A et al (2010) Evaluation of the chemical composition of different non-woody plant fibers used for pulp and paper manufacturing. Open Agric J 4:93–101
Martínez PM, Punt AM, Kabel MA et al (2016) Deconstruction of lignin linked p -coumarates, ferulates and xylan by NaOH enhances the enzymatic conversion of glucan. Biores Technol 216:44–51
Marton J (1964) On the structure of kraft lignin. Tappi 47:713–719
Masle AL, Angot D, Gouin C et al (2014) Development of on-line comprehensive two-dimensional liquid chromatography method for the separation of biomass compounds. J Chromatogr A 1340:90–98
Mattsson C, Andersson SI, Belkheiri T et al (2016) Using 2D NMR to characterize the structure of the low and high molecular weight fractions of bio-oil obtained from LignoBoost™ kraft lignin depolymerized in subcritical water. Biomass Bioenerg 95:364–377
Mbotchak L, Le MC, Duong KL et al (2015) Purification, structural characterization, and modification of organosolv wheat straw lignin. J Agric Food Chem 63:5178
Michailof C, Sfetsas T, Stefanidis S et al (2014) Quantitative and qualitative analysis of hemicellulose, cellulose and lignin bio-oils by comprehensive two-dimensional gas chromatography with time-of-flight mass spectrometry. J Chromatogr A 1369:147–160
Min D, Yang C, Rui S et al (2013) The elucidation of the lignin structure effect on the cellulase-mediated saccharification by genetic engineering poplars (Populus nigra L. × Populus maximowiczii A.). Biomass Bioenerg 58:52–57
Mollica G, Dekhil M, Ziarelli F et al (2015) Quantitative structural constraints for organic powders at natural isotopic abundance using dynamic nuclear polarization solid-state NMR spectroscopy. Angew Chem Int Ed Engl 127:6126–6129
Monteil-Rivera, F, Phuong et al (2013) Isolation and characterization of herbaceous lignins for applications in;biomaterials. Ind Crop & Prod 41:356–364
Montgomery JRD, Lancefield CS, Miles-Barrett DM et al (2017) Fractionation and DOSY NMR as analytical tools: from model polymers to a technical lignin. ACS Omega 2:8466–8474
Morreel K, Ralph J, Kim H et al (2004) Profiling of oligolignols reveals monolignol coupling conditions in lignifying poplar xylem. Plant Physiol 136:3537–3549
Morreel K, Dima O, Kim H et al (2010) Mass spectrometry-based sequencing of lignin oligomers. Plant Physiol 153:1464–1478
Morrow SL, Mascia P, Self KA et al (1997) Molecular characterization of a brown midrib3 deletion mutation in maize. Mol Breeding 3:351–357
Moxley G, Gaspar AR, Higgins D et al (2012) Structural changes of corn stover lignin during acid pretreatment. J Ind Microbiol Biotechnol 39:1289–1299
Müse G, Schindler T, Bergfeld R et al (1997) Structure and distribution of lignin in primary and secondary cell walls of maize coleoptiles analyzed by chemical and immunological probes. Planta 201:146–159
Nakagame S, Chandra RP, Saddler JN et al (2011) The influence of lignin on the enzymatic hydrolysis of pretreated biomass substrates. Acs Symp 1067:145–167

Natalie B, Pandey JL, Lucas D et al (2014) Development of a clickable designer monolignol for interrogation of lignification in plant cell walls. Bioconjug Chem 25:2189–2196

Neish A (1968) Monomeric intermediates in the biosynthesis of lignin. Constitution and biosynthesis of lignin, pp 3–43

Nimz H (2010) Beech lignin? Proposal of a constitutional scheme. Angew Chem Int Ed 13:313–321

Nimz H, Robert D, Faix O et al (1981) Carbon-13 NMR spectra of lignins, 8. Structural differences between lignins of hardwoods, softwoods, grasses and compression wood. Holzforsch-Int J Biol, Chem, Phys Technol Wood 35:16–26

Normark M, Winestrand S, Lestander TA et al (2014) Analysis, pretreatment and enzymatic saccharification of different fractions of Scots pine. BMC Biotechnol 14:20

Novo-Uzal E, Pomar F, Ros LVG et al (2012) Chapter 9—evolutionary history of lignins. Adv Bot Res 61(309):11–50

Nsimba RY, Mullen CA, West NM et al (2013) Structure-property characteristics of pyrolytic lignins derived from fast pyrolysis of a lignin rich biomass extract. Acs Sustain Chem & Eng 1:260–267

Ohra-Aho T, Linnekoski J (2015) Catalytic pyrolysis of lignin by using analytical pyrolysis-GC–MS. J Anal Appl Pyrol 113:186–192

Onorevoli B, Machado ME, Dariva C et al (2014) A one-dimensional and comprehensive two-dimensional gas chromatography study of the oil and the bio-oil of the residual cakes from the seeds of Crambe abyssinica. Ind Crops Prod 52:8–16

Osakabe K, Tsao CC, Li L et al (1999) Coniferyl aldehyde 5-hydroxylation and methylation direct syringyl lignin biosynthesis in angiosperms. Proc Natl Acad Sci USA 96:8955–8960

Ouyang XP, Chen ZL, Qiu XQ (2014) Determination of monophenolic compounds from lignin oxidative degradation using ultra performance liquid chromatography/high resolution mass spectrometry. Chin J Anal Chem 42:723–728

Owen BC, Haupert LJ, Jarrell TM et al (2012) High-performance liquid chromatography/high-resolution multiple stage tandem mass spectrometry using negative-ion-mode hydroxide-doped electrospray ionization for the characterization of lignin degradation products. Anal Chem 84:6000–6007

Papa G, Varanasi P, Sun L et al (2012) Exploring the effect of different plant lignin content and composition on ionic liquid pretreatment efficiency and enzymatic saccharification of Eucalyptus globulus L. mutants. Biores Technol 117:352–359

Passauer L, Salzwedel K, Struch M et al (2016) Quantitative analysis of the etherification degree of phenolic hydroxyl groups in oxyethylated lignins: correlation of selective aminolysis with FTIR spectroscopy. J Inst Brew 4:96–101

Patwardhan PR, Brown RC, Shanks BH (2011) Understanding the fast pyrolysis of lignin. Chemsuschem 4:1629–163

Patil SV, Argyropoulos DS (2017) Stable organic radicals in lignin: a review. ChemSusChem 10: 3284–3303

Perera PN, Martin S, Chiang VL et al (2012) Raman-spectroscopy-based noninvasive microanalysis of native lignin structure. Anal Bioanal Chem 402:983–987

Perras FA, Luo H, Zhang X et al (2016) Atomic level structure characterization of biomass pre and post lignin treatment by dynamic nuclear polarization-enhanced solid state NMR. J Phys Chem A 121:623

Petrik DL, Karlen SD, Cass CL et al (2014) p-Coumaroyl-CoA: monolignol transferase (PMT) acts specifically in the lignin biosynthetic pathway in Brachypodium distachyon. Plant J 77:713–726

Pillinger JM, Cooper JA, Harding CJ (1996) Stable free radical from plant litter decomposing in water. J Chem Ecol 22:1001–1011

Prasad K, Claus F (2010) Characterization of lignin during oxidative and hydrothermal pre-treatment processes of wheat straw and corn stover. Biores Technol 101:3175–3181

Prothmann J, Spégel P, Sandahl M et al (2018) Identification of lignin oligomers in Kraft lignin using ultra-high-performance liquid chromatography/high-resolution multiple-stage tandem mass spectrometry (UHPLC/HRMS n). Anal Bioanal Chem 410:7803–7814

Pu Y, Cao S, Ragauskas AJ (2011) Application of quantitative 31P NMR in biomass lignin and biofuel precursors characterization. Energy Environ Sci 4:3154. https://doi.org/10.1039/c1ee01201k
Purcaro G, Barp L, Beccaria M et al (2016) Characterisation of minor components in vegetable oil by comprehensive gas chromatography with dual detection. Food Chem 212:730–738
Qin L, Li WC, Zhu JQ et al (2015) Ethylenediamine pretreatment changes cellulose allomorph and lignin structure of lignocellulose at ambient pressure. Biotechnol Biofuels 8:174
Rahikainen JL, Martin-Sampedro R, Heikkinen H et al (2013) Inhibitory effect of lignin during cellulose bioconversion: the effect of lignin chemistry on non-productive enzyme adsorption. Bioresour Technol 133:270–278
Ralph J (2010) Hydroxycinnamates in lignification. Phytochem Rev 9:65–83
Ralph J, Landucci L (2010) Lignin and lignans: advances in chemistry. Heitner C:137
Ralph J, Lu F (1998) The DFRC method for lignin analysis. 6. A simple modification for identifying natural acetates on lignins. J Agric Food Chem 46:4616–4619
Ralph J, Lapierre C, Lu F et al (2001) NMR evidence for benzodioxane structures resulting from incorporation of 5-hydroxyconiferyl alcohol into Lignins of O-methyltransferase-deficient poplars. J Agric Food Chem 49:86–91
Ralph J, Lundquist K, Brunow G et al (2004) Lignins: natural polymers from oxidative coupling of 4-hydroxyphenyl- propanoids. Phytochem Rev 3:29–60
Rashid T, Chong FK, Regupathi I et al (2016) Dissolution of kraft lignin using Protic Ionic Liquids and characterization. Ind Crops Prod 84:284–293
Rencoret J, Marques G, Gutierrez A et al (2008) Structural characterization of milled wood lignins from different eucalypt species. Holzforschung 62:514–526. https://doi.org/10.1515/hf.2008.096
Rencoret J, Kim H, Evaristo AB et al (2017) Variability in Lignin Composition and Structure in Cell Walls of Different Parts of Macaúba (Acrocomia aculeata) Palm Fruit. J Agric Food Chem 66:acs.jafc.7b04638
Rex RW (1960) Electron paramagnetic resonance studies of stable free radicals in lignins and humic acids. Nature 188:1185–1186
Río JCD, Gutiérrez A, Hernando M et al (2005) Determining the influence of eucalypt lignin composition in paper pulp yield using Py-GC/MS. J Anal Appl Pyrol 74:110–115
Río JCD, Gutiérrez A, Rodríguez IM et al (2007) Composition of non-woody plant lignins and cinnamic acids by Py-GC/MS, Py/TMAH and FT-IR. J Anal Appl Pyrol 79:39–46
Río JC, Del Jorge R, Gisela M et al (2008) Highly acylated (acetylated and/or p-coumaroylated) native lignins from diverse herbaceous plants. J Agric Food Chem 56:9525–9534
Río JC, Del Jorge R, Pepijn P et al (2012) Structural characterization of wheat straw lignin as revealed by analytical pyrolysis, 2D-NMR, and reductive cleavage methods. J Agric Food Chem 60:5922
Robinson AR, Mansfield SD (2010) Rapid analysis of poplar lignin monomer composition by a streamlined thioacidolysis procedure and near-infrared reflectance-based prediction modeling. Plant Journal 58:706–714
Rodrigo CP, James WH, Zwier TS (2011) Single-conformation ultraviolet and infrared spectra of jet-cooled monolignols: p-coumaryl alcohol, coniferyl alcohol, and sinapyl alcohol. J Am Chem Soc 133:2632–2641
Rodrigues Pinto PC, Borges da Silva EA, Rodrigues AE (2012) Lignin as source of fine chemicals: vanillin and syringaldehyde. In: Baskar C, Baskar S, Dhillon RS (eds) Biomass conversion: the interface of biotechnology, chemistry and materials science. Springer Berlin Heidelberg, Berlin, Heidelberg, pp 381–420. https://doi.org/10.1007/978-3-642-28418-2_12
Rosler J, Krekel F, Amrhein N et al (1997) Maize phenylalanine ammonia-lyase has tyrosine ammonia-lyase activity. Plant Physiol 113:175–179. https://doi.org/10.1104/pp.113.1.175
Saito K, Kaiho A, Sakai R et al (2016) Characterization of the interunit bonds of lignin oligomers released by acid-catalyzed selective solvolysis of Cryptomeria japonica and Eucalyptus globulus woods via thioacidolysis and 2D-NMR. J Agric Food Chem 64:9152–9160

Sangha AK, Loukas P, Smith JC et al (2012) Molecular simulation as a tool for studying lignin. Environ Prog & Sustain Energy 31:47–54
Sannigrahi P, Ragauskas AJ, Tuskan GA (2010) Poplar as a feedstock for biofuels: a review of compositional characteristics. Biofuels, Bioprod Biorefin 4:209–226. https://doi.org/10.1002/bbb.206
Santoni I, Callone E, Sandak A et al (2015) Solid state NMR and IR characterization of wood polymer structure in relation to tree provenance. Carbohyd Polym 117:710–721
Santos RB, Capanema EA, Balakshin MY et al (2011a) Effects of hardwoods characteristics on kraft pulping process: empahsis on lignin structure. BioResources 6:3623
Santos RB, Capanema EA, Balakshin MY et al (2011b) Lignin structural variation in hardwood species. In: 16thInternational Symposium on Wood, Fiber and Pulping Chemistry pp 4923–4930
Sarkanen KV, Ludwig CH (1971) Lignins: occurrence, formation, structure and reactions. Lignins Occurrence Formation Structure & Reactions
Sathitsuksanoh N, Holtman KM, Yelle DJ et al (2014) Lignin fate and characterization during ionic liquid biomass pretreatment for renewable chemicals and fuels production. Green Chem 16:1236–1247
Savy D, Piccolo A (2014) Physical–chemical characteristics of lignins separated from biomasses for second-generation ethanol. Biomass Bioenerg 62:58–67
Schaefer J, Stejskal EOR (1976) Carbon-13 nuclear magnetic resonance of polymers spinning at the magic angle. J Am Chem Soc 98:1848–1854
Schonland DS (1959) On the determination of the principal g-values in electron spin resonance. Proc Phys Soc 73:788
Schuetz M, Smith R, Ellis B (2013) Xylem tissue specification, patterning, and differentiation mechanisms. J Exp Bot 64:11–31
Schulze F (1857) Beitrage zur Kenntniss des lignins. Chem Zentr 21:321–325
Sederoff R, Campbell M, O'Malley D et al (1994) Genetic regulation of lignin biosynthesis and the potential modification of wood by genetic engineering in loblolly pine. In: Ellis BE, Kuroki GW, Stafford HA (eds) Genetic engineering of plant secondary metabolism. Springer US, Boston, MA, pp 313–355. https://doi.org/10.1007/978-1-4615-2544-8_12
Seiji N, Chandra RP, Kadla JF et al (2011) Enhancing the enzymatic hydrolysis of lignocellulosic biomass by increasing the carboxylic acid content of the associated lignin. Biotechnol Bioeng 108:538–548
Selig MJ, Viamajala S, Decker SR et al (2007) Deposition of lignin droplets produced during dilute acid pretreatment of maize stems retards enzymatic hydrolysis of cellulose. Biotechnol Prog 23:1333
Senn HM, Thiel W (2007) QM/MM studies of enzymes. Curr Opin Chem Biol 11:182–187. https://doi.org/10.1016/j.cbpa.2007.01.684
Sevillano R, Mortha G, Barrelle M et al (2001) 19F NMR spectroscopy for the quantitative analysis of carbonyl groups in lignins. Holzforschung 55:286–295
Shi J, Gladden JM et al (2013) One-pot ionic liquid pretreatment and saccharification of switchgrass. Green Chem 15:2579–2589
Shi J, Pattathil S, Parthasarathi R et al (2016) Impact of engineered lignin composition on biomass recalcitrance and ionic liquid pretreatment efficiency. Green Chem 18:4884–4895. https://doi.org/10.1039/c6gc01193d
Simionescu C, Anton I (1970) study of some representative types of reed lignin. Cellul Chem Technol
Sluiter A, Hames B, Ruiz R et al (2008) Determination of structural carbohydrates and lignin in biomass. Lab Anal Proced 1617:1–16
Sofía C, Gaspar AR, Anderson G et al (2008) Propensity of lignin to associate: light scattering photometry study with native lignins. Biomacromol 9:3362–3369
Springer SD, He J, Chui M et al (2016) Peroxidative oxidation of lignin and a lignin model compound by a manganese salen derivative. Acs Sustain Chem Eng 4:3212–3219

Steelink C (1964) Free radical studies of lignin, lignin degradation products and soil humic acids. Geochim Cosmochim Acta 28:1615–1622
Steelink C (1965) Stable phenoxy radicals derived from phenols related to Lignin. J Am Chem Soc 87:2056–2057
Steelink C, Reid T, Tollin G (2002) On the nature of the free-radical moiety in Lignin. J Am Chem Soc 85:4048–4049
Stevenson PG, Tarafder A, Guiochon G (2012) Comprehensive two-dimensional chromatography with coupling of reversed phase high performance liquid chromatography and supercritical fluid chromatography. J Chromatogr A 1220:175–178
Stewart D (2008) Lignin as a base material for materials applications: Chemistry, application and economics. Ind Crops Prod 27:202–207
Studer MH, Demartini JD, Davis MF et al (2011) Lignin content in natural Populus variants affects sugar release. Proc Natl Acad Sci USA 108:6300–6305
Sun R (2010) Cereal straw as a resource for sustainable biomaterials and biofuels: chemistry, extractives, lignins, hemicelluloses and cellulose. Elsevier
Sun M, Lidén G, Sandahl M et al (2016) Ultra-high performance supercritical fluid chromatography of lignin-derived phenols from alkaline cupric oxide oxidation. J Sep Sci 39:3123–3129
Sun M, Sandahl M, Turner C (2018) Comprehensive on-line two-dimensional liquid chromatography × supercritical fluid chromatography with trapping column-assisted modulation for depolymerised lignin analysis. J Chromatogr A 1541:21–30. S0021967318301407
Taboada A, Novo-Uzal E, Flores G et al (2010) Digestibility of silages in relation to their hydroxycinnamic acid content and lignin composition. J Sci Food Agric 90:1155–1162
Takao K, Wakako C, Kaori S et al (2010) Influence of syringyl to guaiacyl ratio on the structure of natural and synthetic lignins. J Agric Food Chem 58:895–901
Tan SSY, Macfarlane DR, Upfal J et al (2009) Extraction of lignin from lignocellulose at atmospheric pressure using alkylbenzenesulfonate ionic liquid. Green Chem 11:339–345
Terrett OM, Dupree P (2019) Covalent interactions between lignin and hemicelluloses in plant secondary cell walls. Curr Opin Biotechnol 56:97–104. https://doi.org/10.1016/j.copbio.2018.10.010
Tessarolo NS, Silva RVS, Vanini G et al (2016) Characterization of thermal and catalytic pyrolysis bio-oils by high-resolution techniques: 1 H NMR, GC × GC-TOFMS and FT-ICR MS. J Anal Appl Pyrol 117:257–267
Tetsuo K, Takashi W, Jun-ichi A (1984) Existence of benzylated carbohydrate moiety in lignin–carbohydrate complex from pine wood. Chem Lett 13:1737–1740. https://doi.org/10.1246/cl.1984.1737
Tobimatsu Y, Schuetz M (2019) Lignin polymerization: how do plants manage the chemistry so well? Curr Opin Biotechnol 56:75–81. https://doi.org/10.1016/j.copbio.2018.10.001
Tobimatsu Y, Elumalai S, Grabber JH et al (2012) Hydroxycinnamate conjugates as potential monolignol replacements: in vitro lignification and cell wall studies with rosmarinic acid. Chemsuschem 5:676–686
Trajano HL, Engle NL, Foston M et al (2013) The fate of lignin during hydrothermal pretreatment. Biotechnol Biofuels 6:110
Upton BM, Kasko AM (2015) Strategies for the Conversion of Lignin to High-Value Polymeric Materials: Review and Perspective. Chem Rev 116:2275–2306 acs.chemrev.5b00345
Van EG, De VR, Dwh M et al (2017) Quantification of lignin and its structural features in plant biomass using (13)C lignin as internal standard for pyrolysis-GC-SIM-MS. Anal Chem 89:10907–10916. acs.analchem.7b02632
Vered T, Galili G (2010) New Insights into the Shikimate and Aromatic Amino Acids Biosynthesis Pathways in Plants. Molecular Plant 3:956–972
Wang G, Chen H (2013) Fractionation and characterization of lignin from steam-exploded corn stalk by sequential dissolution in ethanol–water solvent. Sep Purif Technol 120:402–409

Wang SZ, Xing F, Zheng AL et al (2014) Evaluation of the oxidation of rice husks with sodium hypochlorite using gas chromatography-mass spectrometry and direct analysis in real time-mass spectrometry. Anal Lett 47:77–90

Watanabe T, Koshijima T (1988) Evidence for an Ester Linkage between Lignin and Glucuronic Acid in Lignina Carbohydrate Complexes by DDQ-Oxidation. J Agric Chem Soc Jpn 52:2953–2955

Werner RA, Andreas R, Christine S et al (2004) Biosynthesis of gallic acid in Rhus typhina: discrimination between alternative pathways from natural oxygen isotope abundance. Phytochemistry 65:2809–2813

Wilkerson CG, Mansfield SD, Lu F et al (2014) Monolignol ferulate transferase introduces chemically labile linkages into the lignin backbone. Science 344:90–93

Windt M, Meier D, Marsman JH et al (2009) Micro-pyrolysis of technical lignins in a new modular rig and product analysis by GC–MS/FID and GC × GC–TOFMS/FID. J Anal Appl Pyrol 85:38–46. https://doi.org/10.1016/j.jaap.2008.11.011

Wu S, Argyropoulos DS (2004) An improved method for isolating residual kraft lignin in high yield and purity. J Tianjin Univ Sci Technol 19:108–114

Wu Z, Zhang M, Wang L et al (2013) Biomass digestibility is predominantly affected by three factors of wall polymer features distinctive in wheat accessions and rice mutants. Biotechnol Biofuels 6:183

Xie Y, Yasuda S, Wu H et al (2012) Analysis of the structure of lignin-carbohydrate complexes by the specific ^{13}C tracer method. J Wood Sci 46:130–136

Xin G, Laskar DD, Zeng J et al (2015) A ^{13}C CP/MAS-based nondegradative method for lignin content analysis. Acs Sustain Chem & Eng 3:153–162

Xiong F, Han Y, Wang S et al (2017) Preparation and formation mechanism of renewable lignin hollow nanospheres with a single hole by self-assembly. Acs Sustain Chem & Eng 5:2273–2281

Xu N (2012) Hemicelluloses negatively affect lignocellulose crystallinity for high biomass digestibility under NaOH and HSO pretreatments in Miscanthus. Biotechnol Biofuels 5:58

Xu C, Ferdosian F (2017) Conversion of lignin into bio-based chemicals and materials. Springer

Xu F, Sun RC, Zhai MZ et al (2010) Comparative study of three lignin fractions isolated from mild ball-milled Tamarix austromogoliac and Caragana sepium. J Appl Polym Sci 108:1158–1168

Xu W, Miller SJ, Agrawal PK et al (2012) Depolymerization and hydrodeoxygenation of switchgrass lignin with formic acid. Chemsuschem 5:667–675

Yang NC, Gaoni Y (2002) Charge-transfer interaction in organic polymers. J Am Chem Soc 86:5022–5023

Yang Q, Pan X (2016) Correlation between lignin physicochemical properties and inhibition to enzymatic hydrolysis of cellulose. Biotechnol Bioeng 113:1213–1224

Yang Q, Shi J, Lu L (2012) Characterization of structural changes of lignin in the process of cooking of bagasse with solid alkali and active oxygen as a pretreatment for lignin conversion. Energy Fuels 26:6999–7004

Yang L, Wang D, Zhou D et al (2016) Effect of different isolation methods on structure and properties of lignin from valonea of Quercus variabilis. Int J Biol Macromol 85:417–424

Yann LB, Luc D, Jesus R et al (2015) High resolution solid state 2D NMR analysis of biomass and biochar. Anal Chem 87:843–847

You TT, Mao JZ, Yuan TQ et al (2013) Structural elucidation of the lignins from stems and foliage of Arundo donax Linn. J Agric Food Chem 61:5361–5370

Yu Z, Tanya C, Brigitte P et al (2011) Impact of lignin structure and cell wall reticulation on maize cell wall degradability. J Agric Food Chem 59:10129–10135

Yu Z, Gwak KS, Treasure T et al (2014) Effect of lignin chemistry on the enzymatic hydrolysis of woody biomass. Chemsuschem 7:1942–1950

Yuan TQ, Sun SN, Xu F et al (2011) Characterization of lignin structures and lignin-carbohydrate complex (LCC) linkages by quantitative 13C and 2D HSQC NMR spectroscopy. J Agric Food Chem 59:10604–10614

Zeng JJ, Helms GL, Gao X et al (2013) Quantification of wheat straw lignin structure by comprehensive NMR analysis. J Agric Food Chem 61:10848–10857

Zhang L, Gellerstedt G (2001) NMR observation of a new lignin structure, a spiro-dienoneElectronic supplementary information (ESI) available: 13C, QUAT, HMBC and HSQC NMR spectra. See http://www.rsc.org/suppdata/cc/b1/b108285j. Chem Commun 2744–2745. https://doi.org/10.1039/b108285j
Zhang L, Gellerstedt G (2011) Quantitative 2D HSQC NMR determination of polymer structures by selecting suitable internal standard references. Magn Reson Chem Mrc 45:37–45
Zhang L, Gellerstedt G, Ralph J et al (2006) NMR studies on the occurrence of spirodienone structures in lignins. J Wood Chem Technol 26:65–79
Zhang A-P, Liu C-F, Sun R-C et al (2013) Extraction, purification, and characterization of lignin fractions from sugarcane bagasse. BioResources 8:1604–1614
Zhang J, Zhang H, Zhang J (2014) Evaluation of liquid ammonia treatment on surface characteristics of hemp fiber. Cellulose 21:569–579
Zhang L, Yin R, Mei Y et al (2017) Characterization of crude and ethanol-stabilized bio-oils before and after accelerated aging treatment by comprehensive two-dimensional gas-chromatography with time-of-flight mass spectrometry. J Energy Inst 90:646–659. https://doi.org/10.1016/j.joei.2016.04.009
Zhang B, Yin X, Wu C et al (2012) Structure and pyrolysis characteristics of lignin derived from wood powder hydrolysis residues. Appl Biochem Biotechnol 168:37–46
Zhao BC, Chen BY, Yang S et al (2016) Structural variation of lignin and lignin-carbohydrate complex in Eucalyptus grandis E.urophylla during its growth process. AcS Sustain Chem Eng 5:2273–2281. https://doi.org/10.1021/acssuschemeng.6b02396
Zhen F, Smith Jr RL (2016) Production of Biofuels and Chemicals from Lignin. In Lignin and Its Production, 263–287, Springer
Zobel BJ, van Buijtenen JP (1989) Wood variation: its causes and control. Wood Variation Its Causes & Control, Springer, Berlin
Zoia L, Perazzini R, Crestini C et al (2011) Understanding the radical mechanism of lipoxygenases using 31P NMR spin trapping. Bioorg Med Chem 19:3022–3028

Lignin Synthesis and Degradation

Vishal Ahuja and Raya Roy

Abstract Lignin, an amorphous hetero-polymer, deposited over the secondary cell walls to form an inflexible and impervious layer where it acts as structural barrier to protect the plant cells from chemical and biological invasions. Besides this, it also contributes to nutrients and water transportation through vascular bundles in plants. Lignin is an inseparable but least explored and under-utilized part of renewable biomass. Its recalcitrant nature owns to its phenolic, branched heteropolymeric structure. Therefore, it is necessary to segregate lignin from biomass in order to utilize the available sugars efficiently for commercial purposes. To fulfill the aim, various pretreatment methods have been evaluated, but most of them generate toxic residues and damage the sugar moieties in biomass as well. Biological depolymerization, however, offers a suitable and efficient alternative that improves the economics of lignocellulose-based industries by facilitating delignification under mild conditions, improving biomass fermentations besides generating valuable aromatic compounds as by-products. Ligninolytic enzymes have made significant contribution for developing an efficient bioprocess in various industries including paper and pulp, food, chemical industries and also in pollution control. The present chapter emphasized on the role of enzymes in lignin biosynthesis, modification and degradation along with brief description of cofactors that participate in lignin depolymerization.

Keywords Lignin · Delignification · Ligninolytic enzymes · Value-added products

1 Introduction

Lignin, one of the most important secondary metabolites, encompasses large group of aromatic polymers that are synthesized by oxidative coupling of 4-hydroxyphenyl propanoids in phenylalanine/tyrosine metabolic pathway (Boerjan et al. 2003). It forms an inflexible and impervious layer over the secondary cell walls under the influence of various abiotic and biotic factors. Besides providing mechanical support,

---

V. Ahuja (✉) · R. Roy
Department of Biotechnology, Himachal Pradesh University Summer Hill, Shimla, H.P., India
e-mail: vahuja3@gmail.com

S. Sharma and A. Kumar (eds.), *Lignin*, Springer Series on Polymer and Composite Materials, https://doi.org/10.1007/978-3-030-40663-9_3

it also forms stress response element for and plant defense against microbial decay (Tronchet et al. 2010). In addition, lignin also allows the transportation of nutrients and water through the vascular bundles in plants due to its hydrophobic nature. Earlier, it was proposed that the evolution of lignin is related to the adaptation of plants to a terrestrial habitat. Later, it was also found to protect marine algae-like *Calliarthron* from the pressure of breaking waves. Presence of lignin both in terrestrial and marine life forms reflects strong conserved and convergent evolutionary trend (Weng and Chapple 2010; Vanholme et al. 2012).

Monolignols including 4-hydroxycinnamyl alcohol (H), coniferyl alcohol (G) and 4-hydroxy-3, 5-dimethoxycinnamyl alcohol (S) (Fig. 1) are the major monomeric units for lignin synthesis (Vanholme et al. 2012; Lu et al. 2018) besides some other organic compounds including tricin flavones, hydroxy cinnamaldehydes, hydroxystilbenes, xenobiotics, etc. (Whetten and Sederoff 1995; Liu et al. 2018). While differentiation of tracheary elements (protoxylam) in *Arabidopsis*, lignin monomers can move freely to extracellular space of cells but their polymerization occurs only in secondary cell walls. In 1857 Schulz coined the term lignin after the discovery of an oxidizable incrustant by Payen that accompanies cellulose in wood. The relation between lignin and coniferyl aldehyde was proposed by Klasson in 1897. The elemental analysis revealed the presence of ether linkages between

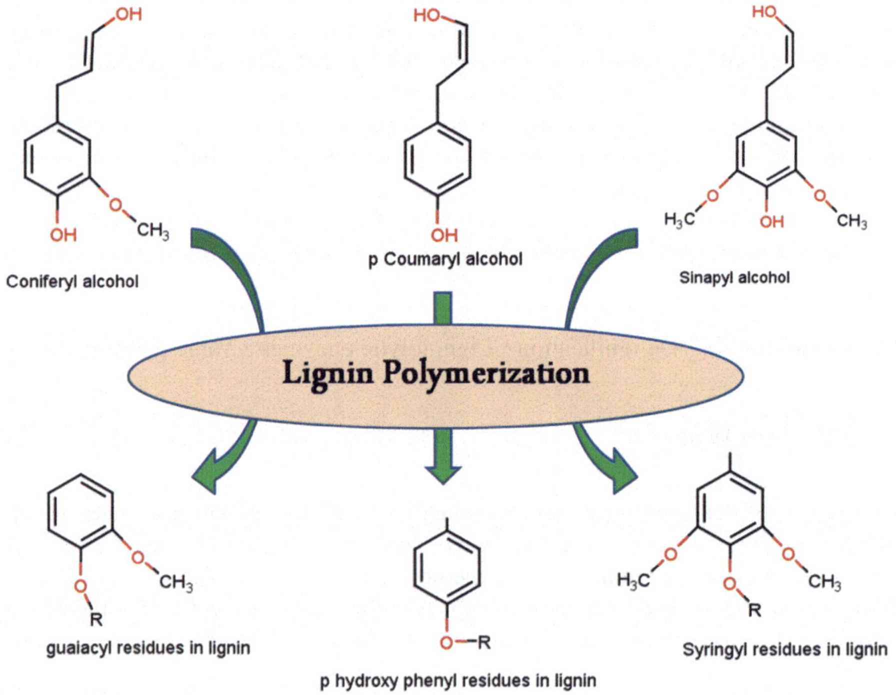

Fig. 1 Lignin monomers (monolignols) and their respective lignin residues

coniferyl alcohol monomers in lignin. In 1933, Professor H. Erdtman suggested that lignin is synthesized from 4-coumaryl alcohol by dehydrogenative polymerization. In the very same year, Freudenberg also reported similar findings from radioactive labeling of coniferin and phenylalanine to track their incorporation into the lignin while polymerization (van Parijs et al. 2010).

2 Lignin Biosynthesis

Lignin is one of the most prolific organic polymers of biological origin, accounts for about 30% of organic carbon content of biosphere (Ralph et al. 2004). The complex structure of lignin comprises three phenyl propanoid monomers: syringyl (S), guaiacyl (G) and p-hydroxyphenyl (H) units derived from sinapyl, coniferyl and 4-Hydroxycinnamyl alcohol, respectively. These monomers are synthesized from phenylalanine via various cinnamic acid derivatives (Taiz and Zeiger 2010; Liu et al. 2018). Besides these, a small fraction of p-coumaryl alcohol is also present, but its composition may vary among taxa, species and also among cell types (Nakashima et al. 2008; Vanholme et al. 2012).

Lignin exhibits enormously versatile nature and structure that vary among plant species, tissues as well as different age groups of same individual. In dicotyledons, lignin encompasses G and S units only, whereas in monocotyledons, it is composed of all three precursors: G, S and H units (Lu et al. 2017). Monolignols including coniferol are produced from phenylalanine via multistep phenylpropanoid pathway. The hydroxycinnamyl alcohol monomers: coniferyl, p-coumaryl and sinapyl alcohol undergo various degree of methylation to produce 'G', 'H' and 'S' phenylpropanoids which further condenses in various proportions to form lignin framework (Boerjan et al. 2003; Vanholme et al. 2012). The complex process of lignin biosynthesis is categorized into three steps (Table 1): (i) Synthesis of lignin monomeric units, (ii) transportation to the site of polymerization and (iii) polymerization.

2.1 Synthesis of Lignin Monomeric Units

Synthesis of lignin monomers (monolignols) is complex multistep process that occurs in cytoplasm (Fig. 2). Based on the origin and intermediate phases involved, the lignin synthesis completed in following three distinct steps:

2.1.1 Classical/Conventional Pathway for Lignin Biosynthesis

The classical model of monolignols biosynthesis is initiated with phenylalanine, produced from shikimate pathway. Monolignol synthesis proceeds via phenylpropanoid pathway catalyzed by series of enzymes. Genetic modulation

Table 1 Enzyme involve in lignin synthesis

Enzyme	EC number	Abbreviation	Gene name	Role in lignin synthesis
Phenylalanine ammonia-lyase	4.3.1.5	'PAL'	At2g37040 (PAL1), At3g53260 (PAL2)	Phenylalanine ammonia-lyase resides within microsomes and participates in deamination of phenylalanine to cinnamate. The conversion of phenylalanine to cinnamate is a committed and vital step in regulation of lignin biosynthesis. In vascular plants, 'PAL' exists in its tetrameric form
Cinnamate 4-hydroxylase/Cytochrome P-450-cinnamate 4-hydroxylase	1.14.14.91	'C4H'	At2g30490 (REF3)	Like PAL, C4H also resides inside microsomes. It catalyzes the hydroxylation of cinnamic acid to p-coumaric acid involving molecular oxygen. Out of two, only one oxygen atom transferred to the aromatic ring, and the second atom is utilized to form a water molecule
p-coumaryl coenzyme A synthetase/4-coumarate: Coenzyme A ligase	6.2.1.12	'4CL'	At1g51680 (At4CL1), At3g21240 (At4CL2)	It utilizes ATP for the generation of CoA thioesters from wide range of phenolic acids and alcohols including condensed tannins, benzoic acid, cinnamyl alcohols and flavonoids
Cinnamaldehyde: NADP+ oxidoreductase	1.2.1.44	'CCR'	At1g15950 (CCR1), At1g80820 (CCR2)	p-Coumaroyl-CoA undergoes two types of reaction: p-coumaroyl-CoA to p-coumarayldehyde and also feruloyl-CoA to coniferaldehyde
Hydroxy cinnamoyl-CoA shikimate/Quinate hydroxy cinnamoyl transferase	2.3.1.133	'HCT'	At5g48930	It has dual role to play: conversion of p-coumaroyl shikimate from p-coumaroyl-CoA and caffeoyl shikimate to caffeoyl-CoA

(continued)

Table 1 (continued)

Enzyme	EC number	Abbreviation	Gene name	Role in lignin synthesis
cinnamyl alcohol:NADP+ oxidoreductase	1.1.1.195	'CAD'	At3g19450 (CAD-C/AtCAD4/AtCAD-C), At4g34230 (CAD-D/AtCAD5/AtCAD-D)	CAD is an indicator of lignin biosynthesis. It catalyzes the production of lignin monomeric alcohol from their respective aldehyde. CAD expression induced in infected plants thus displaying both developmental and environmental regulatory roles
p-coumarate 3-hydroxylase	1.10.3.1	'C3H'	At2g40890 (REF8)	It is a copper-containing oxidase which catalyzes the caffeate synthesis by hydroxylation of p-coumarate. It can oxidize wide variety of substrates but preferably p-coumarate
S-adenosyl-L-methionine:caffeoyl-CoA 3-O-methyltransferase	2.1.1.104	'CCoAOMT'	At4g34050 (CCoAOMT1)	Caffeoyl-CoA O-methyltransferase was first identified enzyme responsible for defense response in several dicot plants which methylates caffeoyl-COA as well as 5-hydroxyferuloyl-COA
ferulate 5-hydroxylase	1.14.-.-	'F5H'	At4g36220 (FAH1)	Ferulate 5-hydroxylase is a kind of cytochrome P-450-linked monooxygenase that participates in hydroxylation of ferulate to 5-hydroxyferulate. Mutants lacking F5H resulted in sinapate-derived residues deficient lignin

(continued)

Table 1 (continued)

Enzyme	EC number	Abbreviation	Gene name	Role in lignin synthesis
caffeic acid O-methyltransferase	2.1.1.68	'COMT'	At5g54160 (COMT1/AtOMT1)	It transfers methyl group from S-adenosyl methionine to coniferaldehyde to form sinapaldehyde. It has direct influence on lignin content and monomer composition. Reduction in syringyl to guaiacyl subunits ratio detected in disrupted OMT gene expression in tobacco
Laccase	1.10.3.2	'LAC'	Lac 1	Laccase catalyzes the bioconversion of 4-(3-hydroxyprop-1-en-1-yl)-2-methoxyphenol, 4-hydroxycinnamyl alcohol and sinapic alcohol to guaiacyl (G), p-hydroxyphenyl (H) and syringyl (S) units, respectively

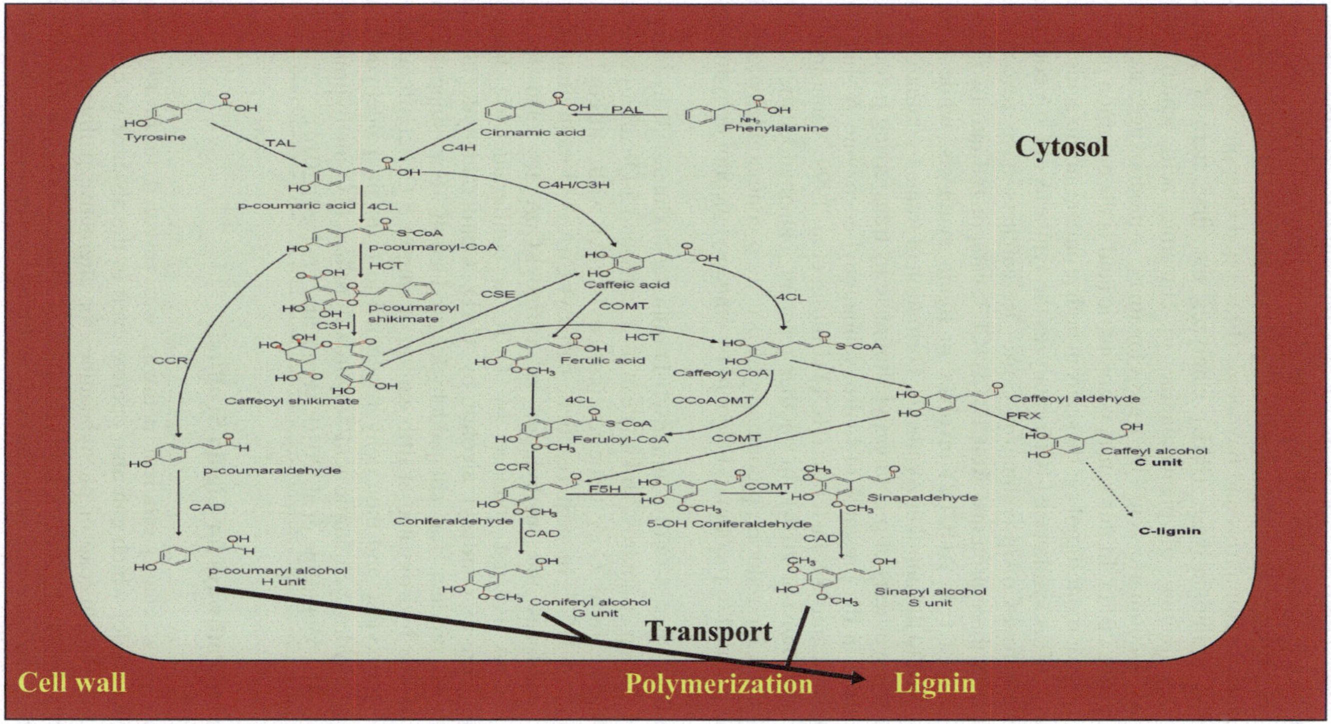

Fig. 2 Lignin biosynthesis from phenylalanine and tyrosine (PAL: phenylalanine ammonia-lyase; 4CL: 4-coumarate:CoA ligase; OMT: O-methyltransferase; CCR: Cinnamoyl-COA reductase; F5H: ferulate 5-hydroxylase; C4H: cinnamate 4-hydroxylase; CAD: Cinnamyl alcohol dehydrogenase)

of relative genes influences the composition and/or accumulation of lignin. This revealed the direct control of phenylpropanoid biosynthesis enzymes over the production of lignin monomers (Wang et al. 2018). Besides lignin monomers, phenylpropanoid pathway also plays significant role in synthesis of many secondary metabolites including coumarins and flavonoids (Fraser and Chapple 2011).

Phenylalanine ammonia-lyase (PAL) catalyzes the first step of pathway, nonoxidative deamination of phenylalanine to cinnamic acid which further hydroxylated to 'p-coumaric acid' by trans-cinnamate 4-monooxygenase (C4H) (Fraser and Chapple 2011). In this sense, p-coumaric acid is the common precursor as well as junction for monolignol synthesis from both aromatic amino acids. In next step, 4-coumaroyl-CoA synthetase (4CL) catalyzes the synthesis of CoA coumarate thioester '4-coumaroyl-CoA/p-coumaroyl-CoA.' Researchers have revealed the presence of at least three proteins: At4CL1-At4CL3, categorized into two classes. Only class I (At4CL1 & At4CL2) involves in lignin synthesis, while class II (At4CL3) is related to flavonoid biosynthesis. Formation of trans-4-coumaroyl-CoA is followed by the synthesis of p-coumaroyl shikimate by transferring p-coumaroyl group to shikimate in the presence of hydroxycinnamoyl-coenzyme A shikimate: quinate hydroxycinnamoyl-transferase (HCT) (Ehlting et al. 1999; Vanholme et al. 2012; Verma and Dwivedi 2014). In another reaction, cinnamaldehyde: $NADP^+$ oxidoreductase (CCR) reduces p-coumaroyl-CoA to p-coumaraldehyde which further dehydrogenated to p-coumaryl alcohol by cinnamyl alcohol dehydrogenase (CAD).

On the other side, 5-O-(4-coumaroyl)-D-quinate/shikimate 3′-hydroxylase (C3′H) hydroxylates p-coumaroyl shikimate at its 3′ carbon to 5-O-caffeoylshikimic acid which further form CoA thioester (caffeoyl-CoA) with caffeoyl moiety in the presence of HCT. In next reaction, caffeoyl-CoA 3-O-methyltransferase (CCoAOMT) catalyzes the methylation of caffeoyl-CoA to form feruloyl-CoA. Feruloyl-CoA is first reduced by CCR to coniferaldehyde followed by dehydrogenation to coniferyl alcohol by CAD. Coniferaldehyde also acts as intermediatory product for sinapaldehyde production. For sinapaldehyde, coniferaldehyde first hydroxyated at fifth carbon of ring to 5-OH coniferaldehyde by ferulate 5-hydroxylase and then dehydrogenated to form sinapaldehyde in the presence of 3,4-Dihydroxycinnamic acid/5-hydroxyferulic acid O-methyltransferase (Schoch et al. 2001; Vanholme et al. 2010; Verma and Dwivedi 2014).

2.1.2 Alternate Pathways for Lignin Synthesis

In recent work, new energy-efficient pathway was revealed in *Brachypodium distachyon* grass model which can also operate even in the absence of cinnamate 4-hydroxylase (C4H). Instead of phenylalanine, for alternative pathway, tyrosine acts as pioneer substrate for monolignols production which undergoes deamination to p-coumarate by bifunctional phenylalanine and tyrosine ammonia-lyase (PTAL) (Barros et al. 2016). The pathway is more efficient in terms energy in comparison with classical pathway as

- It involves fewer steps.
- Tyrosine bypass pathway can utilize available carbon and electrons directly for the biosynthesis of lignin.
- This pathway does not involve cinnamate production (Gregor 1976; Beaudoin-Eagan and Thorpe 1985).

2.1.3 C-Lignin Pathway

In seed coats of most Cactoidae genera and vanilla orchid, lignin is characterized by the presence of an unusual 'catechyl unit (C)' along with G, H and S units. For C-lignin synthesis, caffeic acid reduced to caffeyl alcohols which further undergoes radical coupling reactions and interconnected by benzodioxane bonds in linear fashion. Although its presence has been recorded, the mechanism of caffeyl alcohol synthesis and polymerization is still unclear. The regulation of C-lignin and conventional lignin synthesis is supposed to have different regulatory mechanism (Tobimatsu et al. 2013; Xie et al. 2018). Another interesting fact about C-lignin is its linearity which means there is fewer possibilities for its crosslinking with cellulose and hemicellulose due to rare side chains. It further enhances the biomass digestibility which may offer hope for biofuel production by bioengineering approach.

2.2 *Glycosylation of Monolignols Their Storage and Transport*

Generally, monolignols (coniferyl alcohol, sinapyl alcohol and p-hydroxycinnamyl alcohol) synthesized in cytoplasm are unstable in nature and toxic to cells therefore must not be accumulated in higher concentration. UDP-g1ucose: coniferyl alcohol glucosyltransferase (EC 2.4.1.111) glycosylated phenolic hydroxyl group of monolignols to form their respective 4-O glucosides (p-hydroxycinnamyl alcohol glucoside, coniferin and syringin) which are stable and nontoxic (Liu et al. 2018).

Lignin monomers are stored and transported in the form of glucosides; however, the transport mechanism is not well known. Several models have been proposed earlier to determine the storage and transportation. One of the models proposed, the monolignols are synthesized in the cytoplasm and transported across plasma membrane as 4-O glucosylated forms, i.e., syringin and coniferin to the site of glucosidases where these molecules are deglucosylated (Vanholme et al. 2010). Some plant species including conifers can store these glucosides. Freudenberg and Harkin (1963) extracted 'cambial sap' of Norway spruce and purified all three monolignol glucosides. In a comparative analysis between dicots and gymnosperms (including 19 dicot tree species from 15 genera), the presence monolignol glucoside was reported only from five species belongs to Magnoliaceae and Oleaceae families while all the gymnosperm species have coniferin (Terazawa et al. 1984). Other research groups have also detected monolignol glucosides in differentiating conifer xylem cells (Leinhos and Savidge 1993) and pine xylem cell wall (Leinhos and Savidge 1993; Liu 2012). These facts have revealed the possible role of vacuoles in monolignol glucosides is storage and transportation. But in contrast to this, angiosperms do not have the tendency to store monolignol glucosides and possibly follow some other mechanism for transportation (Terashima and Fukushima 1989). These results suggested that lignin biosynthesis has regulatory role on the storage of monolignol glucosides, and evolved angiosperms may have separate mechanism for monolignol glucoside synthesis, while less evolved angiosperms and gymnosperms have tendency to accumulate glucosides.

2.3 Oxidative Polymerization

Lignin polymerization, multistep process, proceeds via free radical intermediates in the presence of peroxidases and laccases. Erdtman (1933) suggested that lignin is synthesized by polymerization of monolignol precursors by an oxidative combinatorial mechanism. Later, Freudenberg (1965) proposed the two-step polymerization process. In the first step, monolignols oxidized to comparatively stable phenolic radicals in the presence of copper-containing oxygen-dependent oxidase (EC 1.10.3.2; 'laccase') and an H_2O_2-dependent hemoprotein (EC 1.11.1.7; 'peroxidases'). In a subsequent reaction, two monolignol radicals coupled by covalent bond (b-O-4, b-b and b-5) to form dehydro dimers. The mechanism for radical–radical coupling depends upon the nature of monomers as well as the conditions in cell wall (van Parijs et al. 2010).

Recent genetic analysis in *Arabidopsis* revealed that peroxidase PRX17 is decisive for lignin deposition. Knocking out of laccase genes (LAC4 and LAC17) from *Arabidopsis* reduced lignin by 20–40%, while further elimination of LAC11 has resulted in complete elimination of lignin from roots (Cosio et al. 2017; Chou et al. 2018). In addition, hemicellulose can also influence the organization of lignin precursors before polymerization (Atalla and Agarwal 1985; Houtman and Atalla 1995) (Fig. 3).

3 Lignin Degradation

Lignin, an aromatic polymer, comprised up to 30% of woody plant tissues. It is responsible for rigidity and protects cells and cell wall polysaccharides. Due to its complex nature and unavailability of hydrolyzable linkages, depolymerization of lignin is difficult. Although fungi are well-known decomposers, not all fungal genera are keen to degrade lignin. Brown rot fungi have limited lignin-modifying potential, but white-rot fungi have ability to mineralize it completely. Lignin degradation involves many oxidative and reductive extracellular enzymes (Table 2) which

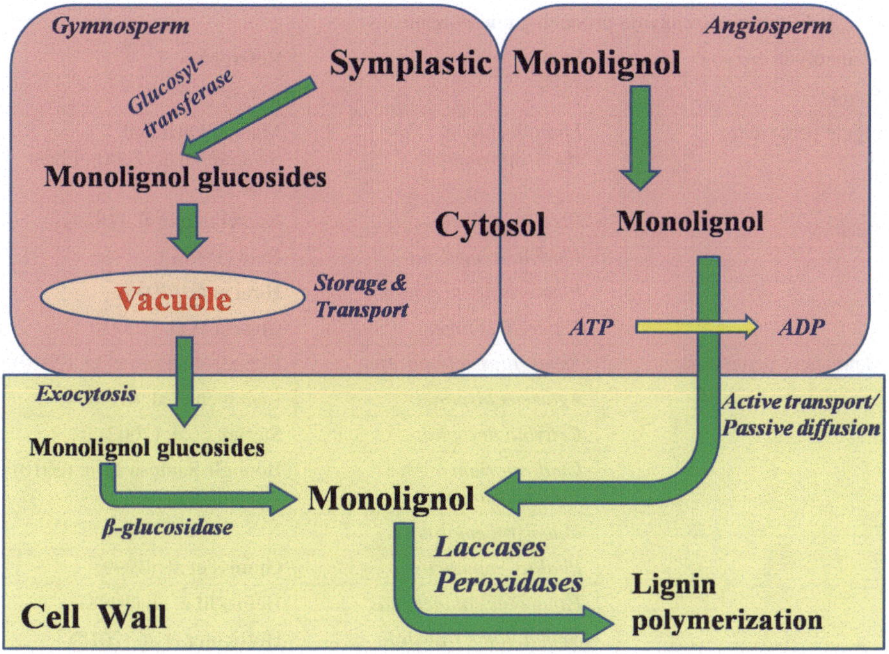

Fig. 3 Storage and transportation of lignin

operates via generation of reactive intermediates. Therefore, efforts are being made to harness the potential of these enzymes for delignification of biomass which further enhances the utilization of available sugar moieties for various industrial products. Besides delignification, ligninolytic microbes also aid to degradation of pollutants from wastewaters and soils (Sticklen 2008; Weng et al. 2008; Mansfield 2009).

Based on the action mechanism, the enzymes involved in lignin depolymerization are categorized into lignin-modifying enzymes (LME) and lignin-degrading auxiliary (LDA) enzymes. LME can modify the structure of polymer in different manner. It cannot degrade lignin by itself but assist LDA in the degradation process. These enzymes are further classified as peroxidases including metal peroxidases and laccase. The second class of enzyme is 'lignin-degrading auxiliary (LDA) enzymes.' As name suggests, LDA refers to the enzymes involved in degradation process. However, LDA is also not capable of degrading the lignin alone but necessary for complete degradation process (Coelho-Moreira et al. 2013).

In recent years, a new group of lignin-degrading metalloenzymes 'Heme-thiolate peroxidases' (HTPs) was introduced. In terms of catalytic potential, this new group has been considered as most versatile among hemeprotein family (Hofrichter and Ullrich 2006).

Table 2 Ligninolytic enzyme-producing microorganisms

Ligninolytic enzyme	Source	References
Fungi		
Lignin peroxidase	*Phanerochaete chrysosporium*	Manimekalai and Swaminathan (2000), Pham et al. (2018)
	Phlebia radiata	Kantelinen et al. (1989)
	Phlebia tremellosa	Reid (1992)
	Phanerochaete sordida	Hirai et al. (2005)
	Aspergillus niger	Conesa et al. (2000)
Manganese peroxidase	*Aspergillus sclerotiorum*	Bonugli-Santosa et al. (2010)
	Agaricus bisporus	Lankinen et al. (2005)
	Cettybia dryophila	Steffen et al. (2002)
	Cladosporium cladosporioides	Bonugli-Santosa et al. (2010)
	Mucor racemosus	
	Phallus impudicus	Gramss et al. (1999)
	Panaeolus sphinctrinus	Heinzkill et al. (1998)
	Nematoloma frowardii	Hofrichter et al. (2015)
	Ceriporiopsis subvermispora	Fernández-Fueyo et al. (2012)
Dye-decolorizing peroxidase	*Auricularia auricular-judae*	Strittmatter et al. (2015)
	Irpex lacteus	Duan et al. (2018)
Laccase	*Aspergillus sclerotiorum*	Bonugli-Santosa et al. (2010)
	Cladosporium cladosporioides	Aslam et al. (2012)
	Mucor racemosus	Balaji et al. (2013)
Versatile peroxidase (VP)	*Physisporinus vitreus*	Kong et al. (2017)
	Pleurotus ostreatus	Salame et al. (2012)
Bacteria		
Lignin peroxidase	*Bacillus megaterium*	Patil (2014)
Dye-decolorizing peroxidase A, B and C	*Amycolatopsis* sp.	Brown et al. (2012)
	Bacillus megaterium	Patil (2014)
	Bacillus subtilis	Min et al. (2015)
	Halobacterium salinarum	Chen and Li (2016)
	Pseudomonas putida	Santos et al. (2014)
	Rhodococcus jostii	Roberts et al. (2011)
	Streptococcus thermophilus	Zhang et al. (2015)
	Thermomonospora curvata	Chen et al. (2015)

(continued)

Table 2 (continued)

Ligninolytic enzyme	Source	References
Lac	*Azospirillum lipoferum, Bacillus halodurans, B. pumilus*	Narayanan and Murugan (2014)
	Bacillus subtilis	Muthukumarasamy et al. (2015)

3.1 Lignin-Modifying Enzymes

Peroxidases: Peroxidases oxidize inorganic as well as organic compounds by utilizing H_2O_2 as the electron acceptor. Based on the presence of heme group, enzymes have been categorized into heme (containing protoporphyrin IX) and nonheme peroxidases. Peroxidases have widely distributed among different organisms including microbes, animals and plants (Battistuzzi et al. 2010). After phylogenetic analysis, heme peroxidases have been classified into two superfamilies:

- Peroxidase belongs to cyclo-oxygenase superfamily and
- Peroxidase belongs to catalase superfamily.

Enzymes of catalase superfamily are further grouped into three different subgroups (Welinder 1992).

Class I: Intracellular peroxidases—These peroxidases are present inside the cells and protect host against toxic peroxides, e.g., ascorbate peroxidase (AP).
Class II: Extracellular fungal peroxidases—These peroxides are secreted by fungi and involve in lignin degradation, e.g., lignin peroxidase (LiP).
Class III: Phyto-peroxidases—These peroxidases are of phyto origin and involve in diverse functions including lignification, substrate oxidation, regulate the free radical generation and stress defense, e.g., horseradish peroxidase (HRP) classical secretory plant peroxidases.

Beside this, a new family of iron-containing peroxidases 'dye-peroxidases/dye-decolorizing peroxidases (DyP)' has been added in recent years which can participate in degradation of lignocellulose along with lignin degradation (Sugano 2009; Lundell et al. 2010). In further sections, second class of peroxidases (fungal peroxidases) has been discussed in detail.

3.1.1 Diaryl Propane Oxygenase (EC: 1.11.1.14)

'Diaryl propane oxygenase/lignin peroxidase (LiP)' is a glycosylated protein with iron protoporphyrin prosthetic group that catalyzes oxidative degradation nonphenolic units in H_2O_2-dependent manner. LiP came into light in 1983 with the discovery of lignin degradation route from *P. chrysosporium*. Later, similar kind of activity was also reported from other fungi including *Tramates versicolor*, *Trichoderma viride*, *Phlebia radiata*, *Phanerochaete sordida*, *Bjerkandera* sp. and *Phlebia tremellosa*. Besides white and brown rot fungi, LiP activity has also been reported from some bacterial groups including protobacteria and actinomycetes (Dashtban et al. 2010; Zamocky and Obinger 2010; Aarti et al. 2015; Falade et al. 2016).

In 1989, Farrell and his colleagues reported six different isoforms of LiP from culture broth of *P. chrysosporium* BKM-F-1767 and designated as H1–H10. Later, another isoform 'Ha' was also added in 1990 by Dass and Reddy. Variation with respect to substrate specificity, sugar composition, protein stability and N-terminal amino acid sequences suggested that these isoforms are expressed from different genes. In proceeding years, around ten LiP isozymes gene sequences have been explored from *P. chrysosporium* strain RP78 (Falade et al. 2016).

LiP is a monomeric, globular protein with a size of 50 × 40 × 40 Å (Piontek et al. 1993) of about 343 amino acid residues. 3D structure of the enzyme is stabilized with the help of four disulfide bridges along with two glycosylation sites and two Ca^{2+} binding sites (Edwards et al. 1993). The folding motif of LiP comprised of eight major and minor helices each and three short antiparallel β sheets. Heme is firmly fixed by H-bond to Asp-183 of apoenzyme and differentiated it into two structurally differentiated domains, i.e., proximal and distal domains. The weaker iron–nitrogen bond between heme and Nε2 atom of the proximal histidine residues makes heme more electrons deficient which is responsible for higher redox potential of LiP (Tien and Kirk 1983; Glumoff et al. 1990; Choinowski et al. 1999; Furukawa et al. 2014).

Catalytic Reactions

LiPs have wide range of organic substrates including phenolic as well as nonphenolic alcohol and other carbonyl compounds such as β-O-4 linkage-type arylglycerol-aryl ethers. LiP catalysis characteristics include one-electron oxidation, hydroxylation, demethylation, side-chain cleavage, intramolecular addition and rearrangements, etc. (Kirk et al. 1986; Harvey and Palmer 1990; Baciocchi et al. 2001; Wong 2009; Furukawa et al. 2014).

The oxidation ability of LiPs is due to its high reduction–oxidation potential (1.2 V at pH 3.0) which becomes much more prominent in the presence of veratryl alcohol (Christian et al. 2005; Sigoillot et al. 2012). The catalytic mechanism of LiP completes in one oxidation reaction followed by two reduction reactions (Abdel-Hamid et al. 2013) which can be summarized as follows (Fig. 4):

Step 1: H_2O_2 oxidizes native enzyme ((LiP)-Fe (III)) to Compound I 'oxo-ferryl intermediate (Fe(IV))';
Step 2: 'oxo-ferryl intermediate (Fe (IV))' reduced to Compound II by oxidizing nonphenolic aromatic substrate (A);
Step 3: In third step, Compound II is again reduced to reform the native ferric state by oxidizing substrate A again.

Enzyme activity is further modulated by its distinct potential to generated tryptophanyl radical from tryptophan residue (Trp171) exposed at the surface of protein (Ivancich et al. 2001).

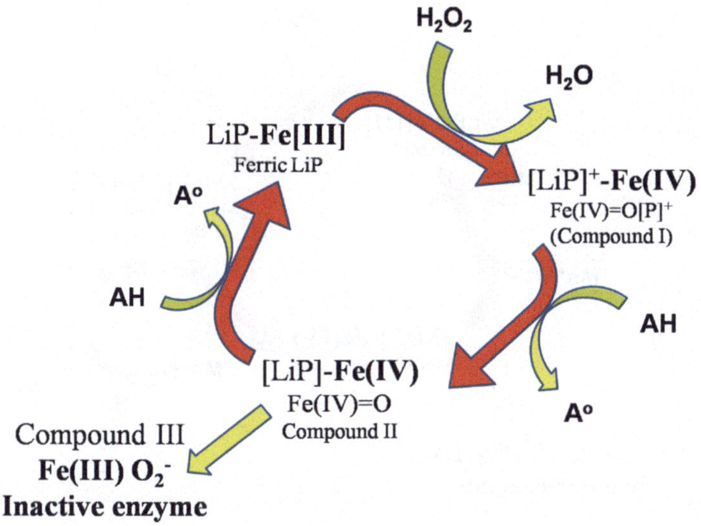

Fig. 4 Schematic diagram for lignin peroxidase catalyzed reactions

3.1.2 Mn(II): Hydrogen-Peroxide Oxidoreductase (EC 1.11.1.13)

Mn(II):hydrogen-peroxide oxidoreductase/manganese Peroxidase (MnP)/Mn-dependent (NADH-oxidizing) peroxidase, an important ligninolytic enzyme, produced by wood-rotting fungi. It causes much more degradation of phenolic lignin then laccase (Perez and Jeffries 1990; Ten Have and Teunissen 2001; Morgenstern et al. 2008). Likewise LiP, MnP is also heme-based glycoproteins which oxidizes substrates in H_2O_2-dependent manner. It was first detected in *P. chrysosporium,* but thereafter, its different isoforms have been detected from culture broth of other lignin-degrading fungi including *Ceriporiopsis subvermispora, Dichomeris squalens, Fusarium, Phanerochaete radiata* and *P. rivulosu* (Chowdhary et al. 2019).

MnPs are larger glycosylated hemeproteins with 47–60 kDa molecular weights. The native enzyme has high spin Fe^{3+} ligated to the proximal histidine by five bonds, five disulfide bridges and two Ca^{2+} ions. In inactive form, heme is sandwiched between two α-helices domains (Sutherland et al. 1997). Crystal structures revealed only one major Mn-binding site, located in proximity to the porphyrin macrocycle having one aspartate and two glutamate γ-carboxylic groups (Sundaramoorthy et al. 1994; Sigoillot et al. 2012).

Catalytic Mechanism

Similar to LiP, MnP uniquely utilizes Mn(II) as a reactant and generates Mn(III). The catalytic cycle is completed in following three steps (Fig. 5):

Step 1: In the first step, heme of MnP oxidized to Fe (IV) from Fe (III) in the presence of H_2O_2 and form compound I.

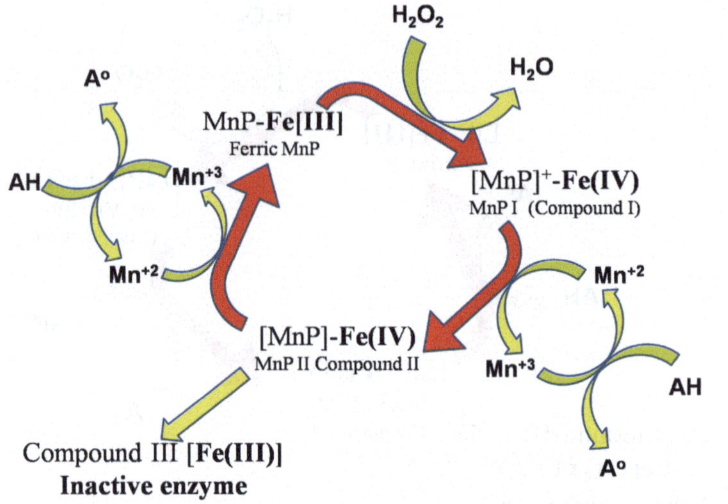

Fig. 5 Schematic diagram of manganese peroxidase catalytic activity

Step 2: Mn (II) reduces compound I to compound II and oxidizes to Mn (III).
Step 3: In the final step, Mn (II) again reduces compound II and oxidizes to Mn (III).

During the oxidation–reduction of compound I and II, phenolic compounds also act as electron donors, but the conversion rate is quite slow as phenolics are not that efficient in transformation of MnP compound II to its native enzyme. Moreover, the presence of Mn(II) is also necessary in the catalytic cycle as it acts as an obligatory redox coupler. Similar to the LiP, native form becomes inactive in the presence of excess of H_2O_2 (Wariishi et al. 1988; Manavalan et al. 2015).

Outside the enzyme system, organic acids including glyoxylate, fumarate, malate and oxalate stabilize Mn(III) by chelation and influence the liberation of Mn (III) resides on enzyme's active site. In addition, MnP also supports the oxidation of organic substrates by enacting as diffusible redox moderator (Chowdhary et al. 2019).

3.1.3 Polyvalent Peroxidase (EC 1.11.1.16)

'Polyvalent peroxidase/hybrid peroxidase/versatile peroxidase (VP)' is an important and least explored enzyme for lignin depolymerization which has the catalytic potential of both Mn-dependent peroxidases and diaryl propane oxygenase. Its diverse catalytic nature is due to its hybrid molecular design which allows the oxidation of wide range of phenolic and nonphenolic substrates with high as well as low redox potentials such as aromatic alcohol and azo-dyes without involving any external mediators (Camarero et al. 1999; Hofrichter and Ullrich 2010; Garcia-Ruiz et al. 2014).

VPs are also heme glycoproteins in nature that are secreted in several isoforms. It was first characterized from *Bjerkandera,* but other fungal genera including *Calocybe, Dichomitous, Lepista, Pleurotus, Panus Spongipellis* and *Trametes* have also been reported for the production of different isoforms of VPs (Ravichandran and Sridhar 2016).

Its molecular weight ranges from 40 to 45 kDa with a pI between 3.4 and 3.9 (Mester and Field 1998). Besides Mn^{2+}-binding site similar to MnPs, its molecular architecture was characterized with around one heme pocket, two structural Ca^{2+} sites, four disulfide bridges and 12 helices (Perez-Boada et al. 2005).

Catalytic Activity

VPs also utilize two transitional compounds I and II to complete its catalytic cycle like other peroxidases, but the mechanism is much more complex (Sigoillot et al. 2012). VPs' diverse catalytic nature can be understood by the presence of two active sites, one is tryptophanyl radical (Trp-164) at enzyme's surface, involved in direct oxidation of wide range of substrate with different redox potential, and other is exposed heme edge that catalyzes Mn^{2+} independent oxidation of amines and phenols (Ravichandran and Sridhar 2016).

3.1.4 p-Diphenol Oxidase (E.C. 1.10.3.2)

'p-Diphenol oxidase/Laccase (Lac)' is a metalloenzyme which oxidizes both inorganic and organic compounds including aromatic and phenolics. It is a metalloprotein, belongs to the family of polyphenol oxidases with copper cluster at catalytic site, and therefore, it is also known as blue multicopper oxidases (Fig. 6) (Shradsha et al. 2011).

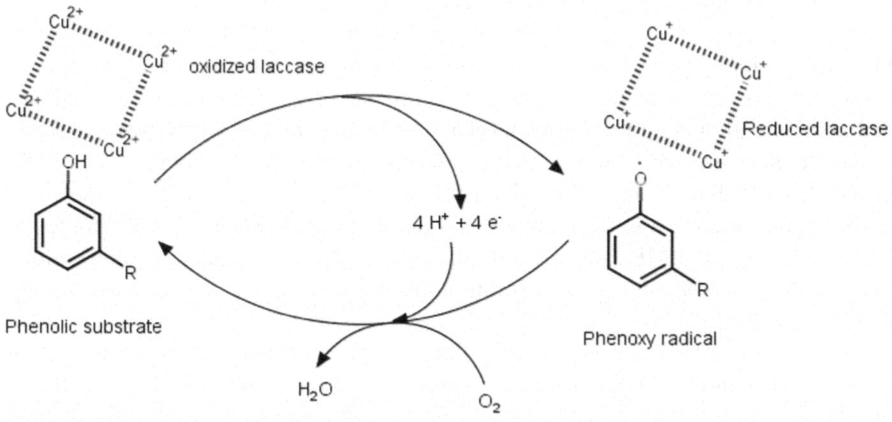

Fig. 6 Catalytic cycle of laccase

Initially, it was suggested that Lacs could oxidize only phenolic compound (Kersten and Cullen 1993), but later, it was found that Lacs can also oxidize wide variety of substance but in the presence of a mediator. Although, enzyme alone is sufficient to utilize even complex substrates, but interaction between active site and large-sized substrate molecules becomes tedious. Therefore, mediators facilitate the interaction between active sites and large-sized substrates (Li et al. 1999). Laccases are widely dispersed in higher plants, brown and white wood-rotting fungi, insects and bacteria such as *Streptomyces cyaneus, Streptomyces lavendulae* and *Marinomonas mediterranea.* However, the occurrence is more profound in fungi including *Theiophora terrestris, Pleurotus ostreatus, Trichoderma T. atroviride, T. harzianum, T. longibrachiatum* (Suderman et al. 2006; Shraddha et al. 2011).

Catalytic Activity
Laccases are the most vital constituent of the ligninolytic enzyme complex which oxidize a large number of aromatic and organo-metallic compounds via single electron transfer and produce free radicles (Brijwani et al. 2010). Lacs have multiple copper atoms exhibiting different spectroscopic characteristics (Shraddha et al. 2011):

Type 1 copper (T1Cu): It is an intense blue-colored enzyme (λ_{max} around 600 nm).
Type 2 copper (T2Cu): It is colorless showing poor absorption within the visible spectrum range.
Type 3 copper (T3Cu): λ_{max} around 330 nm.

In three-dimensional structure, T-1 provides a surface for attachment of reducing substrate, while T2/T3 forms tri-nuclear metal-cluster which facilitates the generation of water from O_2.

3.1.5 Dye-Decolorizing Peroxidase (DyP; EC 1.11.1.19)

'Dye-decolorizing peroxidase/reactive-blue-5:hydrogen-peroxide oxidoreductase' belongs to new family in heme-based peroxidase superfamily that catalyzes radical-mediated oxidation of lignin (Zámocký et al. 2015). Besides lignin, DyPs are also able to oxidize inorganic Mn^{+2} and aromatic compounds (Fernández-Fueyo et al. 2015). Although DyPs have catalytic mechanism relatable to manganese peroxidase and versatile peroxidases, it is phylogenetically distinct to LiP, MnP and VP. Moreover, DyPs and classical LMPs share similarities in catalytic site (heme pocket) which revealed the possible convergent evolution among DyPs and classical LMP family (Linde et al. 2015).

The difference between DyPs and other peroxidases can also be seen in their molecular structure as DyPs distinctly possess $\alpha + \beta$ ferredoxin-like fold. Its three-dimensional architecture of enzyme is differentiated into two structurally distinct domains, i.e., antiparallel β-sheets and α-helices. These structural moieties together form a cavity for binding of heme cofactor (Colpa et al. 2014).

DyPs were first discovered from *Bjerkandera adusta* (earlier known as *Geotrichum candidum*). It showed a tremendous potential to bleach a wide range of dyes, thus named 'dye-decolorizing peroxidase.' Later researches have suggested that DyPs are widely distributed among microorganisms including *Enterobacter lignolyticus*, *Auricularia delicate*, *Trametes versicolor* and *Irpex lacteus* (Fernandez-Fueyo et al. 2015; Shrestha et al. 2017; Duan et al. 2018).

DyPs predominate among prokaryotes like bacteria then fungi. Its activity in bacteria is supposed to be equivalent to fungal lignin-degrading peroxidases (Colpa et al. 2014). Phylogenetic analysis of genome sequence differentiated DyPs into four following subfamilies:

Type-A: DyPs in bacteria containing 'Tat-dependent signal sequence' which allow the enzyme to act outside the cytoplasm.
Type-B: It is mostly reported from the cytoplasm of Ascomycota and *Dictyostelium discoideum*.
Type-C: It encompasses intracellular DyPs similar to type B.
Type-D: This subclass is dominated by a fungal version of the enzyme, especially from Basidiomycota and Ascomycota.

Usually, Types A–C are predominated by bacteria, while Type D mainly belongs to fungi. All DyPs have peroxidase activities along with oxygenase or hydrolase activity (Sugano 2009).

3.2 Supplementary Enzymes for Lignin Degradation (LDEs)

Supplementary enzymes participate in different steps of lignin degradation involving various proteins and cofactors including ethanedial oxidase, veratryl alcohol oxidase, FAD-dependent glucose dehydrogenases, glucose 2-oxidase and cellobiose oxidoreductase.

3.2.1 Heme-Thiolate Haloperoxidases

Heme-thiolate haloperoxidases, the most adaptable and diverse biocatalysts, belong to heme protein family. Its versatile nature can be defined by its common catalytic characteristics with at least three classes of iron-containing oxidoreductases: cytochrome P450 mono-oxygenases, catalase and fungal and plant peroxidases. Post-translationally modified heme or heme-b acts as redox cofactor for heme peroxidases and hydrogen peroxide-mediated one- and two-electron oxidation of target substrate including coniferyl alcohol, tyrosine, Mn^{2+}, ascorbate or halides or even proteins such as cytochrome C (Fig. 7). Usually, four steps are involved in catalytic activity of peroxidases (Zámocký et al. 2015).

In the first step of catalytic cycle, H_2O_2 is reduced to water and oxidizes one-electron donors (AH_2) to respective radicals ($^{\cdot}AH$).

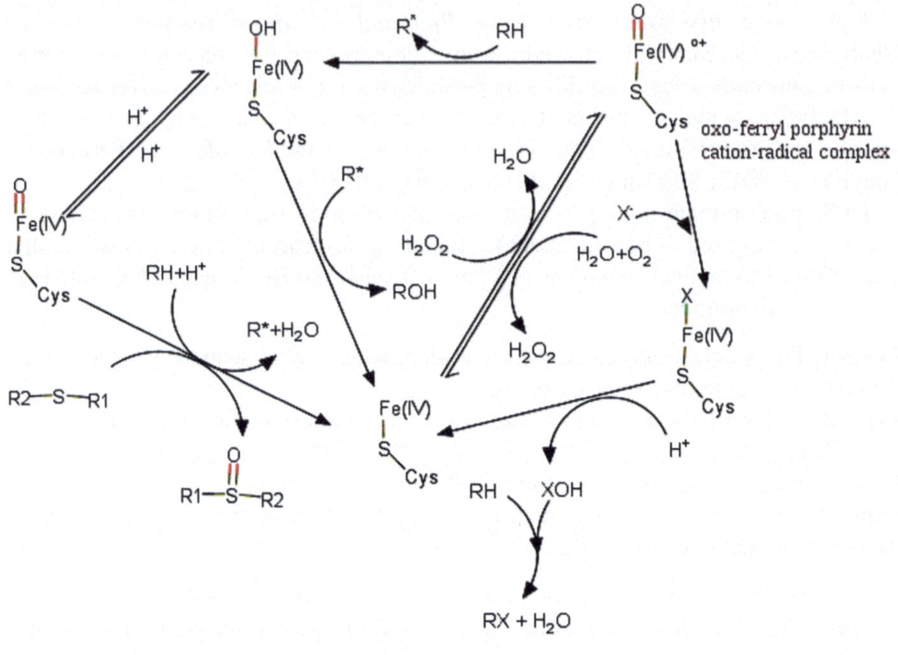

Fig. 7 Catalytic cycle of Heme-thiolate haloperoxidases

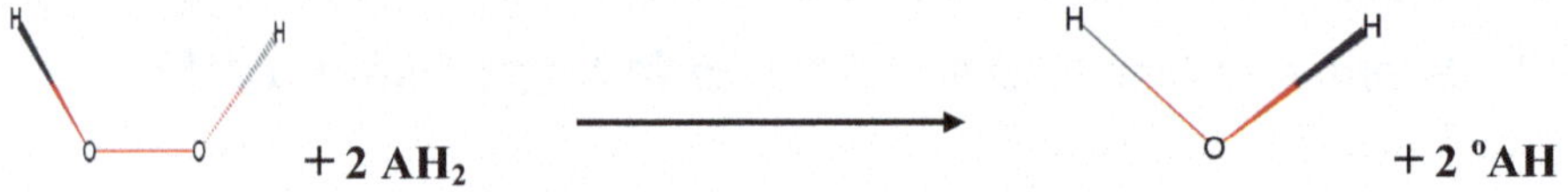

Second step involves the oxidation of two-electron donors like halides (X^-). Here, halide ions are oxidized hypohalous acids (HOX) and water.

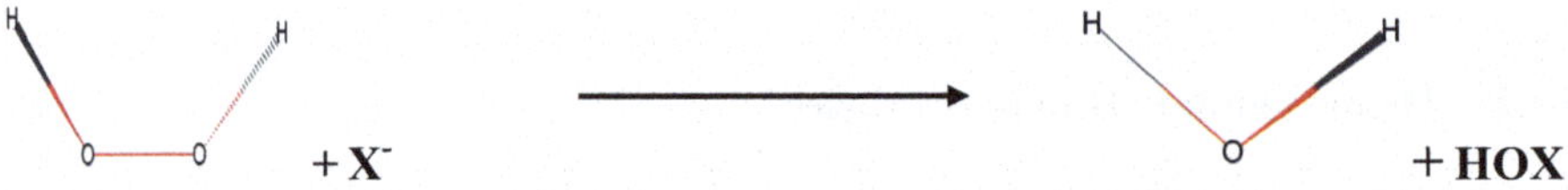

In the third step, peroxidases perform catalytic reaction and use second hydrogen peroxide molecule as electron donor.

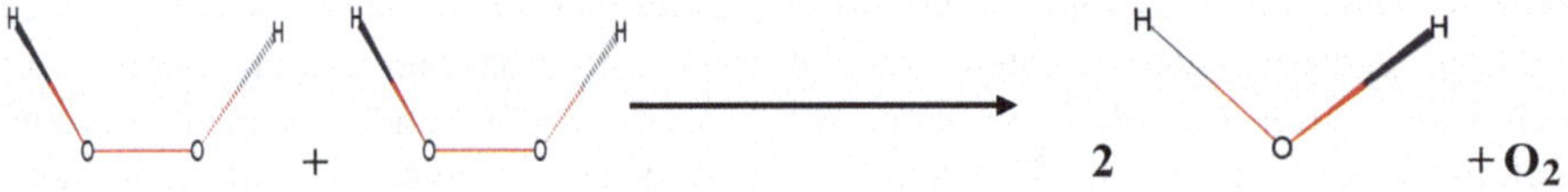

In the last step, organic molecules peroxygenated by the insertion of oxygen and formation of water.

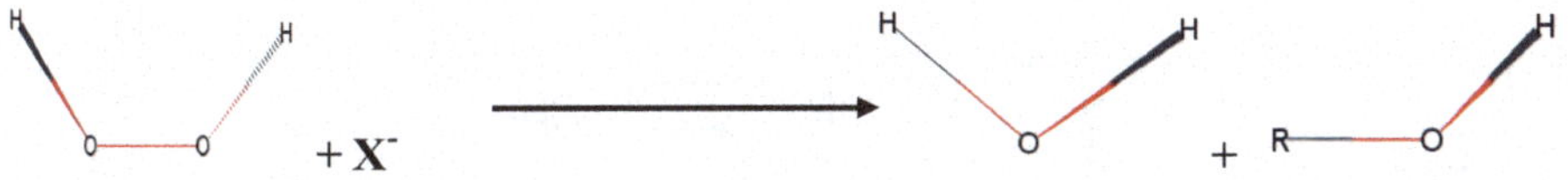

Based on the cations, haloperoxidases are grouped into two subclasses: chloroperoxidases (CPO) and aromatic/unspecific peroxygenases (APO).

Chloroperoxidases (EC 1.11.1.10; CPO): It is most widely studied halogenating enzymes, reported from Ascomycete *Caldariomyces* fumago for the first time by Hager and coworkers. It is carbohydrate-rich protein comprised of about 30% carbohydrate moieties with two N-glycosylation sites (Morris and Hager 1966; Kenigsberg et al. 1987). There are multiple forms 'isoenzymes' of CPO has been reported (Hashimoto and Pickard 1984). Besides chlorination and cleavage of lignin, its catalytic specialization also includes hydroxylation, epoxidation of C–H and double bonds and sulfoxidation (Wever and Barnett 2017).

Catalytic Activity

The basic catalytic cycle of CPO-induced halogenations can be epitomized in the following sections (Hager et al. 1966; Kellner et al. 2014).

- In native form, active site contains ferric protoporphyrin IX (heme Fe^{3+}).
- The cycle begins with heterolytic cleavage of hydrogen peroxide which transfers two electrons from heme.
- It forms CPO compound I 'oxo-ferryl porphyrin cation-radical complex' and liberates H_2O.
- CPO compound I highly unstable transitional state that further produces ferric hypo halite adduct 'compound X' (heme Fe^{3+}–O–X) on reacting with halide (X^-).
- Similarly, compound X also has low stability. It disintegrates into native enzyme and halonium ion (X+) that form hypohalous acid (HOX) on reacting with –OH (present in aqueous solution) as the counterpart (Griffin 1983).
- In presence of excess hypohalous acid at active site, the enzyme catalyzes selective halogenations.

Miscellaneous reactions of CPO: Apart from halogenations, CPO is also known for other halogens independent miscellaneous reactions including oxidation, epoxidation, hydroxylation, N-dealkylation and sulfoxidation. Some of these reactions hold special place in biotechnology as it prefers regio- and enatio-selective oxygenations of substrate molecules (Littlechild 1999) (Fig. 8).

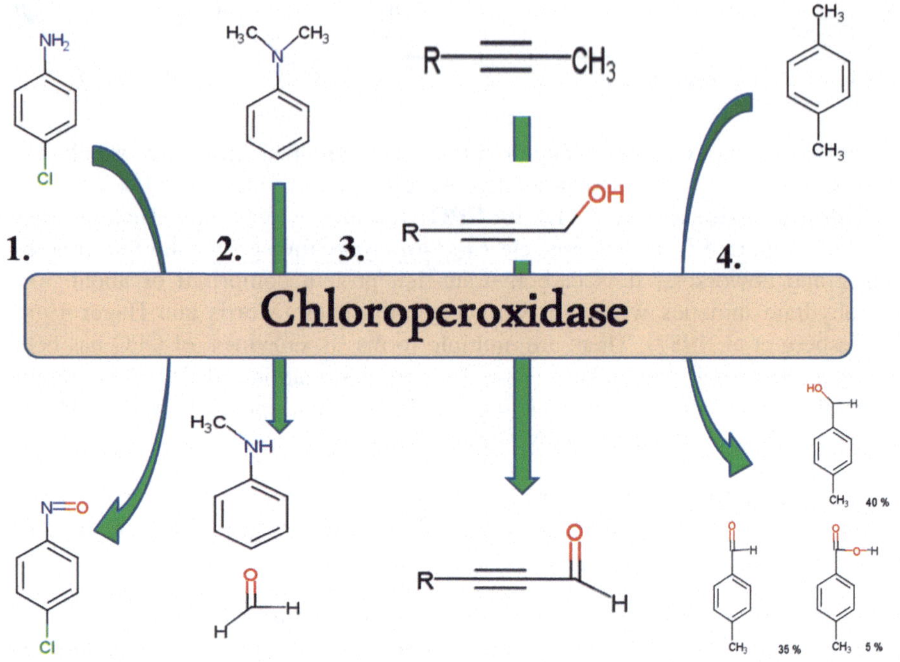

Fig. 8 Nonhalogenating reaction catalyzed by chloroperoxidases (1. Oxidation of amino group of chloro-aniline to nitroso compound, 2. 1 N dealkylation, 3. Propargylic hydroxylation, 4. Selective hydroxylation and oxidation)

Aromatic/unspecific peroxygenases (APO/UPO; EC 1.11.2.1): Peroxygenase is a new class of heme-containing peroxidases, dedicated to oxygen transfer, oxidation of aliphatic and aromatic compounds (Fig. 9). APO was first reported from *Agrocybe aegerita* (MW = 45 kDa) in 2004 for transferring oxygen. Initially, it was known for the hydroxylation of naphthalene, but after some time, its oxygen transfer potential was also revealed which resulted in renaming of enzyme from haloperoxidase to haloperoxidase–peroxygenase (Hofrichter and Ullrich 2010; Piontek et al. 2010; Wang et al. 2018). As new potential of enzyme discovered with the time, the name of the enzyme first changed from haloperoxidase (Ullrich et al. 2004) to haloperoxidase–peroxygenase (Kluge et al. 2007) followed by aromatic peroxygenase (*Agrocybe aegerita;* Ullrich et al. 2008) and finally to unspecific peroxygenase (UPO; Peter et al. 2013). UPOs contain histidine imidazole with heme-coordinated cysteine residue as a fifth ligand. Presence of basic ferryl group made it more reactive toward phenolic and benzylic substrates (Piontek et al. 2013; Wang et al. 2015).

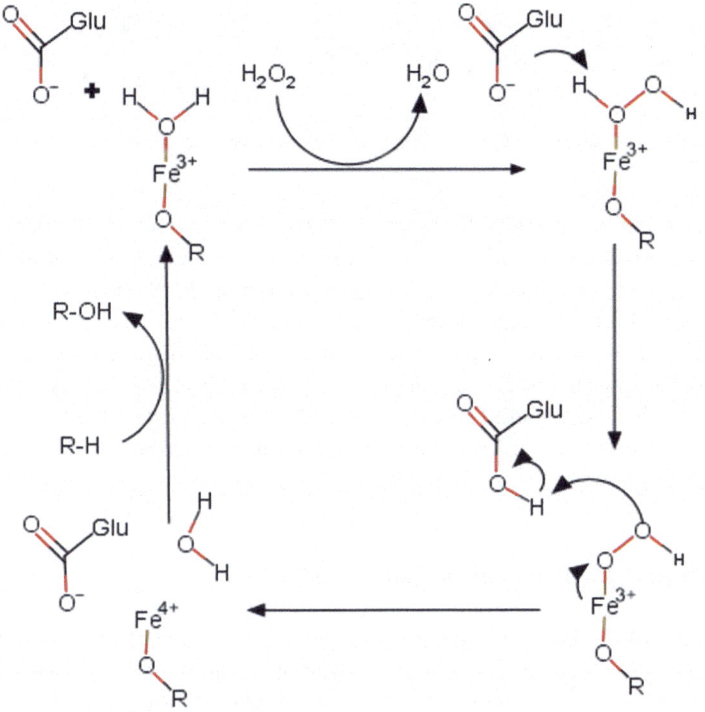

Fig. 9 Catalytic cycle of unspecific peroxygenases

3.2.2 Ethanedial Oxidase (EC 1.2.3.5)

Ethanedial oxidase/glyoxal oxidase (GLOx) is an extracellular copper-containing metalloenzyme that oxidizes a wide range of hydroxycarbonyl and dicarbonyl compound. It contains free radical-coupled copper active site, and therefore, Glox was listed among 'radical-copper peroxidases' (CRO) of 'auxiliary activity family 5 (AA5)' under 'carbohydrate activating enzyme (CAZymes)' (Yamada et al. 2014; Daou et al. 2016). Further research revealed the deviation in function of GLOx and galactose oxidase (1.1.3.9; GAO) belongs to same family AA5 even with similar tertiary structure and active site. This led to bifurcation of family into subfamilies AA5-1 and AA5-2. Former includes GLOx and enzymes with similar sequences, while later includes GAO enzymes (Daou and Faulds 2017). Spectroscopic analysis revealed the presence of phenoxyl radical-copper motif at the active site of both GLOx and GAO at which copper ion is synchronized with one tyrosine residue coupled cysteine residue and two histidine residues (Kadowaki et al. 2018).

Catalytic Activity
Although the exact role of GLOx is not clear, its activity is supposed to be important for the development of hyphal tip, fungal pathogenesis, cell wall stress response and

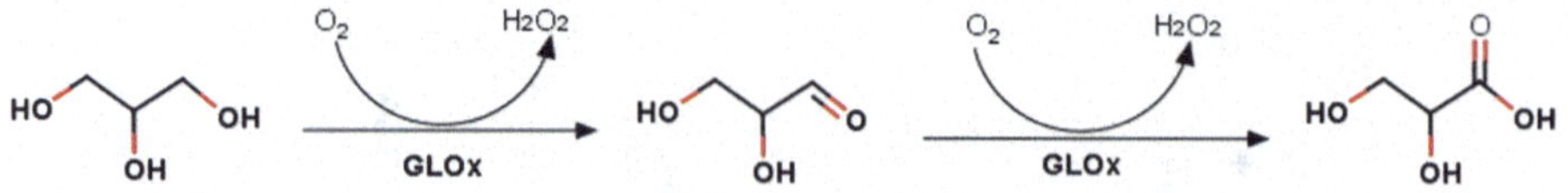

Fig. 10 Catalytic mechanism of glyoxal oxidase (oxidation of glycerol to glyceric acid)

integrity (Kerem et al. 1999; Song et al. 2016). GLOx oxidizes the aldehydes to carboxylic acids using O_2 as electron acceptor via two-electron oxidation, and in allied reaction, O_2 is reduced to H_2O_2 (Kadowaki et al. 2018) (Fig. 10).

Among various alcohol and carbonyl compounds, glyoxal has been reported to be the best substrates for GLOx production during glucose degradation. Besides this, GLOx is also known to have regulatory effect on peroxidase activity. As we know that glycol-aldehydes are potential substrates for GLOx, it suggested that the enzyme can possibly be used for detoxifying the broth by acting on inhibitory aldehydes (Daou and Faulds 2017).

3.2.3 Veratryl Alcohol Oxidase (EC 1.1.3.7)

Veratryl alcohol oxidase/aryl alcohol: oxygen oxidoreductase/aryl alcohol oxidase (AAO) is a flavoenzyme sharing common structural feature with glucose–methanol–choline (GMC) oxidoreductase family, i.e., FAD-binding domain along with ADP binding motif at N-terminal. It was first outlined from fungus *Polystictus versicolor/Trametes versicolor* in 1960. Later on, it was also reported from other microorganisms including *Bjerkandera adusta, Pleurotus eryngii, P. ostreatus, P. pulmonarius, Geotrichum candidum, Brachypsectra fulva, Fusarium solani* and *Botrytis cinerea* (Farmer et al. 1960; Goswami et al. 2013).

AAO is monomeric enzyme with molecular mass ranging from 66.7 to 334 kDa. The three-dimensional molecule has two distinct domains attached to FAD via noncovalent interaction (Guillen et al. 1992; Fernandez et al. 2009).

Catalytic Activity

AAO catalyzes oxidative dehydrogenation of aromatic as well as aliphatic alcohols to respective aldehydes and generates H_2O_2. AAO catalytic cycle divided into two halves: reductive half and oxidative half. Reaction cycle begins with oxidative dehydrogenation of the substrate, and the cycle is completed by the production of hydrogen peroxide from the oxidation of flavin (Ferreira et al. 2009; Goswami et al. 2013) (Fig. 11).

AAO catalysis proceeds transfer of hydride to flavin via electrophilic attack in first phase of the reaction. Simultaneously, molecular oxygen reduced to hydrogen peroxide (Ferreira et al. 2005). His-546 and His-502 present adjacent to FAD, and two aromatic residues are necessary for catalysis of AAOs (Ferreira et al. 2006).

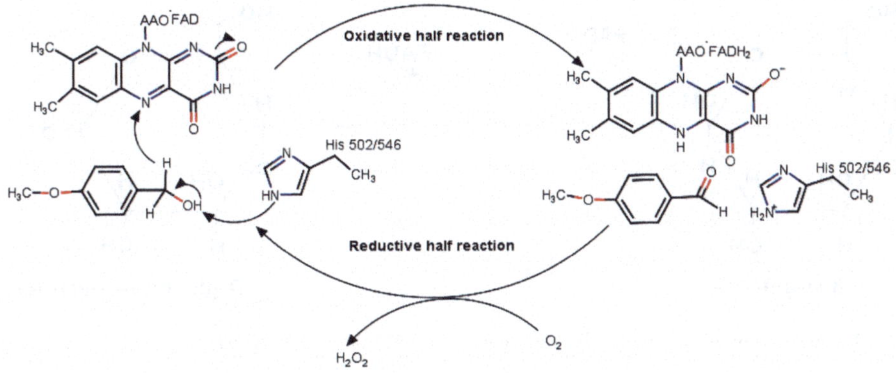

Fig. 11 Catalytic cycle of veratryl alcohol oxidase

3.2.4 FAD-Dependent Glucose Dehydrogenases (FADGDHs/GDH; EC 1.1.99.10)

FADGDHs are an oxidoreductase that utilizes FAD as the primary electron acceptor for the oxidation of first –OH group of sugar molecules. Based on origin, FADGDHs can be of three distinct types (Ferri et al. 2011).

- Fungal FAD-dependent GDHs.
- Bacterial membrane-binding protein (reported from Gram-negative bacteria).
- Insect cytosolic enzymes.

Three-dimensional structure of glucose dehydrogenase (GDH) from *Aspergillus flavus* revealed that the molecule is made up of two major domains. GDH are comparatively larger protein comprised of 18 helices and 20 β strands (Yoshida et al. 2015):

FAD-binding domain: Domain has four prominent structures:

- *Sandwich structure* consists of three layers (β-3, β-5 and α-3): first 'B9–B11,' second (B1–B2, B8, B12 and B20) and third (H1, H5 and H18)
- *Short helices* encompass eight α helices (H2–H4, H6–H9 and H17)
- *Irregular β sheet* composed of 'B3–B4 & B6–B7' and
- *A long loop* containing comprised of short antiparallel β sheet (B17 and B18).

C-terminal domain: It is composed of both types of structures including six-antiparallel β sheet (B5, B13–B16 and B19) surrounded by six α helices (H11–H16) accompanied by short α helix 'H10'.

FADGDH catalyzed the removal of hydrogen from anomeric carbon, oxidizes alcohol to ketone and reduces FAD to $FADH_2$ (Fig. 12). Among these three classes discussed above, fungal FADGDH has gained much attention as it is secreted extracellularly and participates in lignin degradation. Besides this, GDHs also play significant role in reducing quinones, produced by plants and protecting the fungal cell from phenoxy radicals. Generally, GDHs use FAD as external electron

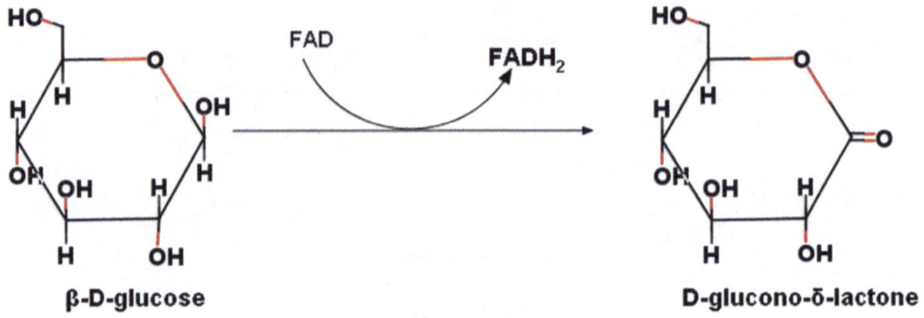

Fig. 12 Oxidation of glucose to glucono-lactone by FAD-dependent GDH

acceptors but can also use iron complexes such as ferricyanide and ferrocenium hexafluorophosphate, phenoxy radicals, quinones, redox dyes but never uses oxygen. Due to its narrow substrate specificity and oxygen insensitivity, FADGDH has been used for blood glucose self-monitoring (SMBG) (Ferri et al. 2011; Piumi et al. 2014; Pongtharangkul et al. 2015).

3.2.5 Glucose 2-Oxidase (EC 1.1.3.10)

Glucose 2-oxidase/pyranose oxidase/pyranose: oxygen 2-oxidoreductase/flavin adenine dinucleotide (FAD)-dependent oxidase (POX) is a hydrogen peroxide generating enzyme that can mineralize lignin completely. It is homotetrameric protein of 300 kDa, localized in periplasmic space in association with membrane-bound vesicles of wood-degrading fungi (de Koker et al. 2004; Prongjit et al. 2009; Pitsawong et al. 2010; Wongnate and Chaiyen 2013). Its three-dimensional structure displayed three domains that are conserved throughout the evolution:

- Prosthetic group-flavin attachment loop,
- substrate-binding region and
- FAD-binding site.

The catalytic cycle of POx completed into two following steps (Fig. 13):

- **Oxidation of carbohydrate by FAD**: In the first step, carbohydrate get oxidized to respective ketone/aldehyde
- **Regeneration of reduced FAD**: In second step, molecular oxygen gets reduced to hydrogen peroxide to generate native FAD.

This is the only enzyme that can reduce pyranoses at C2 position. POx has wide range of substrate specificity which includes D-xylose, L-sorbose and D-glucose (Freimund et al. 1998).

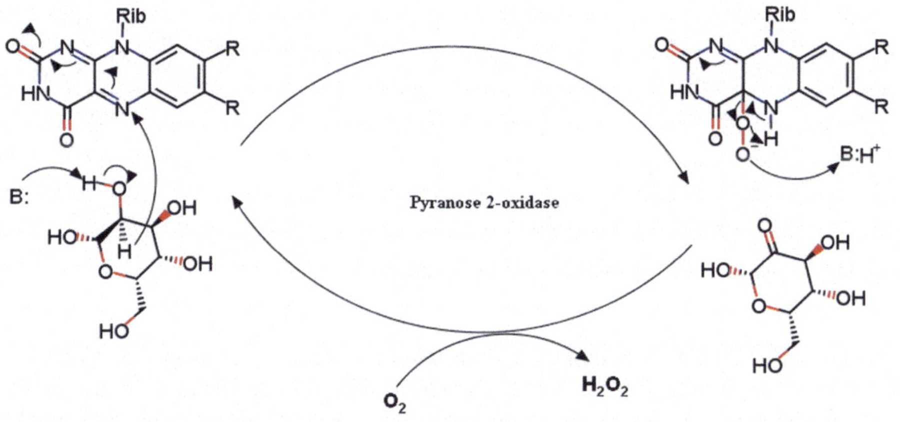

Fig. 13 Proposed mechanism of glucose 2-oxidase activity

3.2.6 Cellobiose Oxidoreductase (CDH-EC 1.1.99.18)

Cellobiose oxidoreductase/cellobiose dehydrogenase/Cellobiose: acceptor 1-oxidoreductase is an extracellular glycoprotein, secreted by wood-degrading fungi which oxidizes lactose, cellodextrins and maltodextrins to their respective lactones using a wide range of electron acceptors including Fe^{3+}, Cu^{2+}, triiodide ion and organic compounds such as phenoxy radicals and quinones via ping–pong mechanism. In 1974, Westermark and Eriksson first reported reduction of quinones by cellobiose from wood rot fungi. CDH is mostly secreted by wood rot fungi, but no bacteria and brown rot fungi except *Coniophora puteana* has been reported to produce CDH. In contrast to celluloses degradation, ligninolysis required Fe^{3+} and O_2 (Henriksson et al. 2000; Harreither et al. 2011; Kracher and Ludwig 2016).

A large number of genomes have been sequenced which categorized CDH sequences into three classes:

Class-I: Its presence was reported from Basidiomycota. Class I enzymes have strong affinity for cellulose due to the presence of cellulose-binding module similar to cellulolytic enzymes along with catalytic domain.

Class-II: These enzymes secreted by Ascomycota can utilize glucose as a substrate and can work efficiently at neutral pH. Based on the presence of CBM, Class II enzymes are categorized into class IIA having type 1 CBM and class IIB lacking the CBM1 domain.

Class-III: No functional sequences have been expressed that belong to class III CDHs.

Molecular structure of CDHs has a bipartite domain organization, encoded by single polypeptide chains (Hallberg et al. 2002; Kracher et al. 2015; Kracher and Ludwig 2016):

C-terminal cytochrome-binding fragment (CYT) domain: β-sandwich-like structural topology of CYT domain (190 amino acid) forms compact globular

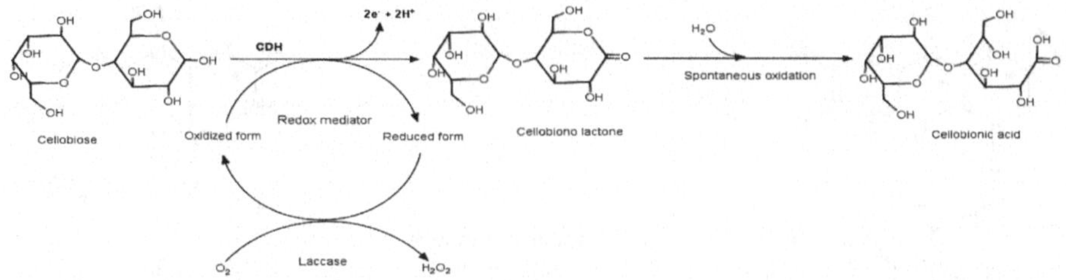

Fig. 14 Catalytic cycle of CDH and oxidation of cellobiose

structure of 35 Å. CYT domain has a hydrophobic pocket on its surface to bind with heme-b. Heme has hexacoordinated interaction with Met & His and propionate-A arm of the heme cofactor can have limited contact with surrounding solvent. In the active three-dimensional structure, CYT along with heme-b oriented toward active site, while propionate-A chain approaches substrate-binding pocket (Kracher and Ludwig 2016).

Catalytic flavo-dehydrogenase (DH) domain: The structure of DH domain resembles with p-hydroxybenzoate hydroxylase (PHBH)-fold, characteristics of flavin-containing enzymes. DH domain is further subdivided into two distinct subdomains:

Substrate-binding domain (S-domain) that interacts with the substrate and flavin-binding domain (F-domain) which contains a βαβ-motif and binds with FAD/NAD by noncovalent interaction (Hallberg et al. 2002).

Catalytic activity: CDH belongs to AA3 group of CAZy leads to generation of $FADH_2$ from FAD. Catalytic cycle of CDH consists of two half reaction phases: oxidative and reductive half reactions (Kracher and Ludwig 2016).

Oxidative half reaction: In the first half of the reaction, C1 position of a saccharide oxidized to lactone from hemiacetal. Generated lactone is further hydrolyzed to carboxylic acid (cellobionic acid). During the course of reaction, electron movement is facilitated by electron acceptors.

Reductive half reaction: The second half of catalytic cycle involves the regeneration of redox mediator. This phase may also involve some other enzymes such as laccase (Fig. 14).

Besides enzymes, small-sized molecules ions/organic molecules (Table 3) play important role in catalytic activity of ligninolytic enzyme such as MnP and LiP (Bourbonnais and Paice 1992; Flournoy et al. 1993; Bourbonnais et al. 1995).

4 Way Forward

Under the light of some sophisticated tools of genetic engineering, polymer remodeling and in silico analysis have opened new possibilities for lignin biology. One of the approaches includes the incorporation of mono/oligomeric molecules

Table 3 Cofactors and mediators for lignin degradation

Cofactor/mediators	Formula	Role in lignin degradation	References
Manganese	Mn^{2+}/Mn^{3+}	• Mn^{2+} used as inducer for MnP production • Mn^{2+} scavenges peroxide radical, reduces oxidative stress • Mn^{3+} aids oxidation of phenolic substrates	Datta et al. (2017), Rothschild et al. (1999)
3,4-Dimethoxybenzyl alcohol	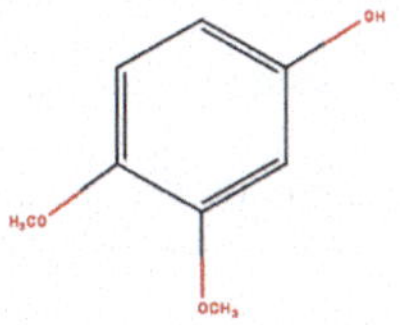	• VA is the physiological substrate of LiP • Its production fro glucose is influenced by secondary lignin metabolites • VA is the key factor for LiP-mediated/catalyzed metabolic reaction	Lundquist and Kirk (1978), Datta et al. (2017)
Oxalate	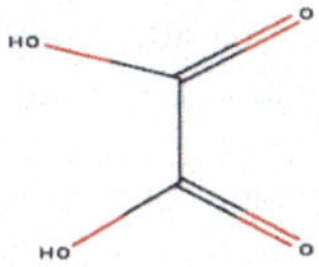	• It is one of the major aliphatic organic acids secreted by white-rot fungi, which further generated super-oxide on decomposition and participates in lignin oxidation	Shimada et al. (1997), Datta et al. (2017)
2-Chloro-1, 4-dimethoxybenzene	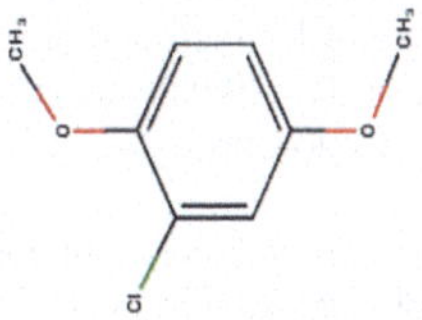	• It is an efficient cofactor cum redox mediator for lignin peroxidase (LiP) for the oxidation 4-methoxybenzyl alcohol • It participates in lignolysis, but its role is not very much clear	Kirk et al. (1978), Teunissen and Field (1998), Datta et al. (2017)

to increase the digestibility of lignin. This proposal ignited the discovery of new lignin monomers that can readily copolymerize with natural lignin monomers but comparatively less resistant to degradation (Pilate et al. 2002; Ralph 2006; Leplé et al. 2007). This very perception has successfully been applied in lignin molding by incorporating coniferyl ferulate with natural monolignols in biomimetic system of maize primary cell walls. Resulted lignin contains breakable ester bonds in the lignin backbone that need less energy intense pretreatments in order to liberate cellulose and hemicellulose from biomass (Grabber et al. 2008).

With additional insight on lignin structure determining factors and Freudenberg's concept of monolignols coupling, a mathematical model has been developed to predict the bond type, polymeric structure and impact of different variables on lignin degradation (Vanholme et al. 2008; van Parijs et al. 2010). After long journey of research, advantages of sophisticated tools have elucidated almost every aspect of lignin biochemistry, but any kind of modification in lignin structure can also be lethal to plants as it alters the plant defense against diverse extreme conditions and pathogens. Only inhouse trails are neither sufficient nor reliable with respect to open environment, and therefore, open field trials are needed against various extreme environment and pathogens. Lignin engineering will only be successful if does not interrupt the expression of other important genes.

References

Aarti C, Arasu MV, Agastian P (2015) Lignin degradation: a microbial approach. Indian J Biol Sci 1:119–127

Abdel-Hamid AM, Solbiati JO, Cann IKO (2013) Insights into lignin degradation and its potential industrial applications. Adv Appl Microbiol 82:1–28

Aslam MS, Aishy A, Samra ZQ et al (2012) Identification, Purification and Characterization of a novel extracellular laccase from *Cladosporiumcladosporioides*. Biotechnol Biotec Eq 26:3345–3350. https://doi.org/10.5504/BBEQ.2012.0107

Atalla RH, Agarwal UP (1985) Raman microprobe evidence for lignin orientation in the cell walls of native woody tissue. Science 227:636–638

Baciocchi E, Gerini MF, Lanzalunga O et al (2001) Isotope-effect profiles in the oxidative N-demethylation of N, N-dimethylanilines catalyzed by lignin peroxidase and a chemical model. Eur J Org Chem 2001:2305–2310

Balaji V, Arulazhagan P, Ebenezer P (2013) Enzymatic bioremediation of polyaromatic hydrocarbons by fungal consortia enriched from petroleum contaminated soil and oil seeds. J Environ Biol 35. https://www.kau.edu.sa/Files/188/Researches/65402_36751.pdf

Barros J, Serrani-Yarce JC, Chen F et al (2016) Role of bifunctional ammonia-lyase in grass cell wall biosynthesis. Nat Plants 2:16050. https://doi.org/10.1038/nplants.2016.50

Battistuzzi G, Bellei M, Bortolotti CA et al (2010) Redox properties of heme peroxidases. Arch Biochem Biophy 500:21–36

Beaudoin-Eagan LD, Thorpe TA (1985) Tyrosine and phenylalanine ammonia lyase activities during shoot initiation in tobacco callus cultures. Plant Physiol 78:438–441

Boerjan W, Ralph J, Baucher M (2003) Lignin biosynthesis. Ann Rev Plant Bio 54:519–546. https://doi.org/10.1146/annurev.arplant.54.031902.134938

Bonugli-Santosa RC, Durrant LR, da Silva M et al (2010) Production of laccase, manganese peroxidase and lignin peroxidase by brazilian marine-derived fungi. Enzyme Microb Technol 46:32–37

Bourbonnais R, Paice M (1992) Demethylation and delignification of kraft pulp by *Trametes versicolor* laccase in the presence of 2,20-azinobis-(3-ethylbenzthiazoline-6-sulphonate). Appl Microbiol Biotechnol 36:823–827

Bourbonnais R, Paice MG, Reid ID et al (1995) Lignin oxidation by laccase isozymes from Trametes versicolor and role of the mediator 2,20-azinobis(3-ethylbenzthiazoline-6-sulfonate) in kraft lignin depolymerization. Appl Environ Microbiol 61:1876–1880

Brijwani K, Rigdon A, Vadlani PV (2010) Fungal laccases: production, function and applications in food processing. Enz Res 2010:149748. https://doi.org/10.4061/2010/149748

Brown ME, Barros T, Chang MCY (2012) Identification and characterization of a multifunctional dye peroxidase from a lignin-reactive bacterium. ACS Chem Biol 7:2074–2081

Camarero S, Sarkar S, Ruiz-Duenas FJ et al (1999) Description of a versatile peroxidase involved in the natural degradation of lignin that has both manganese peroxidase and lignin peroxidase substrate interaction sites. J Biol Chem 274:10324–10330

Chen C, Li T (2016) Bacterial dye-decolorizing peroxidases: biochemical properties and biotechnological opportunities. Phy Sci Rev 2016:20160051. https://doi.org/10.1515/psr-2016-0051

Chen C, Shrestha R, Jia K et al (2015) Characterization of Dye-decolorizing peroxidase (DyP) from *Thermomonospora curvata* reveals unique catalytic properties of A-type DyPs. J Biol Chem 290:23447–23463

Choinowski T, Blodig W, Winterhalter KH et al (1999) The crystal structure of lignin peroxidase at 1.70 Å resolution reveals a hydroxy group on the Cβ of tryptophan 171: a novel radical site formed during the redox cycle. J Mol Biol 286:809–827
Chou EY, Schuetz M, Hoffmann N et al (2018) Distribution, mobility, and anchoring of lignin-related oxidative enzymes in *Arabidopsis* secondary cell walls. J Exp Bot 69:1849–1859
Chowdhary P, Shukla G, Raj G et al (2019) Microbial manganese peroxidase: a ligninolytic enzyme and its ample opportunities in research. SN Appl Sci 1:45. https://doi.org/10.1007/s42452-018-0046-3
Christian V, Shrivastava R, Shukla D et al (2005) Mediator role of veratryl alcohol in the lignin peroxidase-catalyzed oxidative decolorization of remazol brilliant blue R. Enzyme Microb Technol 36:426–431
Coelho-Moreira JS, Maciel GM, Castoldi R et al (2013) Involvement of lignin-modifying enzymes in the degradation of herbicides. In: Herbicides—advances in research. INTECH, pp 165–187
Colpa DI, Fraaije MW, van Bloois E (2014) DyP-type peroxidases: a promising and versatile class of enzymes. J Ind Microbiol Biot 41:1–7
Conesa A, Hondel CAMJJ, Punt PJ (2000) Studies on the production of fungal peroxidases in *Aspergillus niger*. Appl Environ Microbiol 66:3016–3023
Cosio C, Ranocha P, Francoz E et al (2017) The class III peroxidase PRX17 is a direct target of the MADS-box transcription factor AGAMOUS-LIKE15 (AGL15) and participates in lignified tissue formation. New Phytol 213:250–263. https://doi.org/10.1111/nph.14127
Daou M, Piumi F, Cullen D et al (2016) Heterologous production and characterization of two glyoxal oxidases from *Pycnoporus cinnabarinus*. Appl Environ Microbiol 82:4867–4875
Daou M, Faulds CB (2017) Glyoxal oxidases: their nature and properties. World J Microbiol Biotechnol 33:87
Dashtban M, Schraft H, Syed TA et al (2010) Fungal biodegradation and enzymatic modification of lignin. Int J Biochem Mol Biol 1:36–50
Dass SB, Reddy CA (1990) Characterization of extracellular peroxidases produced by acetate-buffered cultures of the lignin-degrading basidiomycete *Phanerochaete chrysosporium*. FEMS Microbiol Lett 69:221–224
Datta R, Kelkar A, Baraniya D et al (2017) Enzymatic degradation of lignin in Soil: a review. Sustainability 9:1163. https://doi.org/10.3390/su9071163
de Koker TH, Mozuch MD, Cullen D et al (2004) Isolation and purification of pyranose 2-oxidase from *Phanerochaete chrysosporium* and characterization of gene structure and regulation. Appl Environ Microb 70:5794–5800
Duan Z, Shen R, Liu B et al (2018) Comprehensive investigation of a dye-decolorizing peroxidase and a manganese peroxidase from *Irpex lacteus* F17, a lignin-degrading basidiomycete. AMB Expr 8:119. https://doi.org/10.1186/s13568-018-0648-6
Edwards SL, Raag R, Wariishi H et al (1993) Crystal structure of lignin peroxidase. Proc Natl Acad Sci 90:750–754
Ehlting J, Bu¨ttner D, Wang Q et al (1999) Three 4-coumarate: coenzyme a ligases in *Arabidopsis* thaliana represent two evolutionarily divergent classes in angiosperms. Plant J 19:9–20
Erdtman H (1933) Dehydrierung in der Coniferylreihe. 1. Dehydrodieugenol und dehydro di isoeugenol. Z Biochem 258:172–180
Falade AO, Nwodo UU, Iweriebor BC et al (2016) Lignin peroxidase functionalities and prospective applications. Microbiologyopen 6:e00394. https://doi.org/10.1002/mbo3.394
Farmer VC, Henderson MEK, Russell JD (1960) Aromatic-alcohol-oxidaseactivity in the growth medium of *Polystictus versicolor*. Biochem J 74:257–262
Farrell RL, Murtagh KE, Tien M et al (1989) Characterization of lignin peroxidase isoenzymes. Enyzme Microb Technol 11:322–328
Fernandez IS, Ruiz-Duenas FJ, Santillana E et al (2009) Novel structural features in the GMC family of oxidoreductases revealed by the crystal structure of fungal aryl-alcohol oxidase. Acta Crystallogr D 65:1196–1205

Fernández-Fueyo E, Linde D, Almendral D et al (2015) Description of the first fungal dye-decolorizing peroxidase oxidizing manganese (II). Appl Microbiol Biotechnol 99:8927–8942
Fernández-Fueyo E, Ruiz-Dueñas FJ, Miki Y et al (2012) Lignin-degrading Peroxidases from genome of selective ligninolytic fungus *Ceriporiopsis subvermispora*. J Biol Chem 287:16903–16916
Ferreira P, Hernandez-Ortega A, Herguedas B et al (2009) Aryl-alcohol oxidase involved in lignin degradation: a mechanistic study based on steady and pre-steady state kinetics and primary and solvent isotope effects with two alcohol substrates. J Biol Chem 284:24840–24847
Ferreira P, Medina M, Guill´en F et al (2005) Spectral and catalytic properties of aryl-alcohol oxidase, a fungal flavoenzyme acting on polyunsaturated alcohols. Biochem J 389:731–738
Ferreira P, Ruiz-Duen˜as FJ, Martı´nez MJ et al (2006) Site-directed mutagenesis of selected residues at the active site of aryl-alcohol oxidase, an H2O2-producing ligninolytic enzyme. FEBS J 273:4878–4888
Ferri S, Kojima K, Sode K (2011) Review of glucose oxidases and glucose dehydrogenases: a bird's eye view of glucose sensing enzymes. J Diabetes Sci Technol. 5:1068–1076
Flournoy DS, Paul JA, Kirk TK et al (1993) Changes in the size and volume of pores in sweetgum wood during simultaneous rot by *Phanerochaete chrysosporium* Burds. Holzforschung 47:297–301
Fraser CM, Chapple C (2011) The phenylpropanoid pathway in *Arabidopsis*. Arabidopsis Book 9:e0152. https://doi.org/10.1199/tab.0152
Freimund S, Huwig A, Giffhorn F et al (1998) Rare keto-aldoses from enzymatic oxidation: substrates and oxidation products of pyranose 2-oxidase. Chem: Eur J 4:2442–2455
Freudenberg K (1965) Lignin: Its constitution and formation from p-hydroxycinnamyl alcohols. Science 148:595–600
Freudenberg K, Harkin JM (1963) The glucosides of cambial sap of spruce. Phytochemistry 2:189–193
Furukawa T, Bello FO, Horsfall L (2014) Microbial enzyme systems for lignin degradation and their transcriptional regulation. Front Biol 9:448–471
Garcia-Ruiz E, Mate DM, Gonzalez-Perez D et al (2014) Directed evolution of ligninolytic oxidoreductases: from functional expression to stabilization and beyond. In: Fessner W (ed) Cascade Biocatalysis. Wiley-VCH Verlag GmbH & Co. KGaA, Weinheim, Germany, pp 1–22
Glumoff T, Harvey P, Molinari S et al (1990) Lignin peroxidase from *Phanerochaete chrysosporium* molecular and kinetic characterization of isozymes. Euro J Biochem 187:515–520
Goswami P, Chinnadayyala SSR, Chakraborty M et al (2013) An overview on alcohol oxidases and their potential applications. Appl Microbiol Biotechnol 97:4259–4275. https://doi.org/10.1007/s00253-013-4842-9
Grabber JH, Hatfield RD, Lu F et al (2008) Coniferyl ferulate incorporation into lignin enhances the alkaline delignification and enzymatic degradation of cell walls. Biomacromol 9:2510–2516
Gramss G, Kirsche B, Voigt KD et al (1999) Conversion rates of five polycyclic aromatic hydrocarbons in liquid cultures of fifty-eight fungi and the concomitant production of oxidative enzymes. Mycol Res 103:1009–1018
Gregor HD (1976) Studies on tyrosine ammonia-lyase activity in castor bean endosperm. Z Pflanzenphysiol 77:372–375
Griffin BW (1983) Mechanism of halide-stimulated activity of chloroperoxidase evidence for enzymatic formation of free hypohalous acid. Biochem Biophys Res Commun 116:873–879
Guillen F, Martinez AT, Martinez MJ (1992) Substrate specificity and properties of the aryl-alcohol oxidase from the ligninolytic fungus *Pleurotus eryngii*. Eur J Biochem 209:603–611
Hager LP, Morris DR, Brown FS et al (1966) Chloroperoxidase: II. utilization of halogen anions. J Biol Chem 241:1769–1777
Hallberg BM, Henriksson G, Pettersson G et al (2002) Crystal structure of the flavoprotein domain of the extracellular flavocytochrome cellobiose dehydrogenase. J Mol Biol 315:421–434
Harreither W, Sygmund C, Augustin M et al (2011) Catalytic properties and classification of cellobiose dehydrogenases from ascomycetes. Appl Environ Microb 77:1804–1815

Harvey PJ, Palmer JM (1990) Oxidation of phenolic compounds by ligninase. J Biotechnol 13:169–179

Hashimoto A, Pickard PA (1984) Chloroperoxidases from Caldariomyces (*Leptoxyphium*) cultures: glycoproteins with variable carbohydrate content and isoenzymic forms. J Gen Microbiol 130:2051–2058

Heinzkill M, Bech L, Halkier T et al (1998) Characterization of laccases and peroxidases from wood-rotting fungi (family Coprinaceae). Appl Environ Microbiol 64:1601–1606

Henriksson G, Johansson G, Pettersson G (2000) A critical review of cellobiose dehydrogenases. J Biotechnol 78:93–113

Hirai H, Sugiura M, Kawai S et al (2005) Characteristics of novel lignin peroxidases produced by white-rot fungus *Phanerochaete sordida* YK-624. FEMS Microbiol Lett 246:19–24

Hofrichter M, Kellner H, Pecyna MJ (2015) Fungal unspecific peroxygenases: heme-thiolate proteins that combine peroxidase and cytochrome P450 properties. monooxygenase, peroxidase and peroxygenase properties and mechanisms of cytochrome P450. Adv Exp Med Biol 851:341–368. https://doi.org/10.1007/978-3-319-16009-2_13

Hofrichter M, Ullrich R (2006) Heme-thiolate haloperoxidases: versatile biocatalysts with biotechnological and environmental significance. Appl Microbiol Biotechnol 71:276–288

Hofrichter M, Ullrich R (2010) New trends in fungalbiooxidation. In: Hofrichter M (ed) Industrial applications, 2nd edn. Springer-Verlag, Berlin, pp 425–449

Houtman CJ, Atalla RH (1995) Celluloses lignin interactions. Plant Physiol 107:977–984

Ivancich A, Mazza G, Desbois A (2001) Comparative electron paramagnetic resonance study of radical intermediates in turnip peroxidase isozymes. Biochem 40:6860–6866

Kadowaki MAS, de Godoy MO, Kumagai PS et al (2018) Characterization of a new glyoxal oxidase from the thermophilic fungus *Myceliophthora thermophila* M77: hydrogen peroxide production retained in 5-hydroxymethylfurfural oxidation. Catalysts 8:476. https://doi.org/10.3390/catal8100476

Kantelinen A, Hatakka A, Viikari L (1989) Production of lignin peroxidase and laccase by *Phlebia radiata*. Appl Microbiol Biotechnol 31:234–239

Kellner H, Luis P, Pecyna MJ et al (2014) Widespread occurrence of expressed fungal secretory peroxidases in forest soils. PLoS ONE 9:e95557. https://doi.org/10.1371/journal.pone.0095557

Kenigsberg P, Fang GH, Hager LP (1987) Post-translational modifications of chloro-peroxidase from *Caldariomyces fumago*. Arch Biochem Biophys 254:409–415

Kerem Z, Jensen KA, Hammel KE (1999) Biodegradative mechanism of the brown rot basidiomycete *Gloeophyllum trabeum*: Evidence for an extracellular hydroquinone-driven fenton reaction. FEBS Lett 446:49–54

Kersten PJ, Cullen D (1993) Cloning and characterization of cDNA encoding glyoxal oxidase, a H_2O_2-producing enzyme from the lignin-degrading basidiomycete *Phanerochaete chrysosporium*. Proc Nat Acad Sci USA 90:7411–7413

Kirk TK, Croan S, Tien M et al (1986) Production of multiple ligninases by *Phanerochaete chrysosporium*: effect of selected growth conditions and use of a mutant strain. Enyzme Microb Technol 8:27–32

Kirk TK, Schultz E, Connors WJ et al (1978) Influence of culture parameters on lignin metabolism by *Phanerochaete chrysosporium*. Arch Microbiol 117:277–285

Kluge MG, Ullrich R, Scheibner K (2007) Spectrophotometric assay for detection of aromatic hydroxylation catalyzed by fungal haloperoxidase-peroxygenase. Appl Microbiol Biotechnol 75:1473–1478

Kong W, Fu X, Wang L et al (2017) A novel and efficient fungal delignification strategy based on versatile peroxidase for lignocellulose bioconversion. Biotechnol Biofuels 10:218. https://doi.org/10.1186/s13068-017-0906-x

Kracher D, Ludwig R (2016) Cellobiose dehydrogenase: Ein essentielles Enzym für den Lignozelluloseabbau in der Natur—Eine Übersicht. Die Bodenkultur: J Land Manag Food Environ 67:145–163

Kracher D, Zahma K, Schulz C et al (2015) Inter-domain electron transfer in cellobiose dehydrogenase: modulation by pH and divalent cations. FEBS J 282:3136–3148

Lankinen P, Hildén K, Aro N et al (2005) Manganese peroxidase of *Agaricus bisporus*: grain bran-promoted production and gene characterization. Appl Microbiol Biotechnol 66:401–407

Leinhos V, Savidge RA (1993) Isolation of protoplasts from developing xylem of *Pinus banksiana* and *Pinus strobos*. Can J For Res 23:343–348

Leplé JC, Dauwe R, Morreel K et al (2007) Downregulation of cinnamoyl-coenzyme A reductase in poplar: multiple-level phenotyping reveals effects on cell wall polymer metabolism and structure. Plant Cell 19:3669–3691

Li K, Xu F, Eriksson KEL (1999) Comparison of fungal laccases and redox mediators in oxidation of a nonphenolic lignin model compound. Appl Environ Microbiol 65:2654–2660

Linde D, Ruiz-Duenas FJ, Fernandez-Fueyo E et al (2015) Basidiomycete DyPs: genomic diversity, structural-functional aspects, reaction mechanism and environmental significance. Arch Biochem Biophys 574:66–74

Littlechild J (1999) Haloperoxidases and their role in biotransformation reactions. Curr Opin Chem Biol 3:28–34

Liu CJ (2012) Deciphering the enigma of lignification: precursor transport, oxidation, and the topochemistry of lignin assembly. Mol Plant 5:304–317

Liu Q, Luo L, Zheng L (2018) Lignins: biosynthesis and biological functions in plants. Int J Mol Sci 19:335. https://doi.org/10.3390/ijms19020335

Lu Y, Lu Y, Hu H et al (2017) Structural characterization of lignin and its degradation products with spectroscopic methods. J Spectroscopy 1–15. https://doi.org/10.1155/2017/8951658

Lundell TK, Makela MR, Hilden K (2010) Lignin-modifying enzymes in filamentous basidiomycetes–ecological, functional and phylogenetic review. J Basic Microbiol 50:5–20

Lundquist K, Kirk TK (1978) De novo synthesis and decomposition of veratryl alcohol by a lignin-degrading basidiomycete. Phytochemistry 17:1676

Manavalan T, Manavalan A, Heese K (2015) Characterization of lignocellulolytic enzymes from white-rot fungi. Curr Microbiol 70:485–498

Manimekalai R, Swaminathan T (2000) Removal of hazardous compounds by lignin peroxidase from *Phanerochaete chrysosporium*. Bioproc Eng 22:29–33

Mansfield SD (2009) Solutions for dissolution—engineering cell walls for deconstruction. Curr Opin Biotechnol 20:286–294

Mester T, Field JA (1998) Characterization of a novel manganese peroxidase-lignin peroxidase hybrid isozyme produced by Bjerkandera species strain BOS55 in the absence of manganese. J Biol Chem 273:15412–15417

Min K, Gong G, Woo HM et al (2015) A dye-decolorizing peroxidase from *Bacillus subtilis* exhibiting substrate-dependent optimum temperature for dyes and β-ether lignin dimer. Sci Rep 5:8245

Morgenstern I, Klopman S, Hibbett DS (2008) Molecular evolution and diversity of lignin degrading heme peroxidases in the agaricomycetes. J Mol Evol 66:243–257

Morris DR, Hager LP (1966) Chloroperoxidase. I. Isolation and properties of the crystalline glycoprotein. J Biol Chem 241:1763–1768

Muthukumarasamy NP, Jackson B, Raj AJ et al (2015) Production of extracellular laccase from *Bacillus subtilis* MTCC 2414 using agroresidues as a potential substrate. Biochem Res Int 2015:765190. https://doi.org/10.1155/2015/765190

Nakashima J, Chen F, Jackson L et al (2008) Multi-site genetic modification of monolignol biosynthesis in alfalfa (*Medicago sativa*): effects on lignin composition in specific cell types. New Phytol 179:738–750

Narayanan PM, Murugan S (2014) Production, purification and application of bacterial laccase: a review. Biotechnology 13:196–205

Patil SR (2014) Production and purification of lignin peroxidase from *Bacillus Megaterium* and its application in bioremidation. CIBTech J Microbiol 3:22–28

Perez J, Jeffries TW (1990) Mineralization of 14C-ring-labeled synthetic lignin correlates with the production of lignin peroxidase, not of manganese peroxidase or laccase. Appl Environ Microbiol 56:1806–1812

Perez-Boada M, Ruiz-Duenas FJ, Pogni R et al (2005) Versatile peroxidase oxidation of high redox potential aromatic compounds: site-directed mutagenesis, spectroscopic and crystallographic investigation of three long-range electron transfer pathways. J Mol Biol 354:385–402

Peter S, Kinne M, Ullrich R et al (2013) Epoxidation of linear, branched and cyclicalkenes catalyzed by unspecific peroxygenase. Enzyme Microb Technol 52:370–376

Pham LTM, Seo H, Kim KJ et al (2018) In silico-designed lignin peroxidase from *Phanerochaete chrysosporium* shows enhanced acid stability for depolymerization of lignin. Biotechnol Biofuels 11:325

Pilate G, Guiney E, Holt K et al (2002) Field and pulping performances of transgenic trees with altered lignification. Nat Biotechnol 20:607–612

Piontek K, Glumoff T, Winterhatter K (1993) Low pH crystal structure of glycosylated lignin peroxidase from *Phanerochaete chrysosporium*. FEBS Lett 315:119–124

Piontek K, Strittmatter E, Ullrich R et al (2013) Structural basis of substrate conversion in a new aromatic peroxygenase: cyto-chrome P450 functionality with benefits. J BiolChem 288:34767–34776

Piontek K, Ullrich R, Liers C et al (2010) Crystallization of a 45 kDa peroxygenase/peroxidase from the mushroom *Agrocybe aegerita* and structure determination by SAD utilizing only the haem iron. Acta Crystallogr, Sect F: Struct Biol Cryst Commun 66:693–698

Pitsawong W, Sucharitakul J, Prongjit M et al (2010) A conserved active-site threonine is important for both sugar and flavin oxidations of pyranose 2-oxidase. J Biol Chem 285:9697–9705

Piumi F, Levasseur A, Navarro D et al (2014) A novel glucose dehydrogenase from the white-rot fungus *Pycnoporus cinnabarinus*: production in *Aspergillus niger* and physicochemical characterization of the recombinant enzyme. Appl Microbiol Biot 98:10105–10118

Pongtharangkul T, Chuekitkumchorn P, Suwanampa N et al (2015) Kinetic properties and stability of glucose dehydrogenase from *Bacillus amyloliquefaciens* SB5 and its potential for cofactor regeneration. AMB Expr. 5:68

Prongjit M, Sucharitakul J, Wongnate T et al (2009) Kinetic mechanism of pyranose 2-oxidase from *Trametes multicolor*. Biochemistry 48:4170–4180

Ralph J (2006) What makes a good monolignol substitute? In: Hayashi T (ed) The science and lore of the plant cell wall: biosynthesis, structure and function. BrownWalker Press, Boca Raton, FL, pp 285–293

Ralph J, Lundquist K, Brunow G et al (2004) Lignins: natural polymers from oxidative coupling of 4-hydroxyphenyl- propanoids. Phytochem Rev 3:29–60. https://doi.org/10.1023/B:PHYT.0000047809.65444.a4

Ravichandran A, Sridhar M (2016) Versatile peroxidases: super peroxidases with potential biotechnological applications-a mini review. J Dairy Vet Anim Res. 4:277–280

Reid ID (1992) Demethylation of the lignin model dimer 1-(3′,4′-dimethoxyphenyl)-2-(o-methoxyphenoxy)-propane-1,3 diol by the white-rot fungus *Phlebia tremellosa*. Can J Bot 70:453–460

Roberts JN, Singh R, Grigg JC et al (2011) Characterization of dyedecolorizing peroxidases from *Rhodococcus jostii* RHA1. Biochemistry 50:5108–5119

Rothschild N, Levkowitz A, Hadar Y et al (1999) Manganese deficiency can replace high oxygen levels needed for lignin peroxidase formation by *Phanerochaete chrysosporium*. Appl Environ Microbiol 65:483–488

Salame TM, Knop D, Levinson D et al (2012) Release of *Pleurotus ostreatus* versatile-peroxidase from Mn^{2+} repression enhances anthropogenic and natural substrate degradation. PLoS ONE 7:e52446

Santos A, Mendes S, Brissos V et al (2014) New dye-decolorizing peroxidases from *Bacillus subtilis* and *Pseudomonas putida* MET94: towards biotechnological applications. Appl Microbiol Biotechnol 98:2053. https://doi.org/10.1007/s00253-013-5041-4

Schoch G, Goepfert S, Morant M et al (2001) CYP98A3 from *Arabidopsis thaliana* is a 3′-hydroxylase of phenolic esters, a missing link in the phenylpropanoid pathway. J Biol Chem 276:36566–36574
Shimada M, Akamtsu Y, Tokimatsu T et al (1997) Possible biochemical roles of oxalic acid as a low molecular weight compound involved in brown-rot and white-rot wood decays. J Biotechnol 53:103–113
Shraddha Shekher R, Sehgal S et al (2011) Laccase: microbial sources, production, purification and potential biotechnological applications. Enz Res 2011:217861. https://doi.org/10.4061/2011/217861
Shrestha R, Huang G, Meekins D et al (2017) Mechanistic insights into dye-decolorizing peroxidase revealed by solvent isotope and viscosity effects. ACS Catalysis 7:6352–6364
Sigoillot JC, Berrin JG, Bey M et al (2012) Fungal strategies for lignin degradation. In: Jouanin L, Lapierre C (eds) Lignins: biosynthesis, biodegradation and bioengineering vol. 61, London, UK: Academic Press, Elsevier. pp 263–308
Song XS, Xing S, Li HP et al (2016) An antibody that confers plant disease resistance targets a membrane-bound glyoxal oxidase in *Fusarium*. New Phytol 210:997–1010
Steffen KT, Hatakka A, Hofrichter M (2002) Degradation of humic acids by the litter-decomposing basidiomycete *Collybia dryophila*. Appl Environ Microbiol 68:3442–3448
Sticklen MB (2008) Plant genetic engineering for biofuel production: towards affordable cellulosic ethanol. Nat Rev Genet 9:433–443
Strittmatter E, Serrer K, Liers C et al (2015) The toolbox of *Auricularia auricula-judae* dye-decolorizing peroxidase—identification of three new potential substrate-interaction sites. Arch Biochem Biophys 574:75–85. https://doi.org/10.1016/j.abb.2014.12.016
Suderman RJ, Dittmer NT, Kanost MR et al (2006) Model reactions for insect cuticle sclerotization: cross-linking of recombinant cuticular proteins upon their laccase-catalyzed oxidative conjugation with catechols. Insect Biochem Mol Biol 36:353–365
Sugano Y (2009) DyP-type peroxidases comprise a novel heme peroxidase family. Cell Mol Life Sci 66:1387–1403
Sundaramoorthy M, Kishi K, Gold MH (1994) The crystal structure of manganese peroxidase from *Phanerochaete chrysosporium* at 2.06-A resolution. J Biol Chem 269:32759–32767
Sutherland GR, Zapanta LS, Tien M et al (1997) Role of calcium in maintaining the heme environment of manganese peroxidase. Biochemistry 36:3654–3662
Taiz L, Zeiger E (2010) Plant Physiol. Sinauer, Sunderland, MA
Ten Have R, Teunissen PJM (2001) Oxidative mechanisms involved in lignin degradation by white-rot fungi. Chem Rev 101:3397–3414
Terashima N, Fukushima K (1989) Biogenesis and structure of macromolecular lignin in the cell wall of tree xylem as studied by microautoradiography. In: Lewis NG, Paice MG (eds) Plant cell wall polymers: biogenesis and biodegradation, ACS Symposium Series 399. American Chemical Society, Washington, DC, pp 160–168
Terazawa M, Okuyama H, Miyake M (1984) Phenolic compounds in living tissue of woods. I. Phenolic P-glucosides of 4-hydroxycinnamyl alcohol derivatives in the cambial sap of woods. Mokuzai Gakkaishi 30:322–328
Teunissen PJM, Field JA (1998) 2-Chloro-1,4-dimethoxybenzene as a novel catalytic cofactor for oxidation of anisyl alcohol by lignin peroxidase. Appl Environ Microbiol 64:830–835
Tien M, Kirk K (1983) Lignin-degrading enzyme from the hymenomycete *Phanerochaete chrysosporium*. Science 221:661–663
Tobimatsu Y, Chen F, Nakashima J et al (2013) Coexistence but independent biosynthesis of catechyl and guaiacyl/syringyl lignin polymers in seed coats. Plant Cell 25:2587–2600. https://doi.org/10.1105/tpc.113.113142
Tronchet M, Balagué C, Kroj T et al (2010) Cinnamyl alcohol dehydrogenases C and D, key enzymes in lignin biosynthesis, play an essential role in disease resistance in *Arabidopsis*. Mol Plant Pathol 11:83–92

Ullrich R, Dolge C, Kluge M et al (2008) Pyridine as novel substrate for regio selective oxygenation with aromatic peroxygenase from *Agrocybe aegerita*. FEBS Lett 582:4100–4106
Ullrich R, Nu¨ske J, Scheibner K et al (2004) Novel haloperoxidase fromthe agaric basidiomycete *Agrocybe aegerita* oxidizesaryl alcohols and aldehydes. Appl Environ Microbiol 70:4575–4581
van Parijs FRD, Morreel K, Ralph J et al (2010) Modeling lignin polymerization. I. simulation model of dehydrogenation polymers. Plant Physiol 153:1332–1344
Vanholme R, Demedts B, Morreel K et al (2010) Lignin biosynthesis and structure. Plant Physiol 153:895–905. https://doi.org/10.1104/pp.110.155119
Vanholme R, Morreel K, Darrah C et al (2012) Metabolic engineering of novel lignin in biomass crops. New Phytol 196:978–1000
Vanholme R, Morreel K, Ralph J et al (2008) Lignin engineering. Curr Opin Plant Biol 11:278–285
Verma SR, Dwivedi UN (2014) Lignin genetic engineering for improvement of wood quality: applications in paper and textile industries, fodder and bioenergy production. S Afr J Bot 91:107–125
Wang JP, Matthews ML, Williams CM (2018) Improving wood properties for wood utilization through multiomics integration in lignin biosynthesis. Nat Commun 9:1579. https://doi.org/10.1038/s41467-018-03863-z
Wang X, Ullrich R, Hofrichter M et al (2015) Heme-thiolate ferryl of aromatic peroxygenase is basic and reactive. P Natl Acad Sci USA 112:3686–3691
Wariishi H, Akileswaran L, Gold MH et al (1988) Manganese peroxidase from the basidiomycete *Phanerochaete chrysosporium*: spectral characterization of the oxidized states and the catalytic cycle. Biochemistry 27:5365–5370
Welinder KG (1992) Superfamily of plant, fungal and bacterial peroxidases. Curr Opin Struct Biol 2:388–393
Weng JK, Chapple C (2010) The origin and evolution of lignin biosynthesis. New Phytol 187:273–285
Weng JK, Li X, Bonawitz ND et al (2008) Emerging strategies of lignin engineering and degradation for cellulosic biofuel production. Curr Opin Biotechnol 19:166–172
Wever R, Barnett P (2017) Vanadium chloroperoxidases: the missing link in the formation of chlorinated compounds and chloroforming the terrestrial environment? Chem Asian J 12:1997–2007
Whetten R, Sederoff R (1995) Lignin biosynthesis. Plant Cell 7:1001–1013
Wong DWS (2009) Structure and action mechanism of ligninolytic enzymes. Appl Biochem Biotechnol 157:174–209
Wongnate T, Chaiyen P (2013) The substrate oxidation mechanism of pyranose 2-oxidase and other related enzymes in the glucose–methanol–choline superfamily. FEBS J 280:3009–3027
Xie M, Zhang J, Tschaplinski TJ et al (2018) Regulation of lignin biosynthesis and its role in growth-defense tradeoffs. Front Plant Sci 9:1427. https://doi.org/10.3389/fpls.2018.01427
Yamada Y, Wang J, Kawagishi H et al (2014) Improvement of ligninolytic properties by recombinant expression of glyoxal oxidase gene in hyper lignin-degrading fungus *Phanerochaete sordida* YK-624. Biosci Biotech Bioch 78:2128–2133
Yoshida H, Sakai G, Mori K et al (2015) Structural analysis of fungus-derived FAD glucose dehydrogenase. Sci Rep 5:13498. https://doi.org/10.1038/srep13498
Zámocký M, Hofbauer S, Schaffner I et al (2015) Independent evolution of four heme peroxidase superfamilies. Arch Biochem Biophy 574:108–119
Zamocky M, Obinger C (2010) Molecular phylogeny of heme peroxidases. In: Torres E, Ayala M (eds) Biocatalysis based on heme peroxidases. Springer, Heidelberg, pp 7–35
Zhang C, Xin Y, Wang Y et al (2015) Identification of a novel dye-decolorizing peroxidase, EfeB, translocated by a twin-arginine translocation system in *Streptococcus thermophilus* CGMCC 7.179. Appl Environ Microbiol 81:6108–6119

Analysis of Lignin Using Qualitative and Quantitative Methods

Kush Kumar Nayak, Piyush Parkhey and Reecha Sahu

Abstract Lignin, the second richest biopolymer on earth, after cellulose and the primary deposition on the secondary cell wall structure of plants is composed of complex phenolic compounds. Its composition makes it an important natural renewable raw material for production of a number of commodity products. Lignin is also commercially produced as kraft lignin during paper manufacturing. Naturally occurring lignin plays significant biological role affecting plant growth, development, stress tolerance and adaptation. Keeping in view their commercial significance and biological role, much research has been directed towards determining the biosynthetic pathways of lignin and its qualitative and quantitative analysis. The present chapter discusses the structural features of lignin, the associated condensation and biosynthetic steps for lignin accumulation in plants, its biological functions and commercial significance. Additionally, the qualitative analysis techniques such as staining and microscopy along with the qualitative methods of both soluble and insoluble lignin determination, such as spectroscopy, have also been discussed. In the end, future scope of development in lignin application and associated challenges has also been deliberated.

Keywords Lignin · Biosynthesis · Biodegradation · Kraft lignin · Qualitative analysis · Quantitative analysis

1 Introduction

Lignin is the major structural element of plant cell wall and is characterized by high molecular weight, complex phenolic structure and composition. The large polymeric group represented by lignin is derived from 4-hydroxyphenylpropanoids as a result of its oxidative combinatorial coupling (Vanholme et al. 2010). Lignin is the primary deposition on the secondary cell wall structure, thus imparting them rigidity and water resistance. Over the years, there has been an increase in scientific interests

---

K. K. Nayak (✉) · P. Parkhey · R. Sahu
Amity Institute of Biotechnology, Amity University Chhattisgarh, Raipur 493225, India
e-mail: nayak23bt@gmail.com

S. Sharma and A. Kumar (eds.), *Lignin*, Springer Series on Polymer and Composite Materials, https://doi.org/10.1007/978-3-030-40663-9_4

towards understanding the details and intricacies of lignin biosynthesis in plants. The primary reason is the recalcitrant nature of lignin which acts as limiting factor in utilization of plant biomass for biofuel production. Therefore, engineered plants are being developed with low lignin contents (Mansfield 2009).

1.1 Lignin

Lignin is derived from the polyphenolic monolignols or hydroxy cinnamyl alcohols, viz. coniferyl alcohol, sinapyl alcohols and *p*-coumaryl alcohol (Fig. 1), which in turn are synthesized from phenylalanine derived from shikimate biosynthetic pathway (Rippert et al. 2009). The process involves biosynthesis of lignin monomers in cytoplasm by a series of chemical reactions such as deamination, hydroxylation, methylation and reduction. These monomers are transported to apoplast where they are polymerized along with the three monolignols into plant cell walls by the enzymes peroxidase and laccase (Liu et al. 2018). These units upon polymerization are called as guaiacyl (G) (from coniferyl alcohol), syringyl (S) (from sinnapyl alcohol) and p-hydroxyphenyl (H) (*p*-coumaryl alcohol) units. The composition of the units varies across different taxa and species of plants. For example, gymnosperms mostly have G units in their lignin with small amount of H units, while angiosperm lignin consists of mainly G and S units (Vanholme et al. 2010). Grasses on the other hand have somewhat higher levels of H units (Vanholme et al. 2010). It has been suggested that in softwood species, lignin amounts to approximately 30% of the total mass, while in hardwood species, it is about 20–25% (Calvo-Flores and Dobado 2010). The polymerization of lignin is catalysed by a peroxidase-mediated dehydrogenation of monolignols which results in an optically inactive heterogenous structure (Abreu

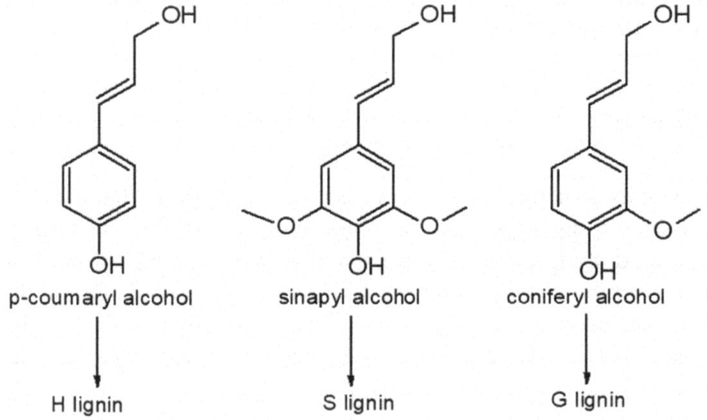

Fig. 1 Three basic structures of lignin monomers: pcoumaryl alcohol [p-hydroxyphenol (H)], sinapyl alcohol [syringyl (S)] and coniferyl alcohol [guaiacyl (G)]

et al. 1999; Calvo-Flores and Dobado 2010). The greater understanding of the intermolecular meshwork of the lignin structure has come from the degradative methods which can provide lignin fragments for GC-MS analysis by controlled bond cleavage (Duval and Lawoko 2014). Lignin, isolated from different sources, has its molecular mass in the range of 1000–20,000 g/mol. However, the challenge is to correctly determine the degree of polymerization. This is mostly because the extraction methods fragmentize lignin yielding various repeated substructures in a random fashion (Thakur et al. 2014). However, based on model studies, two types of lignin structures have been predicted. Type I lignin contains glucomannan-lignin-xylan complex directly bonded to cellulose fibrils through hydrogen bridges, while type II lignin contains xylan-lignin-glucomannan complex embedded thin type I lignin of cellulose fibrils (Kun and Pukánszky 2017).

1.2 Biological Role and Commercial Significance

It is a well-accepted notion that evolution of lignin structure occurred as a result of plant adaptation for a terrestrial life for an erect growth (Vanholme et al. 2010). However, lignin and secondary wall structure have also been reported in marine algae which are a proof of convergent evolution (Martone et al. 2009). In addition, it has also been proposed that lignin biosynthesis can be induced by various biotic and abiotic stresses (Caño-Delgado et al. 2003; Tronchet et al. 2010). Thus, it would be within limits to conclude that lignin biosynthesis and their biological functions are co-related. The primary biological role that lignin plays in plants is discussed below.

1.2.1 Affect on Plant Growth and Development

Probably, the most important role of lignin in plant structure is to provide mechanical strength, water resistance and protection of plant cell polysaccharides against microbial degradation. However, lignin also plays an important role during various stages of plant growth and development. Mutant studies have revealed that lower content of lignin in seed coat drastically reduces the seed germination rate (Liang et al. 2006), stunted growth and less number of seeds (Mir Derikvand et al. 2008; Vanholme et al. 2012). Few studies have also demonstrated loss in apical dominance and male sterility (Schilmiller et al. 2009; Thévenin et al. 2011). The exact mechanism of how lignin affects the developmental processes in plants has still not been well understood, but it is believed that defects in cell wall affect transcriptional processes and signalling pathways, thereby leading to stunted growth (Bonawitz et al. 2014; Liu et al. 2018).

Another important aspect in plant development where lignin plays an important role is in lodging resistance. Lodging is the bending of the crops near the ground which makes them difficult to harvest, thereby seriously affecting the crop growth and yield (Berry et al. 2004). Several studies have been undertaken, wherein the effects of lignin deposition on lodging resistance have been studied. In one of the study, Peng et al. (Berry et al. 2004) studied the effects of paclobutrazol on lignin deposition and resulting lodging tolerance. They concluded that the drug, while decreasing the internode length, increased lignin deposition and improved wheat lodging tolerance. Other factors such as crop density (Zheng et al. 2017) and activity of lignin biosynthesis enzymes (Hu et al. 2017) also play an important role in lignin deposition and lodging resistance. Nutrients like silicon improve lignin accumulation increasing the stalk length and lodging resistance (Hu et al. 2017), while higher nitrogen fertilizers decreased the lignin accumulation and reduced the lodging resistance (Wang et al. 2015; Zhang et al. 2017).

1.2.2 Stress Tolerance and Adaptation

Lignin accumulation in plant cell wall is also associated with imparting resistance to plants against various stresses such as drought, salt, temperature, diseases and heavy metals. Drought and salt stresses induce osmotic stress which leads to dehydration causing plant death. The presence of lignin in the cell wall prevents water loss through transpiration and protects membrane integrity (Monties and Fukushima 2005). Under drought conditions, the amount of lignin in the stem was shown to increase which suggested a strong correlation between lignin biosynthetic genes and drought stress tolerance (Walker et al. 2003; Hu et al. 2009; Srivastava et al. 2015).

Temperature is also an important regulatory factor that governs plant growth. Extremely high temperatures can adversely affect the metabolic pathways of the plant by causing irreversible damage to proteins, enzymes and other biomolecules (Christensen and Christensen 2007; Bita and Gerats 2013). A possible reason to this may be damage to vascular tissues due to reduced lignin assimilation, thereby leading to excessive transpiration and overheating of plant biomolecules (Guy et al. 2007). On the contrary, low temperatures may affect plant photosynthesis inhibiting plant growth and development (Guy et al. 2007). In a study by Guy et al. (2007), heat and low-temperature conditioning downregulated the HSF gene expression which led to reduced lignification. In another study, a post-harvest high-temperature induction of lignin deposition led to development of environmental stress resistance in pericarp of fruit (Guy et al. 2007).

Cell wall also plays a major role in regulating plant heavy metal tolerance. Synthesis of phenolic material and lignin content has been reported to increase in case of heavy metal stress (Liu et al. 2018). It has been proposed that lignin, containing multiple functional groups, can bind to various heavy metal ions and therefore limit the entry of such metals in the cytoplasm (Bhuiyan et al. 2007; Kováčik and Bačkor 2007). Tea plants are known to grow well in high aluminium concentration soils. The enzymes phenylalanine ammonia-lyase and cell wall peroxidase level showed considerable decrement under high aluminium concentration along with reduced lignin concentration. This is suggestive of the fact that tea plants exist in high aluminium concentration (Ghanati et al. 2005). Copper and cadmium have also been reported increase lignin content in plants. The accumulation of phenolics and lignin was reported to increase in ginseng at higher copper concentrations due to enhanced enzymatic activity (Akgül et al. 2007), while cadmium treatment resulted in increased lignin content due to upregulation of POD and LAC activity (Bhuiyan et al. 2007).

It is also proposed that lignin deposition in cell wall of root endoderm alters the transport of heavy metal into xylem (Liu et al. 2018). Lignin biosynthetic genes in *Thlaspi caerulescens* (Zn/Cd hyper accumulator) demonstrated elevated expressions which are directly associated with higher accumulation of Zn and Cd. Another significant role that lignin plays in promotion of plant health is in providing pest and disease resistance to plants. It is believed that pathogen-infected plants synthesize large amount of lignin with higher H unit which might act as a fence the pathogen spread and penetration of toxins into plant cell wall. Also, the other related compounds of lignin may cause the infecting pathogen to lose their pathogenicity against the host and their spread (Liu et al. 2018). Specific studies reveal up-regulation of lignin-producing genes in brown plant hopper-infested insect-resistant rice which is suggestive of the fact that the genes imparting pest resistance to plant work together with lignin-synthesizing genes (Jannoey et al. 2017). In another study, it was demonstrated that enhanced expression of lignin genes in chrysanthemum increased its tolerance levels to attack against aphids. Similarly, the mechanism of action of Sclareol, an antimicrobial molecule that enhanced root-node nematode resistance in *Arabidopsis thaliana,* was increased by ethylene-based lignin accumulation in the roots.

The above reports, along with many others, signify the critical role lignin plays in plants. It is a predominant structural character that is utmost necessary for better plant growth and development and is thus biologically significant. In the subsequent sections, we would study the methods of lignin estimation in plant cells.

2 Lignin Analysis

Analysis of lignin compound and its quantification in a lignocellulosic material is done either to characterize the material or for the selection of method to be utilized for the removal of lignin from the native material. This is routinely done to get cellulose from lignocellulosic biomass or to characterize lignocellulosic material to be used in paper and pulp industries (Dence 1992; Brinkmann et al. 2002). Apart from this, owing to the innumerable novel application lignin molecule possesses, it is emerging as an important biopolymer. The application of lignin offers ranges from controlled drug release (Li et al. 2018), lignin-based activated carbon (Fu et al. 2013), adsorption of heavy metals (Ge and Li 2018), lignin-based jet fuel (Huang et al. 2018) to other biological and pharmaceutical applications (Spiridon 2018). The first step to this is the comprehensive biomass analysis, which includes detection and quantification of lignin compound in the biological samples like lignocellulosic materials. Advances in spectrophotometric analysis, specifically used for in situ analysis of lignocellulosic materials, have made lignin investigation rapid, routine and more accurate (Dence 1992). Such techniques are taking over the older techniques in terms of using the knowledge of biochemistry, molecular biology and computer modelling to probe lignin structure to deliver thorough information about this significant biopolymer. In this section, authors will be discussing the methods and techniques employed and reported for the lignin analysis.

Analysis of lignin components in a lignocellulosic material can be divided into two categories, one which includes the analytical methods/techniques used to analyse lignin present as a component of a solid mixture (qualitative as well as quantitative methods), and another category, lignin content is estimated in solution (quantitative methods).

2.1 Qualitative Methods of Lignin Analysis

2.1.1 Staining and Microscopy

Staining and microscopy are used in visualization/localization of lignin in the sample material. Exact estimation of lignin content cannot be achieved by such methods. Lignin gives bright blue fluorescence under UV light; however, it does not possess colour of its own. About 200 chromogenic reagents or dyes are known to be used for staining lignin to ascertain the presence of lignin in a biomaterial. But, staining using dyes are usually not taken as the confirmatory qualitative test for the presence of lignin as the phenolic compounds interfere the test and give false-positive results. Such test should be supported by analytical techniques utilizing spectrophotometric, NMR, etc. (Dean 1997). However, there have been few reports about effective colour tests for lignin. A comprehensive report by Dean summarizes methods of lignin detection like Maule's test and Cross-Bevan's tests (Dean 1997). Most of the tests are based on chemical degradation of lignin and producing coloured products by subsequent reaction. Chemical compounds like aromatic amines and diazo compounds turn out stable chromophores and absorptive patterns definite for different lignin subunits and have been used in the past as an effective and specific qualitative tool for lignin detection (Dean 1997). New ways for lignin detection were opened with the advent of advanced microscopy. Electron microscopy emerged as effective tool-specific lignin detection at microscopic level using stains like potassium permanganate so that the electron density of lignin can be increased. Potassium permanganate was used as a primary fixative for plant tissues and also as an effective contrasting agent for lignin when used as a stain in glutaraldehyde-osmium fixed tissues (Hepler et al. 1970; Bland et al. 1971; Sant'Anna et al. 2013). Recently, phloroglucinol–HCl was used for staining sections of stone cells obtained from the fruit of two *Pyrus* species (*Pyrus bretschneideri* cv. '*Jingaisu*' and *Pyrus pyrifolia* cv. '*Kousui*') to detect lignin distribution by SEM, TEM and autofluorescence analysis (Tao et al. 2009; Irizarry and White 2018; Jang and Lee 2019). Dyes generally have a specific excitation and emission wavelength which changes with the change in chemical environment or presence of an interfering molecule or a quencher may lead to changes in fluorescence emission. Increase in dye–dye or dye–substrate interactions resulting

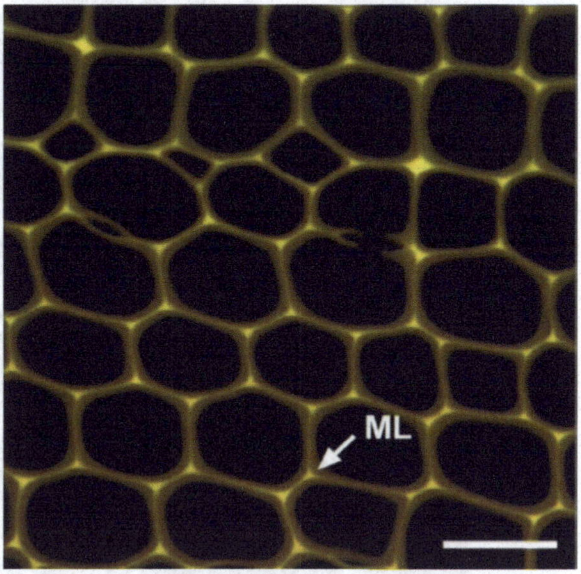

Fig. 2 Autofluorescence image of radiata pine normal wood. High fluorescence intensity corresponds to high lignin content. ML, middle lamella. Scale bar = 40 mm. Reproduced from Bond et al. (2008), with permission from Taylor & Francis

in quenching and/or Forster resonance energy transfer (FRET) brings about changes in fluorescence. Fluorescence behaviour of wood cell walls stained with safranine was studied to identify various biomolecules of a lignocellulosic biomass. Owing to the change in fluorescence emission, safranine was able to differentiate high and low levels of lignin. Regions of high lignin gave red/orange fluorescence, while regions with low lignin showed green/yellow fluorescence as shown in Figs. 2 and 3 (Reproduce with permission from Taylor & Francis). Similar study and staining techniques has also been reported by other research groups (Bond et al. 2008; Simon et al. 2018; Duan et al. 2019; Martin et al. 2019).

Recently, Choong et al. have developed new, spectral-based methods for non-destructive determination and visualization of the chemical composition of lignocellulosic biomass on the basis of structure-responsive optical features of luminescent oligothiophenes. (Choong et al. 2016). They used pentamer hydrogen thiophene ethyl amine (p-HTEA) as a novel probe to detect the structure of cellulose in numerous forms and to assess cellulose purity in conjunction with lignin and hemicellulose detection. It was seen that lignin showed a powdery appearance under phase contrast microscope and displayed an inherent fluorescence regardless of the presence of p-HTEA (Fig. 4) (reproduced with permission). p-HTEA possesses the high affinity to lignin which resulted in suppression of the inherent fluorescence of the probe at 386 nm. Analysis of the associated excitation spectra suggested the presence of lignin (Choong et al. 2016).

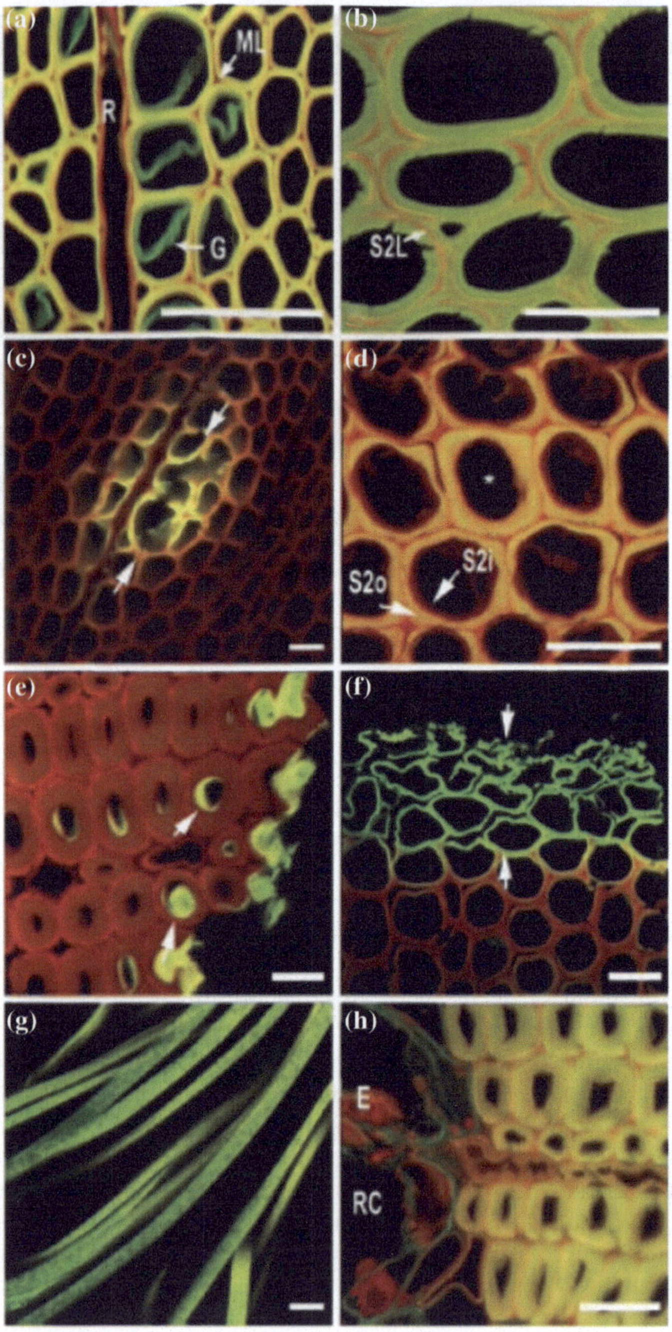
(a)
ML
R
G
(b)
S2L
(c)
(d)
S2i
S2o
(e)
(f)
(g)
(h)
E
RC

◀**Fig. 3** Safranine staining of wood types with varying levels of lignin and polysaccharides. **a** Poplar tension wood. The gelatinous layer (G) is predominantly made up of cellulose. The ray cell (R) and ML are high in lignin; **b** radiata pine compression wood. The S2L region of the secondary cell wall has high lignin content; **c** brown rot-degraded radiata pine. Arrows indicate a region of partially degraded wood containing crystalline cellulose, whereas the surrounding cells have little or no cellulose; **d** trametes versicolor-degraded radiata pine; non-selective white rot. The inner cell wall (S2i) has altered chemistry as shown by the deep red fluorescence; the outer S2 (S2o) appears unaffected. The asterisk shows an undegraded cell; **e** Phellinus pini-degraded Douglas fir; white rot, which selectively removes lignin. Arrows show lignin degraded regions of the cell wall; **f** UV-degraded (artificially weathered) wood. Arrows show the exposed region at the wood surface where lignin has been removed by artificial weathering; **g** cotton fibres; **h** a resin canal in radiata pinewood (RC). Resin canal cell walls do not contain lignin and fluoresce green when stained with safranine while extract (E) fluoresces red. Scale bars =40 mm. Reproduced from Bond et al. (2008), with permission from Taylor & Francis

2.1.2 Ultraviolet Microscopy and Microspectroscopy

Ultraviolet (UV) microscopy is a type of light microscopy that utilizes UV light to generate a magnified image of the sample being analysed. As a result of the shorter wavelength of UV light than visible light, it is possible to view samples with greater magnification and resolution. And, microspectroscopy involves the combination of microscope and spectroscope optics for the purpose of investigating the spectra of small samples. The general methods of microspectroscopy rely on absorption measurements in the ultraviolet, visible, and infrared regions, emission spectroscopy and other applications. Lignin is the predominant material in a lignocellulosic biomass which has UV-absorbing property, and this property has been widely harnessed for localizing and to roughly quantitate lignin. Different biomolecules/materials present in a lignocellulosic material have a characteristic and different absorption spectrum. Taking advantage of this property, microspectrometry has emerged as a rapid and sensitive technique to specifically detect the presence of lignin. Fergus and co-workers harnessed this technique to localize lignin in spruce wood (Fergus et al. 1969). Dean et al. performed UV microspectroscopy of the G and S type lignin subunits, guaiacol (G) and syringol (S), along with their para-methylated derivatives 4-methylguaiacol (4-MG) and 4-methylsyringol (4-MS). The results showed there is a discrimination of G and S chromophores in lignin oligomers and demonstrated the potential for site-selective absorption (Dean et al. 2013). Lu et al. have summarized in detail various spectroscopic methods for qualitative estimation of lignin. One such technique is FTIR spectroscopy. Lignin is characterized by the presence of functional groups like hydroxyl, carbonyl, methoxyl, carboxyl and aromatic and aliphatic C-H, etc. These functional groups can be very well characterized by starching and bending patterns in the FTIR spectrum (Lu et al. 2017). The principal advantage of UV spectroscopy for lignin detection lies in its ability of lignin analysis in situ. Disadvantages associated with the technique are the low specificity in terms of tendency of UV absorption of lignocellulosic biomaterials

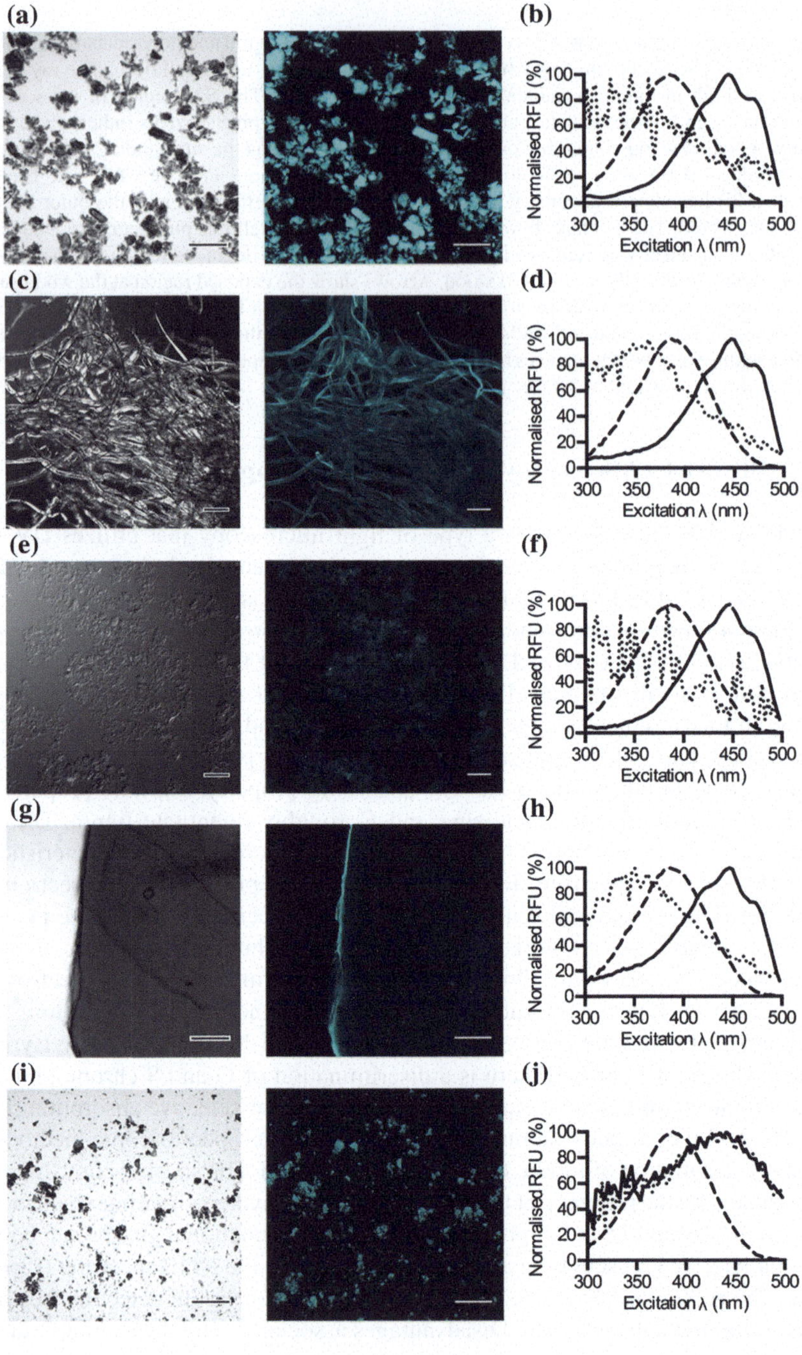
(a)
(b)
(c)
(d)
(e)
(f)
(g)
(h)
(i)
(j)
Normalised RFU (%)
100
80
60
40
20
0
300
350
400
450
500
Excitation λ (nm)

◀**Fig. 4** p-HTEA enables combined fluorescence imaging and spectrometric analysis of cellulose materials. Phase contrast (left) and fluorescence confocal microscopy (right) of **a** M. cellulose, **c** pulpcellulose, **e** cellulose nanofibrils, **g** paper made of cellulose nanofibrils, and **i** lignin stained with p-HTEA. In confocal microscopy, excitation at 473 nm and bandwidth filters detecting 490–530 nm were applied. Normalized excitation spectra of p-HTEA mixed with **b** M. cellulose, **d** pulp cellulose, **f** cellulose nanofibrils, **h** paper made of cellulose nanofibrils, and **j** lignin (solid). Spectra of the pure materials (dotted) and of p-HTEA (dashed) are shown for comparison. Reproduced from Choong et al. (2016), with permission from Taylor & Francis

because it varies not only with structural component changes, but also with changes in lignin structure (Dean 1997).

2.1.3 Immunocytochemical Detection

Immunocytochemical detection of lignin in lignocellulosic biomaterials can be achieved by using antibodies and polyclonal antibodies against lignin molecule. The presence of lignin in bryophytes was detected using two polyclonal antibodies against homoguaiacyl (G) and guaiacyl/syringyl (GS) synthetic lignin-like polymers that were specifically labelling lignified cell walls in tracheophytes and to cell walls in bryophytes (Ligrone et al. 2008). This technique appeared to be effective for selective detection of lignin molecules. Joseleau and co-workers further developed a better technique under same principle of immunocytochemical detection of lignin and semi-quantification using immune gold probes directed especially for the motif structure of lignin (Joseleau et al. 2004). This was visualized by transmission electron microscopy using cryo-fixation–freeze-substitution. The immunological probes exhibited high specificity for condensed and non-condensed guaiacyl (G) and syringyl (S) interunit linkages of lignin. High sensitivity of the probe and the technique employed helped in rapid and selective detection of lignin epitopes of definite chemical structures in the G-layer of tension wood fibres (Joseleau et al. 2004).

2.2 Quantitative Methods of Lignin Analysis

There have been methods for lignin quantification, but earlier methods that have been reported have some or the other disadvantages associated with the method. Factors that contribute to such disadvantages, like the total lignin content of the material, are taken into account. There are different types of lignins (majorly two types G and S) which contain different amounts of the three monomer residues as discussed earlier. Another point which adds up to this is the presence of interfering aromatic and phenolic compounds. Removal of these contaminants always leads to removal of some lignin, but when left in place, the contaminants generally lead to overestimates of lignin content. Another issue is the removal or alteration of lignin molecule during physical treatments of lignocellulosic materials like milling, extraction or any other physiochemical process. This results in significant variation in the final estimation of lignin. However, there have been methods which have overcome certain shortcoming and came up with a better analysis technique for specific quantitative estimation of lignin.

2.2.1 Estimation of Insoluble Lignin

Klason method is the oldest and reliable method for quantitative estimation of lignin in solid materials. It was originally developed by Peter Klason in 1923. He used 61–68 wt% H_2SO_4 at a low temperature and used this method to obtain an acid-insoluble lignin (AIL) content of 28% for European spruce wood. AIL values in the range of 28% for European spruce are now widely accepted. The 72% sulphuric acid method was later published by Ritter et al. in 1932. In 2013, Nicholson et al. modified the protocol and got better results in terms of lesser amount of non-lignin components in AIL (Nicholson et al. 2014). The traditional Klason method of lignin estimation is still most widely used technique and being used routinely for quantifying lignin; recently, it was used for estimation of lignin in *P. elliottii* (Bortoluz et al. 2019).

2.2.2 Estimation of Soluble Lignin

Spectroscopic techniques are the most widely used high-throughput technique for qualitative as well as quantitative estimation of lignin. These techniques are routinely used for detailed qualitative and quantitative information on structural features of lignin, including functional groups, types of chemical bonds and states of atoms (Dean 1997; Lu et al. 2017).

UV Spectroscopy and Fourier-Transform Infrared Spectroscopy

According to Lambert–Beer's Law, UV spectroscopy can be used for the semi-quantitative estimation of lignin purity, its amount in lignocellulosic material and its degradation products by using extinction coefficient. Due to the cross-linking of lignin with various biomolecules like carbohydrates, cellulose and hemicelluloses, its isolation becomes extremely difficult. The low value of extinction coefficient represents the high non-lignin substance content in the isolated lignin (Lu et al. 2017). Using a standard calibration of wood lignin or polyphenols as used in IR spectroscopy, the quantification of soluble lignin can be done. Estimation of lignin quantity using IR spectroscopy was initially reported in 1991 for *Eucalyptus grandis* using wood lignins and native polyphenols for calibration standard (Morais et al. 1991). Huang et al. investigated calibrated visible and near-infrared (NIR) (400–2500 nm) spectroscopy to predict the Klason lignin content of *Moso* bamboo (*Phyllostachys pubescens*), Chinese fir (*Cunninghamialanceolata* (Lamb.) Hook.),

Paulownia (*Paulownia elongata*) and Poplar (*Populusnigra var.*). It was found that NIR can predict the Klason lignin content of the lignocellulosic materials obtained from the plants under study. Fourier-transform infrared spectroscopy (FTIR) has also emerged as an important and sensitive technique to quantitatively and qualitatively estimate lignin. The advantage of FTIR is that it can specifically detect the change in the structure and chemical functional group of lignin. Recently, Horikawa et al. developed a high-throughput analysis method to investigate the relative lignin contents of *Cryptomeria japonica* samples over a broad concentration array (3–73%). The method proposed by the group was simple as compared with the type of chemical pretreatment because it was developed using FTIR spectroscopy. The documented simple and rapid method for quantifying lignin by calculating the intensity of the key band of lignin at 1508 cm^{-1} was calculated using the absorbance at 2900 cm^{-1} as a reference using a reliable calibration curve with an R^2 of 0.968. The calibration curve reported was independent of the chemical treatment performed (Horikawa et al. 2019). In another study, FT-Raman spectroscopy was used for specific and rapid estimation of syringyl units (S) in wood lignins using 70 cm^{-1} Raman band-area intensity (370-area) (Agarwal et al. 2019).

Nuclear Magnetic Resonance Spectroscopy

As discussed earlier, due to the complex structure of lignin polymer, it is difficult to specifically and correctly estimate lignin content. A sequence of chemical analyses has been performed as an attempt to disclose the complex and diverse structure of lignin. Chemical analyses include thioacidolysis (TA), nitrobenzene oxidation (NBO), and derivatization followed by reductive cleavage (DFRC). Recent progresses in nuclear magnetic resonance (NMR) technology make it most widely used technique for lignin analysis and estimation due to its flexibility in demonstration of lignin structural features and structural transformations. Quantitative estimation can be done by quantitative ^{13}C NMR spectrum of non-acetylated lignin. For effective quantification of lignin using quantitative ^{13}C NMR spectrum, there are certain points to be taken care of which are very useful in lignin characterization. Before collecting quantitative ^{13}C-NMR spectra of lignin, there are certain conditions that must be fulfilled. First, contaminants such as carbohydrates or extractives should be removed from lignin sample. Second, to minimize signal-to-noise and to maximize baseline and phasing distortions, lignin sample solution must be made as concentrated as possible. Finally, the inverse-gated-decoupling sequence (i.e. C13IG pulse) is used which involves turning off the proton decouple during the recovery between pulses so that the Nuclear Overhauser Effect (NOE) effect is avoided (Capanema et al. 2004; Wen et al. 2013; Fu et al. 2015; Balakshin and Capanema 2015). Quantitative ^{13}C NMR is used only for the estimation of some specific moieties present in lignin. NMR is mostly for estimation of (S) lignins like FTIR.

Another development in the latest NMR techniques is the two-dimensional heteronuclear single quantum coherence (2D-HSQC) NMR. This NMR technique I emerging as a rapid and powerful tool for lignin characterization and quantification provides richer, detailed and less explicit information (Wen et al. 2013).

Till date, lots of progresses have made in studying lignin qualitatively and quantitatively, which were discussed neatly by Lupoi et al. (2015). Instrumental methods like wet chemistry, chromatography, thermochemical methods, vibrational spectroscopy, nuclear magnetic resonance, electronic spectroscopy, atomic force and electron microscopy and several miscellaneous analytical methods have been applied from long time for lignin analysis (Lupoi et al. 2015). Each method has its own merits and demerits, which are summarized in Table 1. There are array of imaging and analysis techniques available for the analysis of lignin content in lignocellulosic materials (Lu et al. 2017; Ge and Li 2018). Depending on the application, one can choose the technique to be employed for detection and estimation of lignin (Spiridon 2018). However, purification of lignin is an important step which is prior to analysis. The data obtained from the combination of an array of techniques available for lignin estimation and analysis can provide biologically and technologically relevant information on lignin formation and degradation for a variety of applications as discussed below.

Wet chemical methods include detection of carbonyl, carboxyl, aliphatic and phenolic hydroxyl, methoxy, condensed units and free phenolic groups (Lupoi et al. 2015; Serrano et al. 2018). Although the method has been in practice for past several years, the complex structural integrity of lignin has made the reproducible estimation of specific function challenging. The method used to isolate lignin can change its original structure which causes incorrect estimation and/or functional characterization (Lupoi et al. 2015; Kim et al. 2017). Despite the fact that chromatographic methods are often engaged in concurrence with wet chemical methodology, a novel gas chromatography technique was developed to quantify the methoxyl content of isolated lignin (Li et al. 2012). This method is then improved by addition of iodometry method, which engages complex and laborious elementary steps.

Aromatic lignin compounds, especially derived from bio-originated and artificially aged paper, were studied using capillary zone electrophoresis (CZE) (Lupoi et al. 2015; Abdelaziz et al. 2016). Detection for lignin compound by CZE is relatively inadequate but has advantages over gas chromatography since no derivatization is required and the separation times are shorter.

The tremendous amount of information about lignin structure from infrared spectroscopy (IRS) analysis provides both qualitative and quantitative results (Xu and Wang 2015). IRS techniques promise fast, accurate and low-cost analytical methods. Statistical tools are important companions of IRS techniques, which provide robust models to predict the structural information of biomass. IRS technique would be a better option for lignin study as earlier reported that the traditional composition analysis using wet chemical methods is labour-intensive and time-consuming and thus does not meet industrial requirements for providing quick measurements (Xu et al. 2013). The advancement in the IRS technique is under development, and trails have been continued to promote its application in lignin processing. For the analyses

Table 1 Advantages and disadvantages of analytical methods used for lignin analysis

Method	Advantages	Disadvantages
Fluorescence spectroscopy	1. 2–3 orders of magnitude more sensitive than UV/vis 2. Can probe relative lignin content based on brightness in imaging techniques 3. Capable of localization of lignin in plant cell wall 4. Lignin reveals strong fluorescence 5. Cheap instrumentation 6. More selective than absorbance 7. Non-destructive	1. Requires fluorophore (intrinsic or through labelling) 2. Has not led to robust predictive modelling
Gas chromatography	1. Better spectral resolution than HPLC 2. Selectively analyze only volatile species 3. Headspace sampling limits clean-up requirements	1. Destructive 2. Requires volatile molecules, or must derivatize to promote into gas phase
High-performance liquid chromatography	1. Diverse separation configurations and molecular polarities 2. Non-destructive; can collect fractions 3. Separation of non-volatile or thermally unstable molecules	1. Long analysis times 2. May require sample clean-up
Mass spectrometry	1. Can be used for MW determination 2. Rapid analysis 3. Selectivity 4. Sensitivity 5. Variety of different ionization strategies	1. Destructive
Mid-infrared spectroscopy	1. Use of attenuated total reflectance can 2. remove water contribution to spectra 3. More intense spectra than NIR or Raman 4. Non-destructive 5. Little-to-no sample preparation 6. High-throughput	1. May require specific sample preparation (i.e. KBr pellet) 2. Spectral sensitivity to water may require extensive sample drying

(continued)

Table 1 (continued)

Method	Advantages	Disadvantages
Near-infrared spectroscopy	1. Cheaper instrumentation 2. Field-portable 3. High-throughput 4. Non-destructive 5. Little-to-no sample preparation 6. Water-tolerant	1. Requires multivariate analysis 2. Measures predominantly C–H, N–H, O–H and S–H bonds 3. Physical differences in samples detrimental to spectra
Nuclear magnetic resonance	1. Un-paralleled structural information 2. Available 3. Multi-dimensional NMR provides enhanced spectral resolution 4. Non-destructive 5. Non-invasive 6. Recent advances probing higher-throughput capabilities	1. Lignin quantitation is difficult due to spectral overlap 2. Most techniques considered low-throughput 3. Lack of spectral resolution, especially in lignin region 4. Lack of sensitivity 5. Costly instrumentation
Pyrolysis	1. Eliminates need for lignin isolation high-throughput (MBMS) 2. Facile sample preparation 3. Enables detailed structural characterization of pyrolysis products 4. Small sample amounts	1. Destructive 2. Potentially complex data analysis (i.e. which fragments should be included in monomer quantitation
Raman spectroscopy	1. Complimentary to mid-IR 2. Field-portable 3. High-throughput 4. Multiple excitation sources available non-destructive 5. Little-to-no sample preparation 6. Strong signal from lignin 7. Surface enhancement capabilities 8. Water-tolerant	1. Self-absorption from coloured analytes such as lignin 2. Relatively weak phenomenon, compared to IR 3. Cell-wall phenolics can interfere with lignin measurement; need to extract biomass containing high-extractables 4. Fluorescence can mask signal
Size exclusion chromatography	1. Conventional instrument 2. Expansive array of detectable masses in single step 3. Fast analysis 4. Coupling w/MALLS enabled higher MW aggregates to be detected (UV and RI detection do not measure these species)	1. Association of lignin molecules complicates measurement 2. Complicated by polydispersity 3. Condensation effects

(continued)

Table 1 (continued)

Method	Advantages	Disadvantages
Thermogravimetric analysis	1. High-throughput 2. Ideal for polymers like lignin	1. Destructive 2. Temperature ramping must cause mass changes in analyte
UV/Vis spectrophotometry	1. Cheap and simple instruments 2. Convenient 3. Selective and sensitive 4. Lignin has UV electronic transition	1. Difficulty in determining extinction coefficient for lignin quantitation 2. Ubiquitous absorbance in UV for extraneous molecules requires removal prior to acid soluble lignin measurement
Wet chemistry	1. Well-established techniques 2. Selective for specific moieties	1. Tedious, costly sample preparation and analysis 2. May alter/destroy the sample 3. May only detect specific bonds in lignin toxicity

Reproduced from Lupoi et al. (2015), with permission from Elsevier

of lignins, other characterization IRS could be combined with X-ray diffraction and nuclear magnetic resonance for better understanding of lignin structure.

Mass spectrometry, in its numerous configurations, has been employed to characterize the structure of lignin (Lupoi et al. 2015; Prothmann et al. 2018). Time-of-flight (ToF) secondary ion mass spectrometry (SIMS) is one such MS design that provides enhanced sensitivity to surface measurements of solid analytes with high mass spectral and spatial resolution, requires limited sample preparation and yields rapid analyses over wide mass ranges. Goacher and colleagues developed an improved library of secondary ions for the differentiation between lignin and polysaccharide mass fragments, which allowed the segregation of mass spectral data as either lignin or polysaccharide peaks from pine (Goacher et al. 2011). In addition, Lupoi et al. (2015) identified 64 new spectral peaks for polysaccharide and lignin characterization beyond the standard fragments reported in the literature reported by Goacher et al. (2011).

A new instrument for the chemical analysis of lignin volatilization products from raw sources was constructed by combining pyrolysis or laser ablation with resonance-enhanced multiphoton ionization (REMPI) reflectron ToF-SIMS (Mukarakate et al. 2011; Lupoi et al. 2015). Using this instrument, only lignin peaks could be selectively measured without measuring any background peak of polysaccharide entity. In this approach, various monomeric molecules along with some dimer molecules of lignin were prepared by thermal treatment (Lupoi et al. 2015), which allows selective measurement of regions of interest with high selectivity and sensitivity, and with fewer spectral peaks compared to more vigorous ionization techniques.

Matrix-assisted laser desorption/ionization (MALDI) ToF mass spectrometry has provided another constructive method for lignin study (Richel et al. 2012). When used in positive ion mode, this method was found competent for surveying lignins according to "fingerprint" regions (100–800 Da) and discriminating between lignins extracted with ammonia or formic/acetic acid. MALDI mass spectral imaging (MSI) has recently enabled the in situ analysis of *Eucalyptus globulus* and *E. grandis* lignin (Richel et al. 2012).

2.3 Future Challenge and Development

One of the main constituents in wood is lignin, which is a raw material potentially available in large volumes. The value-added products developed from lignin have potential for several possible benefits with respect to the environment and country economy. Lignin-based products can replace product derived from conventional fossil resources at the same time as energy efficiency methods are optimistic in the pulp and paper sector, since an excess of energy can be generated from the mills as lignin. Economically, lignin would be the foundation of new innovation and energy systems in the underdeveloped countries.

The foremost sources of lignin are cooking liquors generated in wood pulping processes, e.g. black and red liquor (AL-Kaabi et al. 2018). In the producing plant, black liquor is predominantly used for energy conversion but not for the generation of value-added product. But with Logno-boost technology (a new Swedish patented method), value-added product can be produced along with energy conversion (Fredrik et al. 2008). This gives us an opportunity to overcome equipment bottle necks when planning for capacity increases in the pulp mills. This opens up for development of the potential of lignin range of application, as discussed below:

(a) Activated Carbon: It can be made from many different sources, out of which lignin is the one. It is used in a wide range of applications, e.g. cleaning of environmentally hazardous emissions to air and water.
(b) Binders: In small-scale industries and particularly in research area, lignin can serve as binder agents, like construction materials, road making and feed pellets.
(c) Carbon Fibres: Oil-based carbon fibres would be replaced by lignin-derived fibres because of its potential mechanical strength.
(d) Phenol: Lignin can be used to prepare phenolic compounds due to its complex structure; lignin-derived phenols can be used as building blocks in a wide range of chemicals.
(e) Plastic materials: Lignin-derived plastics are mainly used in blend with synthetic polymers, which increases biodegradability of the conventional plastics. Polyethylene terephthalate, poly-urethane, etc., can be produced from lignin, and their demand can be met for a significant time in future. Lignin-based hyperbranched polymers developed by lignin valorization exhibit unique properties, such as high functionality and high solubility.
(f) Sorbents: In terms of water and soil purification, lignin can be act as bio-sorbents. By the hydrolysis of lignin, medical enterosorbents can be synthesized,

like sequestration of bile acids. Moreover, ion exchangers derived from lignin express high sorption activity towards heavy metals such as Cu, Cd and Pb as well as organic pollutants (oil products, pesticides, etc.). However, production has so far mainly been tested in laboratory scale, and the production of bio-sorbents requires large amounts of chemicals.

(g) Dispersants: In dyes, construction industry, jet printing ink, oil well drilling muds, etc.
(h) Emulsifiers: It is added in asphalt to make temperature-stable asphalt emulsions.
(i) Aromatics: Production of vanillin, benzene, toluene, xylene and styrene, etc.
(j) With advanced biorefinery approach, an improved and purified lignin can be isolated with their greater applications in the pharmaceuticals, fragrances and flavouring agent.
(k) Lignin hydrogels possessed mechanically responsive and self-healing properties and, due to no cytotoxic effects, have potential biomedical applications.

The future markets for lignin-based industries are difficult to predict because it depends on various factors. Lignin-based phenols are one of the great interests as lignin is the only known natural biological aromatic polymer. But its technology is in native stage and needs improvements to reduce cost and at the same time needs to fulfil the current quality demands for a profitable market in the coming years. Possibly, this will change with a high increase in oil prices, or legal restrictions. Activated carbon is much easier to produce, and it is already possible to make activated carbon from lignin. Due to the existing range of qualities and precursor alternatives, further studies are needed to see whether there exists a profitable niche market for lignin-based alternatives, for example, use in supercapacitors. Carbon fibres are expected to see a large market growth if the prices of the material can be reduced. Another benefit of lignin-based carbon fibres is that it is used in high-value products, making it an interesting application for investments.

3 Conclusion

The present discussion provides a broad synopsis of recent, cutting-edge, research in the effort to cultivate a more thorough comprehension of lignin, qualitatively and quantitatively. In the past few decades, advancements in analytical techniques such as NMR, mass spectroscopy and vibrational spectroscopy have led to development of specific tools and technologies to explore different lignin applications. However, no single method can be considered as superlative alone. There are many laborious wet chemical techniques still at work, and perhaps necessary for certain purposes like achievement of high throughput, screening of biomass that have high-grade biofuel traits like low lignin content or high S/G ratios, which enable scientific community to survey more plants in shorter time of span with minimum investment. Probing of lignin structure and carbohydrates linkage along with inspection of specific functionalities in lignin using two-dimensional nuclear magnetic resonance was one of the predominant methods introduced in the last decade. Further advancements in these analytical tools, as well as in future methods yet to be conceived, can only build upon the foundation laid by the research as reviewed by Lupoi et al. (2015). The recent interests of the lignin researcher's towards environmentally friendly production methods and technologies for the efficient usage of the abundant lignocellulosic biomass are responsible to give birth of biorefinery concept, which has already made a significant technical progress in recent past. Therefore, in near future, lignocellulosic biomass-derived fuels, chemical and its polymers-based biomaterial would be distinctive certainty.

References

Abdelaziz OY, Brink DP, Prothmann J et al (2016) Biological valorization of low molecular weight lignin. Biotechnol Adv 34:1318–1346. https://doi.org/10.1016/J.BIOTECHADV.2016.10.001

Abreu HDS, Do Nascimento AM, Maria MA (1999) Lignin structure and wood properties. Wood Fiber Sci 31:426–433

Agarwal UP, Ralph SA, Padmakshan D et al (2019) Estimation of syringyl units in wood lignins by FT-Raman spectroscopy. J Agric Food Chem 67:4367–4374. https://doi.org/10.1021/acs.jafc.8b06707

Akgül M, Çöpür Y, Temiz S (2007) A comparison of kraft and kraft-sodium borohydrate brutia pine pulps. Build Environ 42:2586–2590. https://doi.org/10.1016/j.buildenv.2006.07.022

AL-Kaabi Z, Pradhan R, Thevathasan N et al (2018) Beneficiation of renewable industrial wastes from paper and pulp processing. AIMS Energy 6:880–907. https://doi.org/10.3934/energy.2018.5.880

Balakshin MY, Capanema EA (2015) Comprehensive structural analysis of biorefinery lignins with a quantitative ^{13} C NMR approach. RSC Adv 5:87187–87199. https://doi.org/10.1039/C5RA16649G

Berry P, Sterling M, Spink J et al (2004) Understanding and reducing lodging in cereals. Adv Agron 84:217–271. https://doi.org/10.1016/S0065-2113(04)84005-7

Bhuiyan NH, Liu W, Liu G et al (2007) Transcriptional regulation of genes involved in the pathways of biosynthesis and supply of methyl units in response to powdery mildew attack and abiotic stresses in wheat. Plant Mol Biol 64:305–318. https://doi.org/10.1007/s11103-007-9155-x

Bita CE, Gerats T (2013) Plant tolerance to high temperature in a changing environment: scientific fundamentals and production of heat stress-tolerant crops. Front Plant Sci 4:273. https://doi.org/10.3389/fpls.2013.00273

Bland DE, Foster RC, Logan AF (1971) The mechanism of permanganate and osmium tetroxide fixation and the distribution of lignin in the cell wall of *Pinus radiata*. Holzforschung 25:137–143. https://doi.org/10.1515/hfsg.1971.25.5.137

Bonawitz ND, Kim JI, Tobimatsu Y et al (2014) Disruption of Mediator rescues the stunted growth of a lignin-deficient Arabidopsis mutant. Nature 509:376–380. https://doi.org/10.1038/nature13084

Bond J, Donaldson L, Hill S, Hitchcock K (2008) Safranine fluorescent staining of wood cell walls. Biotech Histochem 83:161–171. https://doi.org/10.1080/10520290802373354

Bortoluz J, Cemin A, Bonetto LR et al (2019) Isolation, characterization and valorization of lignin from Pinus elliottii sawdust as a low-cost biosorbent for zinc removal. Cellulose 26:4895–4908. https://doi.org/10.1007/s10570-019-02399-9

Brinkmann K, Blaschke L, Polle A (2002) Comparison of different methods for lignin determination as a basis for calibration of near-infrared reflectance spectroscopy and implications of lignoproteins. J Chem Ecol 28:2483–2501

Calvo-Flores FG, Dobado JA (2010) Lignin as renewable raw material. Chemsuschem 3:1227–1235. https://doi.org/10.1002/cssc.201000157
Caño-Delgado A, Penfield S, Smith C et al (2003) Reduced cellulose synthesis invokes lignification and defense responses in Arabidopsis thaliana. Plant J 34:351–362
Capanema EA, Balakshin MY, Kadla JF (2004) A comprehensive approach for quantitative lignin characterization by NMR spectroscopy. https://doi.org/10.1021/jf035282b
Choong FX, Bäck M, Steiner SE et al (2016) Nondestructive, real-time determination and visualization of cellulose, hemicellulose and lignin by luminescent oligothiophenes. Sci Rep 6:35578. https://doi.org/10.1038/srep35578
Christensen JH, Christensen OB (2007) A summary of the PRUDENCE model projections of changes in European climate by the end of this century. Clim Change 81:7–30. https://doi.org/10.1007/s10584-006-9210-7
Dean J (1997) Lignin analysis. In: Dashek WV (ed) Methods in plant biochemistry and molecular biology. CRC Press, Inc
Dean JC, Navotnaya P, Parobek AP et al (2013) Ultraviolet spectroscopy of fundamental lignin subunits: Guaiacol, 4-methylguaiacol, syringol, and 4-methylsyringol. J Chem Phys 139:144313. https://doi.org/10.1063/1.4824019
Dence CW (1992) The determination of Lignin. In: Lin SY, Dence CW (eds) Methods in lignin chemistry. Springer, Berlin, Heidelberg, pp 33–61
Duan A-Q, Feng K, Wang G et al (2019) Elevated gibberellin enhances lignin accumulation in celery (*Apium graveolens* L.) leaves. Protoplasma 256:777–788. https://doi.org/10.1007/s00709-018-01341-3
Duval A, Lawoko M (2014) A review on lignin-based polymeric, micro- and nano-structured materials. React Funct Polym 85:78–96. https://doi.org/10.1016/j.reactfunctpolym.2014.09.017
Fergus BJ, Procter AR, Scott JAN, Goring DAI (1969) The distribution of lignin in sprucewood as determined by ultraviolet microscopy. Wood Sci Technol 3:117–138. https://doi.org/10.1007/BF00639636
Fredrik O, Hans T, Tomani P, Peter A (2008) Method for separating lignin from black liquor, a lignin product, and use of a lignin product for the production of fuels or materials
Fu K, Yue Q, Gao B et al (2013) Preparation, characterization and application of lignin-based activated carbon from black liquor lignin by steam activation. Chem Eng J 228:1074–1082. https://doi.org/10.1016/J.CEJ.2013.05.028
Fu L, McCallum SA, Miao J et al (2015) Rapid and accurate determination of the lignin content of lignocellulosic biomass by solid-state NMR. Fuel 141:39–45. https://doi.org/10.1016/J.FUEL.2014.10.039
Ge Y, Li Z (2018) Application of lignin and its derivatives in adsorption of heavy metal ions in water: a review. ACS Sustain Chem Eng 6:7181–7192. https://doi.org/10.1021/acssuschemeng.8b01345
Ghanati F, Morita A, Yokota H (2005) Deposition of suberin in roots of soybean induced by excess boron. Plant Sci 168:397–405. https://doi.org/10.1016/J.PLANTSCI.2004.09.004
Goacher RE, Jeremic D, Master ER (2011) Expanding the library of secondary ions that distinguish lignin and polysaccharides in time-of-flight secondary ion mass spectrometry analysis of wood. Anal Chem 83:804–812. https://doi.org/10.1021/ac1023028
Guy C, Kaplan F, Kopka J et al (2007) Metabolomics of temperature stress. Physiol Plant 0:071124124159002. https://doi.org/10.1111/j.1399-3054.2007.00999.x
Hepler PK, Fosket DE, Newcomb EH (1970) Lignification during secondary wall formation in coleus: an electron microscopic study. Am J Bot 57:85. https://doi.org/10.2307/2440382
Horikawa Y, Hirano S, Mihashi A et al (2019) Prediction of lignin contents from infrared spectroscopy: chemical digestion and lignin/biomass ratios of cryptomeria japonica. Appl Biochem Biotechnol 1–11. https://doi.org/10.1007/s12010-019-02965-8
Hu Y, Li WC, Xu YQ et al (2009) Differential expression of candidate genes for lignin biosynthesis under drought stress in maize leaves. J Appl Genet 50:213–223. https://doi.org/10.1007/BF03195675

Hu D, Liu XB, She HZ et al (2017) The lignin synthesis related genes and lodging resistance of Fagopyrum esculentum. Biol Plant 61:138–146. https://doi.org/10.1007/s10535-016-0685-4

Huang Y, Duan Y, Qiu S et al (2018) Lignin-first biorefinery: a reusable catalyst for lignin depolymerization and application of lignin oil to jet fuel aromatics and polyurethane feedstock. Sustain Energy Fuels 2:637–647. https://doi.org/10.1039/C7SE00535K

Irizarry I, White JF (2018) *Bacillus amyloliquefaciens* alters gene expression, ROS production and lignin synthesis in cotton seedling roots. J Appl Microbiol 124:1589–1603. https://doi.org/10.1111/jam.13744

Jang JH, Lee OR (2019) Overexpression of ginseng patatin-related phospholipase pPLAIIIβ alters the polarity of cell growth and decreases lignin content in Arabidopsis. J Ginseng Res. https://doi.org/10.1016/J.JGR.2019.01.004

Jannoey P, Channei D, Kotcharerk J et al (2017) Expression analysis of genes related to rice resistance against brown planthopper, Nilaparvata lugens. Rice Sci 24:163–172. https://doi.org/10.1016/J.RSCI.2016.10.001

Joseleau J-P, Imai T, Kuroda K, Ruel K (2004) Detection in situ and characterization of lignin in the G -layer of tension wood fibres of Populus deltoides. Planta 219:338–345. https://doi.org/10.1007/s00425-004-1226-5

Kim H, Padnakshan D, Li Y et al (2017) Characterization and elimination of undesirable protein residues in plant cell wall materials for enhancing lignin analysis by solutionstate nuclear magnetic resonance spectroscopy. Biomacromol 18:4184–4195. https://doi.org/10.1021/acs.biomac.7b01223

Kováčik J, Bačkor M (2007) Phenylalanine ammonia-lyase and phenolic compounds in chamomile tolerance to cadmium and copper excess. Water Air Soil Pollut 185:185–193. https://doi.org/10.1007/s11270-007-9441-x

Kun D, Pukánszky B (2017) Polymer/lignin blends: interactions, properties, applications

Li H, Chai X-S, Liu M, Deng Y (2012) Novel method for the determination of the methoxyl content in lignin by headspace gas chromatography. J Agric Food Chem 60:5307–5310. https://doi.org/10.1021/jf300455g

Li Y, Yang D, Lu S et al (2018) Modified lignin with anionic surfactant and its application in controlled release of avermectin. J Agric Food Chem 66:3457–3464. https://doi.org/10.1021/acs.jafc.8b00393

Liang M, Davis E, Gardner D et al (2006) Involvement of AtLAC15 in lignin synthesis in seeds and in root elongation of Arabidopsis. Planta 224:1185–1196. https://doi.org/10.1007/s00425-006-0300-6

Ligrone R, Carafa A, Duckett JG et al (2008) Immunocytochemical detection of lignin-related epitopes in cell walls in bryophytes and the charalean alga Nitella. Plant Syst Evol 270:257–272. https://doi.org/10.1007/s00606-007-0617-z

Liu Q, Luo L, Zheng L (2018) Lignins: biosynthesis and biological functions in plants. Int J Mol Sci 19:335. https://doi.org/10.3390/ijms19020335

Lu Y, Lu Y-C, Hu H-Q et al (2017) Structural characterization of lignin and its degradation products with spectroscopic methods. J Spectrosc 2017:1–15. https://doi.org/10.1155/2017/8951658

Lupoi JS, Singh S, Parthasarathi R et al (2015) Recent innovations in analytical methods for the qualitative and quantitative assessment of lignin. Renew Sustain Energy Rev 49:871–906. https://doi.org/10.1016/j.rser.2015.04.091

Mansfield SD (2009) Solutions for dissolution—engineering cell walls for deconstruction. Curr Opin Biotechnol 20:286–294. https://doi.org/10.1016/j.copbio.2009.05.001

Martin A., Chandra RP, Dogu D, van Velzen STJ (2019) analytical staining of cellulosic materials: a review. Dept. of Wood and Paper Science, College of Natural Resources, North Carolina State University

Martone PT, Estevez JM, Lu F et al (2009) Discovery of lignin in seaweed reveals convergent evolution of cell-wall architecture. Curr Biol 19:169–175. https://doi.org/10.1016/j.cub.2008.12.031

Mir Derikvand M, Sierra JB, Ruel K et al (2008) Redirection of the phenylpropanoid pathway to feruloyl malate in Arabidopsis mutants deficient for cinnamoyl-CoA reductase 1. Planta 227:943–956. https://doi.org/10.1007/s00425-007-0669-x

Monties B, Fukushima K (2005) Occurrence, function and biosynthesis of lignins. In: Steinbüchel A, Hofrichter M (eds) Biopolymers online. Wiley-VCH Verlag GmbH & Co. KGaA, Weinheim, Germany

Morais SAL, Nascimento EA, Piló-Veloso D (1991) Studies of eucalyptus grandis lignin. part I: estimation of lignin and polyphenols content in eucalyptus grandis by infrared spectroscopy. J Braz Chem Soc 2:129–132

Mukarakate C, Scheer AM, Robichaud DJ et al (2011) Laser ablation with resonance-enhanced multiphoton ionization time-of-flight mass spectrometry for determining aromatic lignin volatilization products from biomass. Rev Sci Instrum 82:033104. https://doi.org/10.1063/1.3563704

Nicholson DJ, Leavitt AT, Francis RC (2014) A three-stage klason method for more accurate determinations of hardwood lignin content. Cellul Chem Technol 48:53–59

Prothmann J, Spégel P, Sandahl M, Turner C (2018) Identification of lignin oligomers in Kraft lignin using ultra-high-performance liquid chromatography/high-resolution multiple-stage tandem mass spectrometry (UHPLC/HRMSn). Anal Bioanal Chem 410:7803–7814. https://doi.org/10.1007/s00216-018-1400-4

Richel A, Vanderghem C, Simon M et al (2012) Evaluation of matrix-assisted laser desorption/ionization mass spectrometry for second-generation lignin analysis. Anal Chem Insights 7:79–89. https://doi.org/10.4137/ACI.S10799

Rippert P, Puyaubert J, Grisollet D et al (2009) Tyrosine and phenylalanine are synthesized within the plastids in arabidopsis. Plant Physiol 149:1251–1260. https://doi.org/10.1104/pp.108.130070

Sant'Anna C, Costa LT, Abud Y et al (2013) Sugarcane cell wall structure and lignin distribution investigated by confocal and electron microscopy. Microsc Res Tech 76:829–834. https://doi.org/10.1002/jemt.22235

Schilmiller AL, Stout J, Weng J-K et al (2009) Mutations in the cinnamate 4-hydroxylase gene impact metabolism, growth and development in Arabidopsis. Plant J 60:771–782. https://doi.org/10.1111/j.1365-313X.2009.03996.x

Serrano L, Esakkimuthu ES, Marlin N et al (2018) Fast, easy, and economical quantification of lignin phenolic hydroxyl groups: comparison with classical techniques. Energy Fuels 32:5969–5977. https://doi.org/10.1021/acs.energyfuels.8b00383

Simon C, Lion C, Biot C et al (2018) Lignification and advances in lignin imaging in plant cell walls. Annual plant reviews online. John Wiley & Sons Ltd, Chichester, UK, pp 1–32

Spiridon I (2018) Biological and pharmaceutical applications of lignin and its derivatives: a mini-review. Cellul Chem Technol 52:543–550

Srivastava S, Vishwakarma RK, Arafat YA et al (2015) Abiotic stress induces change in Cinnamoyl CoA Reductase (CCR) protein abundance and lignin deposition in developing seedlings of Leucaena leucocephala. Physiol Mol Biol Plants 21:197–205. https://doi.org/10.1007/s12298-015-0289-z

Tao S, Khanizadeh S, Zhang H, Zhang S (2009) Anatomy, ultrastructure and lignin distribution of stone cells in two Pyrus species. Plant Sci 176:413–419. https://doi.org/10.1016/J.PLANTSCI.2008.12.011

Thakur VK, Thakur MK, Raghavan P, Kessler MR (2014) Progress in green polymer composites from lignin for multifunctional applications: a review. ACS Sustain Chem Eng 2:1072–1092. https://doi.org/10.1021/sc500087z

Thévenin J, Pollet B, Letarnec B et al (2011) The simultaneous repression of CCR and CAD, two enzymes of the lignin biosynthetic pathway, results in sterility and dwarfism in Arabidopsis thaliana. Mol Plant 4:70–82. https://doi.org/10.1093/mp/ssq045

Tronchet M, Balagua C, Kroj T et al (2010) Cinnamyl alcohol dehydrogenases-C and D, key enzymes in lignin biosynthesis, play an essential role in disease resistance in Arabidopsis. Mol Plant Pathol 11:83–92. https://doi.org/10.1111/j.1364-3703.2009.00578.x

Vanholme R, Demedts B, Morreel K et al (2010) Lignin biosynthesis and structure. Plant Physiol 153:895–905. https://doi.org/10.1104/pp.110.155119

Vanholme R, Storme V, Vanholme B et al (2012) A systems biology view of responses to lignin biosynthesis perturbations in Arabidopsis. Plant Cell 24:3506–3529. https://doi.org/10.1105/tpc.112.102574

Walker TS, Sharp RE, Durachko DM et al (2003) Root exudation and rhizosphere biology. Plant Physiol 132:44–51. https://doi.org/10.1104/pp.102.019661

Wang C, Hu D, Liu X et al (2015) Effects of uniconazole on the lignin metabolism and lodging resistance of culm in common buckwheat (Fagopyrum esculentum M.). F Crop Res 180:46–53. https://doi.org/10.1016/J.FCR.2015.05.009

Wen J-L, Sun S-L, Xue B-L, Sun R-C (2013) Recent advances in characterization of lignin polymer by solution-state nuclear magnetic resonance (NMR) methodology. Mater (Basel, Switzerland) 6:359–391. https://doi.org/10.3390/ma6010359

Xu F, Wang D (2015) Analysis of lignocellulosic biomass using infrared methodology. Pretreat Biomass 7–25. https://doi.org/10.1016/B978-0-12-800080-9.00002-5

Xu F, Yu J, Tesso T et al (2013) Qualitative and quantitative analysis of lignocellulosic biomass using infrared techniques: a mini-review. Appl Energy 104:801–809. https://doi.org/10.1016/j.apenergy.2012.12.019

Zhang W, Wu L, Ding Y et al (2017) Nitrogen fertilizer application affects lodging resistance by altering secondary cell wall synthesis in japonica rice (Oryza sativa). J Plant Res 130:859–871. https://doi.org/10.1007/s10265-017-0943-3

Zheng M, Chen J, Shi Y et al (2017) Manipulation of lignin metabolism by plant densities and its relationship with lodging resistance in wheat. Sci Rep 7:41805. https://doi.org/10.1038/srep41805

Chemical Modification of Lignin by Polymerization and Depolymerization

Karen Lopez-Camas, Muhammad Arshad and Aman Ullah

Abstract Over the past few years, lignin is seen as the remaining part after paper production or biofuels extraction; the most common method to get rid of it by giving it a profitable use has been burning it in a recovery container for steam production which provides enough energy to power their processes. However, the increase in production has left a higher volume of residual lignin than energy demand. So this vast amount of lignin produced as a by-product has gained increasing interest and attention of researchers to use it as a bio-sourced to develop value-added materials. Lignin has a complex structure with phenolic and aliphatic hydroxyl groups, which make it a promising alternative for the production of several chemical compounds. Different chemical modifications have been applied to transform lignin into new materials, either by polymerization which uses lignin as a monomer in the synthesis of polymers or either by creating new reactive sites or by structural modification of the functional groups already existing in the lignin. In addition to that, depolymerization of lignin is another option which involves its fragmentation into smaller molecules having higher reactivity that can be utilized for the synthesis of biobased polymers and biofuels.

Keywords Lignin · Chemical modifications · Polymerization · Depolymerization

1 Introduction

Due to the high consumption of polymeric materials, the demand has increased considerably generating solid waste. Most polymeric materials are manufactured from fossil fuel, which needs hundreds of years to completely decompose. Besides environmental issues, there are economic concerns such as the increase of fossil fuel prices due to the decline of reserves; it has been predicted that the rate of oil extraction will not provide the needs of the world market causing food, water, and security crises

K. Lopez-Camas · M. Arshad · A. Ullah (✉)
Department of Agricultural, Food and Nutritional Science, University of Alberta, Edmonton, AB, Canada
e-mail: ullah2@ualberta.ca

S. Sharma and A. Kumar (eds.), *Lignin*, Springer Series on Polymer and Composite Materials, https://doi.org/10.1007/978-3-030-40663-9_5

(Wang et al. 2001; Mohanty et al. 2005; Höök and Tang 2013; Capellán-Pérez et al. 2014; Abas et al. 2015). As an alternative to fossil fuels, there has been a shift toward using more renewable resources; the problems that agro-based industries are facing with waste disposal, residues can be considered as a substitute for fossil fuel feedstock due to their large-scale availability, easy accessibility, and many of these agro-wastes are currently discarded or used inefficiently (Sanchez-Vazquez et al. 2013).

One of the biowastes produced in high amounts is lignin. It is the smallest fraction contained in the lignocellulosic biomass, about 15–30%; this bio-based polymer occupies the second place in abundance only after cellulose, but it is the highest natural source with aromatic building blocks. The function of lignin in plants is to provide rigidness and strength. It has a very complex chemical structure that is hyperbranched phenolic polymer which is composed of three essential block units or monolignols called *p*-coumaryl, coniferyl, and sinapyl alcohol which produces the aromatics *p*-hydroxyphenyl, guaiacyl, and syringyl units, respectively, with different methoxylation degrees on the aromatic ring and mainly linked by ether and condensed linkages (Calvo-Flores and Dobado 2010; Xu and Ferdosian 2017; Fernández-Rodríguez et al. 2017; Espinoza-Acosta et al. 2018).

The most common application of lignin during past years is a source of energy predominantly in paper-pulping mills and ethanol cellulose-based industries in which lignin has the most significant percentage of all the residues. Approximately, the lignin generation worldwide every year is 45 million tons (Mtons) with no value-added, and it is expected to increase at least five times more by 2030 due to an increase in biofuels production. There is an excellent opportunity to convert lignin into renewable-based materials such as for the synthesis of valuable chemicals or polymers, while unfortunately there is only 1–2% of lignin used for this purpose (Cotana et al. 2014; Santos et al. 2015).

One of the main reasons why lignin has not been exploited to its high potential is due to its complex nature; the monolignols contain methyls groups that increase the steric hindrance and consequently reduces the reactivity, limiting the use of lignin as feedstock for value-added products (Xu and Ferdosian 2017; Park et al. 2018). This chapter gives a brief description of two of the main approaches for lignin applications. The first part is focused on the lignin polymerization including diverse techniques reported to improve reactive sites on lignin structure as well as the ways of using lignin as a macromonomeric unit to produce most common polymers such as polyesters, polyurethanes, epoxide resins and phenolic resins. Furthermore, the next alternative is the depolymerization of lignin, which consists of breaking its linkages, converting the complex polymer into oligomers or aromatic monomers. Through the lignin depolymerization process, it is possible to convert lignin into diverse compounds of low molecular weight such as aromatics and aliphatic compounds cyclohexane (Kleinert and Barth 2008; Ma et al. 2014; Cronin et al. 2017b; Dhar and Vinu 2017).

2 Lignin Polymerization

Lignin in its natural form is constituted by different functional groups such as hydroxyl, carboxyl, carbonyl, and methoxyl groups; these moieties provide a reactive site that allows the synthesis of polymers whether using lignin without chemical modification or improving the existing functional groups or by creating new active sites (Fig. 1) (Laurichesse and Avérous 2014). The following section concentrates on the more common methods used to increase the reactive sites on lignin structure by various chemical modifications and the applications in polymeric materials as well as the use of lignin without modifications in polymers production.

2.1 Methods for the Synthesis of New Reactive Sites in Lignin Structure

Different changes in lignin chemical structure have been studied over the years to increase its reactivity to increase the applications field. The addition of new active sites in the lignin complex network could improve properties in the final product such as mechanical, thermal, physicochemical, and so on. The most studied reactions are sulfomethylation/sulfonation, nitration, amination, alkylation/dealkylation, and hydroxyalkylation, which are briefly discussed in the following sections.

2.1.1 Sulfomethylation and Sulfonation of Lignin

Sulfomethylation of lignin is the addition of methylene sulfonate groups, which provides an improvement in the tanning capacity of lignin products; this reaction involves the addition of methylene sulfonate group ($–CH_2SO_3^-$). In this reaction, the sulfonate groups are introduced by using sulfuric acid or sodium sulfite ($–SO_3^-$) (Meister 2002; Figueiredo et al. 2018).

The most mentioned application for sulfomethylated lignin is as a dye dispersant, due to an improvement in the dispersibility properties, while sulfonated lignin which has a high zeta potential value is mostly used as a dispersing agent in cement matrix due to the presence of electrostatic repulsion forces between the cement particles (Figueiredo et al. 2018). The principal industry that uses this type of modified lignin is the cement and mining industry in which it is applied to obtain higher metal content extracted from the ore (Meister 2002). Other applications of sulfonated lignin due to its new properties are fillers modifiers, flocculants, dust suppressant, dispersants, in composites materials, and others (Oveissi and Fatehi 2015; Aro and Fatehi 2017). Pretreatments such as oxidation, hydrolysis, and hydroxymethylation have been used to increase the active sites of lignin to achieve a higher sulfonation degree (Ouyang et al. 2009; Aro and Fatehi 2017). Ouyang et al. reported a sulfonation

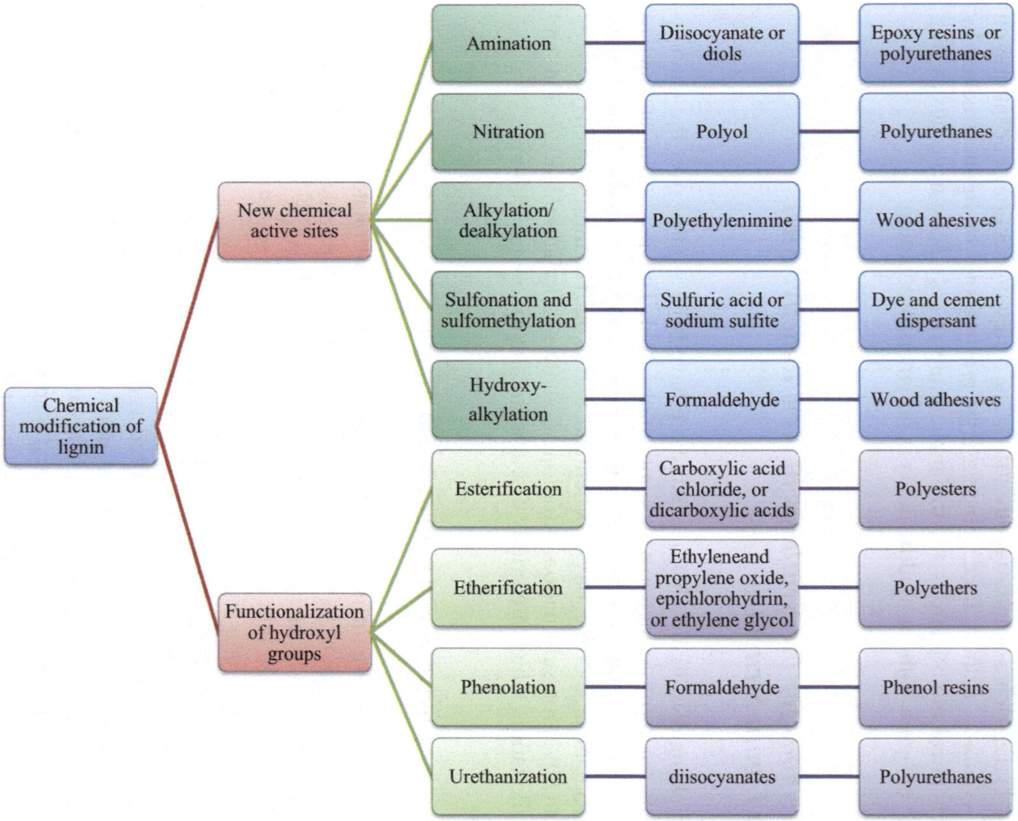

Fig. 1 Types of chemical modifications used for lignin to obtain different polymers

degree of 96.9% when using a combination pretreatment of hydroxymethylation-oxidation reaction; also the zeta potential absolute value increased allowing to use the sulfonated lignin as dispersant on cement matrix (Ouyang et al. 2009).

2.1.2 Nitration of Lignin

Nitration is another method to create new active sites in lignin; this is achieved in non-aqueous solvents with the addition of nitrating agents. The final product is an amorphous brownish powder with a molecular weight between 600 and 2000 Da, and 6–7% of nitrogen content (Meister 2002; Laurichesse and Avérous 2014; Figueiredo et al. 2018). An example of this modification was showed in a polymeric network prepared by mixing lignin modified by nitration with castor oil-based polyurethane prepolymer; films were prepared by casting method and compared with ordinary polyurethane film to study the performance of the material. The results exhibited that the mechanical properties were improved with the addition of 1.4–6% of modified lignin due to an enhancement of cross-linking density (Zhang and Huang 2001). Khabarov et al. reported a nitration procedure of sulfate lignin using nitric acid in an aqueous-dioxane medium under the homogeneous condition which resulted in high nitration degree at 50 °C using a low amount of nitrating agent (Khabarov et al. 2016).

2.1.3 Amination of Lignin

Another modification of lignin is an amination, which introduces an amine functional group into the lignin. The amination of lignin is represented in Fig. 2. The principal and straightforward pathway for amination are by using the Mannich reaction, which consists of charging amines (primary, secondary, or ammonia) with formaldehyde

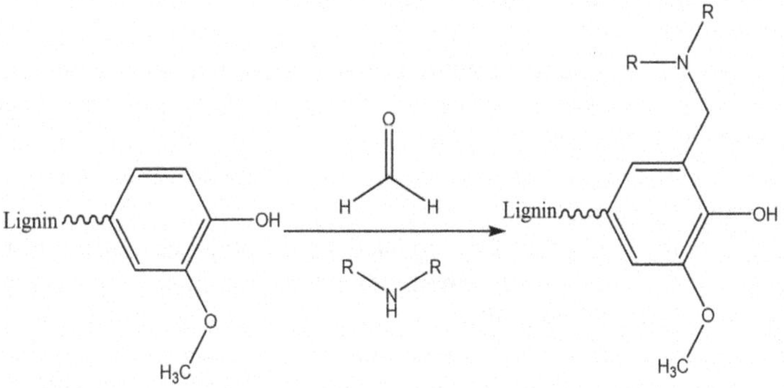

Fig. 2 Reaction scheme showing lignin modification with an amine and formaldehyde. Adapted from Figueiredo et al. (2018)

($H_2C{=}O$) under different pH environments (Figueiredo et al. 2018). The reactions are carried out between a carbon with high electron density and an immonium ion ($CH_2{=}N^+R_2$, where R is the alkyl group) produced from $H_2C{=}O$ and an amine. The importance of Mannich reaction is the possibility of forming a new single C–C bond (Laurichesse and Avérous 2014). Yue et al. applied amination reaction in soda lignin using $H_2C{=}O$ and diethylenetriamine [$HN(CH_2CH_2NH_2)_2$] to apply it as an interfacial modifier in PVC/wood-flour mixtures; the amination process increased the tensile and impact strength of the final material; however, the incorporation of more than 2 wt% of aminated lignin has a counterproductive effect on the mechanical properties (Yue et al. 2011).

Studies have reported the effect of lignin pretreatment as a way to increase the reactive sites and consequently the amination degree. Some of these reactions are phenolation, depolymerization, and demethoxylation (Wang et al. 2018). Jiao et al. produced aminated lignin to use it as an alternative of nitrogen fertilizer of slow-release that indeed showed an improvement compared to urea the conventional fertilizer. The aminated lignin was achieved by applying a phenolation pretreatment followed by Mannich reaction with ethanediamine at 70 °C for 3 h (Jiao et al. 2019). Similarly, a lignin model compound, 1-guaiacyl-1-p-hydroxyphenylethane, was reacted with dimethylamine and H2C=O at 60 °C to increase reactivity by using phenolation pretreatment (Matsushita and Yasuda 2003). Du et al., also reported that the reaction of aminated lignin is enhanced by a phenolation pretreatment process, due to the formation of a new reactive site in the ortho position of the phenolic hydroxyl besides the 5-position of guaiacyl (Du et al. 2014).

2.1.4 Alkylation/dealkylation

Alkylation is the reaction in which an alkyl group is introduced into an organic compound by substitution or addition, and the opposite for dealkylation in which the alkyl group is removed (Speight and Speight 2017). The removal of a methyl group is the most recognized dealkylation method in lignin since representing a by-product of dimethyl sulfoxide (DMSO), which is an organosulfur liquid that dissolves in non-polar and polar compounds; DMSO is essential for polymer synthesis. Figure 3 shows the reaction that takes place between lignin and molten sulfur in basic media; firstly, the methyl groups are shifted to sulfur to produce dimethyl sulfide (DMS), and then by an oxidation process with nitrogen dioxide, DMS is converted to DMSO (Laurichesse and Avérous 2014).

One approach for using the demethylated lignin obtained is the production of wood adhesives which consist of demethylated kraft lignin and a polyethylenimine in an optimum ratio of 1:1 that allowed an enhancement of samples by increasing the strength and hydrophobicity. This combination also serves as a formaldehyde-free wood adhesive. Formaldehyde is a non-renewable essential ingredient of wood adhesives, but through many studies, it has been concluded that it causes skin and respiratory problems and is classified as a human carcinogen, increasing the importance of finding alternatives (Liu and Li 2006).

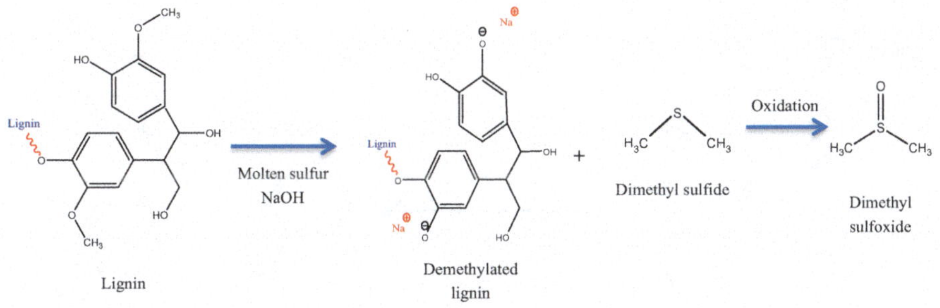

Fig. 3 Scheme representing the preparation of dimethyl sulfoxide by lignin dealkylation. Adapted from Laurichesse and Avérous (2014)

2.1.5 Hydroxyalkylation

As mentioned above, finding options to replace the toxic and non-renewable raw materials commonly used for the synthesis of phenol-formaldehyde resins are of high importance as a way to reduce environmental impact and also an alternative to fossil fuels. Lignin was proposed many years ago as a promising natural resource with the possibility to react with formaldehyde that leads to the formation of wood glues (Laurichesse and Avérous 2014). Different authors quoted two types of reactions that can be applied to obtain lignin-phenol-formaldehyde (LPF) adhesives: by reacting lignin with formaldehyde; and by chemical modifications being the most common reactions used methylolation, phenolation, and demethylation (Vázquez et al. 1997; Hu et al. 2011; Laurichesse and Avérous 2014). Modification of the functional groups has been evaluated more often due to the low quality of the final product when only crude lignin is used; the problematic relies on the steric hindrance of the monolignols, which diminish the reactivity with formaldehyde (El Mansouri et al. 2007; Cheng et al. 2012).

Methylolation, also known as hydroxymethylolation, of lignin, is a reaction in high pH medium in which hydroxymethyl groups are inserted in the fifth carbon in guaiacyl units after reaction between formaldehyde with lignin (Hu et al. 2011; El Mansouri et al. 2011a; Cheng et al. 2012). Mansouri et al. reported that lignin after undergoing a methylolation increased the number of –OH groups which turn out into new reactive sites for the synthesis of the resins; the authors made a substitution of formaldehyde with glyoxal to produce an environmentally friendly material (El Mansouri et al. 2011a). A different way to increase lignin reactivity is by phenolation or phenolysis; this reaction is carried out under organic solvents in low pH medium by reacting phenol with lignin (Hu et al. 2011); Fig. 4 shows part of the lignin structure before and after phenolation. Final phenolated lignin presents better properties that allow its use in thermosetting resins; a successful phenolation reaction highly depends on the parameters used such as reaction time, amount of solvent, temperature, catalyst and most important the proportion of phenol/lignin (Podschun et al. 2015). More examples of lignin phenolation will be described further in this chapter.

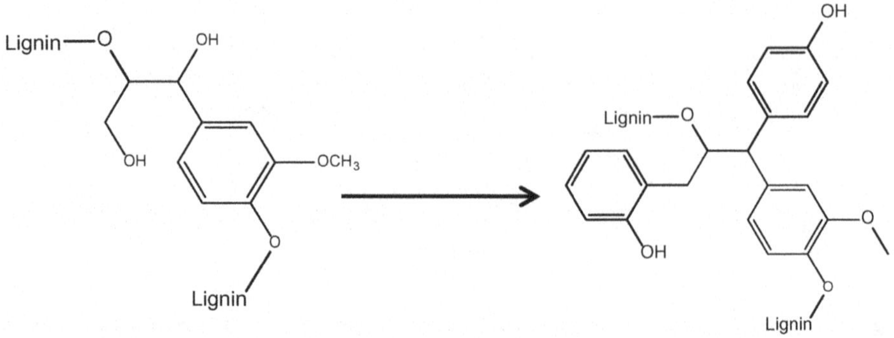

Fig. 4 Representation of the modification of lignin after phenolation process (Podschun et al. 2015)

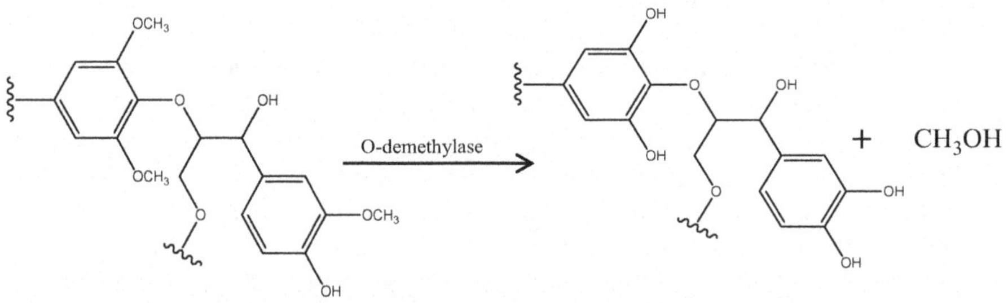

Fig. 5 Demethylation of lignin (Venkatesagowda and Dekker 2019; Venkatesagowda 2019)

Contrary, the demethylation method increases the reactivity by eliminating the methoxyl groups (–O–CH_3) present on the ortho position of the monolignols units (Hu et al. 2011; Venkatesagowda 2019). Some biological demethylation processes have been reported in the literature, in recent publications. For example, *O*-demethylases enzymes are capable of cleaving the methyl groups (–CH_3) from lignin structure without modifying the other bonds existing (see Fig. 5) (Venkatesagowda and Dekker 2019; Venkatesagowda 2019). Zou et al. found that the fungi culture of *Cylindrocladium* ssp. and *Aspergillus* sp. after the third week of incubation were able to eliminate about 40% of the total methyl groups present in lignin structure and keeping about 90% the rest of the lignin structure undamaged (Zou et al. 2015).

2.2 Modification of Lignin's Hydroxyl Groups

Among functional groups present in lignin structure, the phenolic hydroxyl groups and aliphatic hydroxyl groups at C-α and C-γ positions on the side chain are the most reactive and play a crucial role in the chemical reactivity of the newly developed products. The functionalization of hydroxyl groups can lead to the creation of polyol-lignin derivatives; the methodologies reported in the literature at the moment for the modification of hydroxyl groups are esterification, etherification, phenolation, and urethanization (Laurichesse and Avérous 2014; Figueiredo et al. 2018).

2.2.1 Esterification

Esterification is considered one of the easiest methods to modify the hydroxyl groups present on lignin; this process involves the reaction of the hydroxyl groups with either a carboxylic acid or an acyl chloride or an acid anhydride. The bifunctional property of these chemical reagents allows the production of lignin-based polyester networks. The process is done by a condensation reaction or ring-opening polymerization, and depending on the reagent used, a condensed molecule arises such as water, hydrochloric acid, and carboxylic acid. (Laurichesse and Avérous 2014; Duval and Lawoko 2014; Matsushita 2015; Figueiredo et al. 2018).

Recent work on successful esterification of lignin has shown how this chemical modification improves lignin reactivity with other polymers such as PHBV, PP, HDPE, and others (see Table 1). Luo et al. demonstrated that increasing the carbon chain length of the regent used for esterification decreases the glass transition temperature (T_g), and increases the hydrophobicity of lignin which rise its affinity with the polymer used (PHBV and PP), resulting in an improvement of the tensile elongation of the final products when compared with unmodified lignin (Luo et al. 2017). The same authors (Luo et al. 2018a) reported that the incorporation of double bonds could improve the reactivity of esterified lignin. Technical lignin was esterified with anhydrides of the same carbon chain length differing in the unsaturation (crotonic and butyric anhydride); both esterified lignins were thermally processed with a cross-linking agent, dicumyl peroxide, to promote the formation of cross-linking structures. The presence of double bonds increased the T_g; however, no comparison was made against no modified lignin (Luo et al. 2018a). Organosolv lignin from two different sources, spruce and eucalyptus wood, were esterified with dodecanoyl chloride. Both presented an increase on the hydrophobicity, M_n and Mw, and a decrease of T_g after esterification which was related to the addition of the aliphatic chain, and lignin source had no significant effect (Gordobil et al. 2017). Likewise, Koivu et al. by using acyl chlorides esterified kraft lignin successfully.

Compound ε-Caprolactone has been widely used in esterification reactions to synthesize lignin-based polyesters using chemically modified lignin, which showed improvements in thermal properties and increased the reactivity (Matsushita 2015). Luong and collaborators (Luong et al. 2012) polyesterified lignin extracted from black liquor with sebacoyl chloride to synthesize lignin-based copolyesters that showed to be a good alternative as an eco-friendly material with properties almost similar to the commercial lignin-copolyester.

Table 1 Summary of some polyesterification reactions of different lignin sources

Lignin type	Reagents	Reaction parameters			Co-monomer	Applications	Ref.
		T (°C)	Time (h)	Catalyst			
Indulin AT-softwood kraft lignin and Protobind 1000 lignin from agricultural fiber soda	Crotonic and butyric anhydride	120	24	1-methylimidazole	Dicumyl peroxide as initiator	Increase lignin reactivity for a further reaction with other polymers	Luo et al. (2018a)
	Acetic, propionic, butyric, and hexanoic anhydride	65, 95, 120, and 140, respectively	24	1-methylimidazole	PHBV PP	Lignin-based polymers	Luo et al. (2017)
Organosolv lignin from spruce wood and eucalyptus wood	Dodecanoyl chloride	20	2	Pyridine	N.A.	Protective agent on wood products	Gordobil et al. (2017)
Industrial hardwood kraft lignin, softwood lignin, organosolv lignin, soda lignin, and hydrolysis lignin	Acetic, propionic, butyric, and hexanoic anhydride	50	3	*N*-methylimidazole	HDPE	Lignin-based polyethylene	Dehne et al. (2016)

2.2.2 Phenolation

Phenolation is one of the most promising chemical methods that can be applied to lignin to increase its reactivity. As it was mentioned in the hydroxyalkylation section, phenolation consists of the reaction of phenol with the lignin structure; this occurs in an acid environment. Lignin modified by phenolation improves its reactivity to use it as a phenol replacement which can efficiently react with formaldehyde due to phenolic free *ortho* and *para* units (Laurichesse and Avérous 2014). Phenolation of lignin is of high interest as an alternative in the manufacturing of phenol-formaldehyde resins (Jiang et al. 2018). Different reaction conditions have been studied to find the best parameters for a feasible phenolation of lignin. In an industry of wood-based composites, the adhesive is an essential constituent in the final properties of the product. Phenol-formaldehyde resins are the adhesives used in concentrations between 5 and 15% w/w; the final price of the wood product highly depends on the resin cost; thus, finding the cheaper options is the main attraction for wood industries. Lignin-based resins have been proved capable of replacing up to 30% of the phenol in phenol-formaldehyde resins and used as a wood adhesive for particle board (Cetin and Ozmen 2002).

Jiang and collaborators reported phenolation of pine kraft lignin and sweet gum biorefinery lignin under optimized conditions of 90 °C temperature for 2 h with a ratio of lignin/phenol (L/P) 3/5 with 5% acid charge resulting in a 30% phenol insertion, and L/P of 2/5 with 5% of acid charge at 110 °C for 2 h provided an introduction of 60% phenol, respectively. The phenolated lignin showed more free reactive sites which improve its reactivity to use it in thermoset applications (Jiang et al. 2018). The reaction parameters play an essential role in the final degree of phenolations, and it has been observed that at higher concentrations of acid, a lower degree of phenolation was obtained, while the temperature gives a positive effect on the yield depending on the acid charge used. If high concentrations are used, then the temperature should be lower to avoid excessive severe reaction conditions. Moreover, the humidity of lignin itself could reduce the degree of phenolation with water concentrations of about 17% in the reaction, which means giving 50% less degree of phenolation (Podschun et al. 2015; Jiang et al. 2018).

2.2.3 Etherification

Lignin-based polyether is primarily obtained by the etherification method; some of the polymerization reactions applied are based on using: alkaline oxides (such as ethylene and propylene oxides), glycidyl chloride, cross-linking using diglycidyl ethers, solvolysis with ethylene glycol. However, the reaction with alkaline oxides has shown to be a promising technique to achieve the etherification of lignin, predominantly by using propylene oxides. This method is most well-known as "oxypropylation"(Laurichesse and Avérous 2014; Matsushita 2015; Mai and Elder 2016; Figueiredo et al. 2018) (Fig. 6).

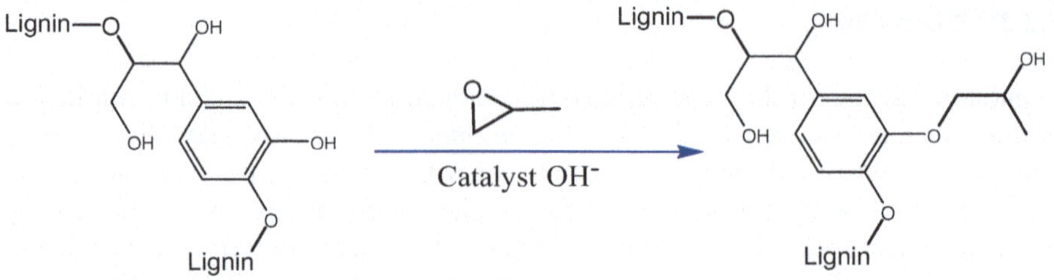

Fig. 6 Oxypropylation of lignin with propylene oxide. Adapted from Laurichesse and Avérous 2014 and Lee et al. 2017

Oxypropylation has been applied to different lignin sources; this process helps to increase the accessibility to hydroxyl groups which are blocked by others bonding, hydroxyl groups are released from steric hindrance and/or electronic constrains. This reaction inserts multiple ether moieties on the lignin structure turning into liquid polyols (Cateto et al. 2009; Bernardini et al. 2015). Subsequently, the polyols obtained are used in the synthesis mostly in polyurethane foams formulations (Bernardini et al. 2015). However, there are few reports focused on using oxypropylated lignin for the production of polyesters, through reacting lignin-based polyols with dicarboxylic acids (Lee et al. 2017).

Different approaches have been used to optimize the oxypropylation to obtain polyols with higher functionality to produce flexible or rigid polyurethanes; Cateto and coworkers prepared rigid polyurethane foams (RPUF) by using oxypropylated lignin of four different sources, Alcell, Indulin AT, Curan 27-11P, and Sarkanda. This reaction was optimized by carefully selecting the proportion of lignin/propylene oxide ratio/the concentration of catalyst (potassium hydroxide) for each type of lignin (Cateto et al. 2009). Later on, these RPUF were evaluated to determine the biodegradability; results showed that polyurethanes using lignin-based polyol were resistant to biodegradation when they are prepared with the proportion of 30/70/2 (Cateto et al. 2014). Bernadini et al. produced flexible polyurethane foams (FPUF) using soda lignin with and without oxypropylation to evaluate if this chemical modification of lignin can improve the FPUF final properties; the foams were prepared with polypropylene glycol triol and castor oil as chain extenders and mixed with polymeric diphenylmethane diisocyanate. Oxypropylated lignin increased the flexibility of the final product, and due to the oxypropylation reaction, part of the lignin structure was deteriorated, causing lighter foams (Bernardini et al. 2015). Additionally, modified lignin can be dissolved easier in organic solvents, which facilitate characterization (D'Souza et al. 2015).

2.2.4 Urethanization

Last but not least, urethanization is also an important hydroxyl functionalization process on lignin. Hydroxyl groups react with isocyanate groups, giving active

RNC=O sites to form a urethane link via an exothermic reaction. This reaction makes the lignin an alternative raw material for the synthesis of polyurethanes, which is based on two main steps. The first one is a one-step reaction by mixing lignin diisocyanate with another diol co-monomer, and the second one is a two-step reaction procedure which includes the synthesis of an isocyanate-based prepolymer followed by the polymerization reaction with the prepolymer and lignin as the polyol chain extender. Studies have shown that lignin can be utilized in the synthesis of polyurethanes with a concentration of up to 30%, while the use of excess amount makes the material brittle (Laurichesse and Avérous 2014; Figueiredo et al. 2018).

2.3 Lignin-Based Polymers

As it was explained in previous sections, lignin can be chemically modified to increase the reactivity and its applications, the addition or improvement of functional groups on lignin allows using it as a macromonomer for the synthesis of different polymers. Due to the functional groups attached to lignin can be utilized for the synthesis of polymers such as polyurethanes, polyesters, epoxide resins, and phenolic resins (Upton and Kasko 2016).

However, lignin not necessarily needs to be modified for its application to synthesize polymers; naturally, lignin contains both phenolic and aliphatic hydroxyl groups that can act as reactive sites for polymerizations. However, in this case, the lignin structure selected to obtain higher yields of polymerizations will depend on the isolated methods and lignin sources. Thus, functionalities have to be adequately characterized to use it as a macromonomer (Laurichesse and Avérous 2014; Figueiredo et al. 2018).

2.3.1 Polyesters

Polyesters are characterized for the ester linkage formed during the reaction either by carboxylic acid with diols or dialdehydes, or by self-polymerization reactions of hydroxycarboxylic acid, or by ring-opening polymerization. Lignin-based polyesters are mainly obtained by a condensation reaction (Figueiredo et al. 2018). As it was mentioned in the first sections of this chapter, lignin possesses a low reactivity, which is the main reason why the synthesis of polyesters using lignin as itself has not reached a great potential. However, there are some publications about the use of lignin in combination with other polymers or chemicals compounds to explore the possibilities of creating new bio-based polymers. Luong et al. synthesized a copolyester using lignin and sebacoyl chloride with a molecular weight of 31,800 equivalents with four to five repeating units of lignin macromonomers (Luong et al. 2012). One of the main parameters to consider in polyester synthesis is the ratio of hydroxyl groups from lignin with the other functional group of the monomer used for the synthesis. However, most of the time, measuring the OH groups is quite complicated due to the complex arrangement of the lignin structure. Kraft lignin was synthesized with sebacoyl chloride using acyl chloride/OHratio of 1, 2, and 3; the copolymer showed a higher molecular weight with a ratio of 2 and with a reaction temperature of 120 °C; this improvement of the molecular weight helps to increase the degradation temperature (Thanh Binh et al. 2009).

Park and collaborators synthesized lignin from black liquor with Ɛ-Caprolactone copolymers by using mass ring-opening polymerization. The resulting copolymer showed an improvement in thermal properties (Park et al. 2019). Likewise, Kai et al. developed lignin-based copolyester with Ɛ-caprolactone and L-lactic acid by using ring-opening polymerization, the main objective of this resulting polymer was to use it as an antioxidant nanofiber destined for healthcare industry (Kai et al. 2017).

2.3.2 Phenolic Resins

Phenolic resins are thermosets materials produced by the acid or based catalyzed step-growth polymerization of phenols and aldehydes that can be added in different ratios which determine the cross-linking degree in the final polymer. The characteristic phenolic resin is the adhesive phenol-formaldehyde resin (PFR). As it was mentioned before, lignin can be an alternative for the synthesis of more environment-friendly phenol-formaldehyde resins, lignin without chemical modifications could act as a monomeric phenol part (Upton and Kasko 2016).

The production of phenolic resins started since 1911 with the bakelite, the first polymer developed from synthetic compounds. Nowadays, the wide applications of phenolic resins and its excellent properties have increased its consumption to about 6 million tons per year (Foyer et al. 2016). Until a couple of years ago, many researchers worked on the incorporation of lignin in the phenol-formaldehyde as a replacement of the non-biobased and highly toxic compounds, but with some drawbacks on the final product. However, a new sustainable phenolic-lignin-based resin for the industry of wood adhesives has been developed. CatLignin technology is an ideal and potential replacement for the toxic phenol in the standard adhesives used in wood industries. The resin was developed by the scientist of VTT Technical Research Center of Finland Ltd., who affirm that the CO_2 footprint of resins can be reduced up to 80% compared to commonly used phenol compounds. Besides this, use lignin also reduces the amount of formaldehyde needed (Wikberg et al. 2018).

Diverse studies have been focused on the methodologies to improve the use of lignin in PFR. Vithanage et al. studied the viability of using formatted assisted fast pyrolysis of lignin oil and its extractives as a replacement of phenol in resins. The resulting product showed that lignin is a potential renewable source to substitute synthetic phenols (Vithanage et al. 2017). Some researchers used modified lignin through some of the chemical reactions explained in the previous paragraphs; however, the importance of replacing phenol in PFR with unmodified lignin lies in decreasing the number of other chemicals required and saving energy and time meaning also the reduction of final costs that makes these approaches more attractive to industries. Kalami and collaborators evaluated the replacement of 100% phenol in resorcinol formaldehyde adhesive for lignin-based resin. Both materials were tested and compared; the results showed no significant difference between the shear strengths of plywood samples elaborated with the 100% lignin-based adhesive (Kalami et al. 2017). The same author reported the effect of lignin source, similarly to other types of chemical reactions applied on lignin; those are highly dependent on its structure; in the study, the authors evaluated the sources: softwood, hardwood, wheat straw, and corn stover and lignin obtained by different isolation process: kraft, organosolv, soda, sulfite, and enzymatic hydrolysis. The results showed that lignin with high concentrations of *p*-hydroxyphenyl or *p*-coumaric acid (mainly found in annual crops) has higher reactivity toward formaldehyde, thus the best candidates to replace phenol in phenol-formaldehyde formulations (Kalami et al. 2018).

2.3.3 Epoxide Resins

Epoxide resins are a type of thermoset polymer, which is composed of monomers with at least one epoxide group; the leading applications are coatings, adhesives, composites, and electronic materials. The most commercialized epoxide resin is prepared by the synthesis of bisphenol A with epichlorohydrin; nevertheless, some studies claim that bisphenol A due to its similarity with the hormone estrogen can cause impotence and cancer. Thus, there is a significant interest and an urgent need to find alternatives to avoid the use of bisphenol A in the synthesis of epoxide resins (Asada et al. 2015; Upton and Kasko 2016).

Before explaining how lignin can be an alternative source in the synthesis of epoxide resins, it is vital to mention that no epoxide groups are present on lignin naturally; thus, lignin needs to be chemically modified to obtain lignin-based epoxide resins (Upton and Kasko 2016).

The easiest way to introduce epoxide groups into lignin structure is through the functionalization of phenolic hydroxyl groups with epichlorohydrin in a basic medium (Fig. 7). Ding et al. studied the partial replacement of bisphenol A in epoxide resins by modified lignin by Mannich reaction which resulted in an improvement in thermal stability of the final product compared with a neat epoxy resin (Ding et al. 2014). The addition of 12.9 wt% modified lignin in the synthesis of epoxide resin improved the mechanical properties compared with the commercial epoxy resins cured with aromatic agents (Nikafshar et al. 2017).

Lignin has been modified through different chemical reactions to enhance its potential application as feedstock for biobased epoxide resins. LignoBoost™ softwood kraft lignin isolated from black liquor was modified with epichlorohydrin in water-organic solvents media with KOH as the catalyst, and resin epoxides were prepared, which resulted in higher charcoal yield and lower degradation rates at high-temperature making it a promising flame-resistant epoxide resin (Jablonskis et al. 2018). Kraft lignin recovered from pulping liquor was modified by methylolation using formaldehyde and glyoxal, and then, methylolated lignin was used for the synthesis of lignin-based epoxy resins (El Mansouri et al. 2011b).

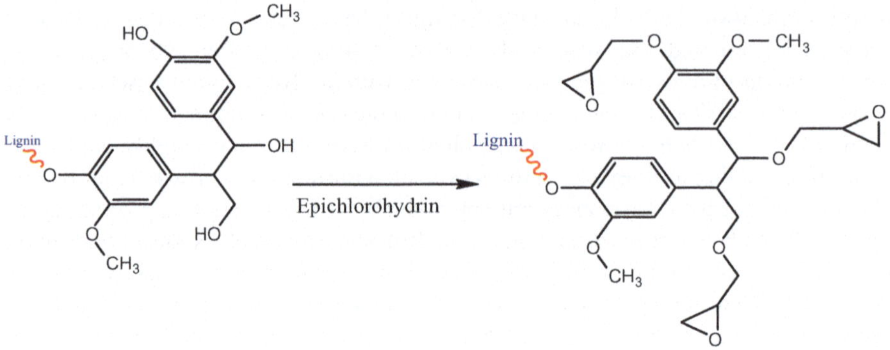

Fig. 7 Epoxide functionalization via a condensation reaction between lignin and epichlorohydrin. Adapted from Upton and Kasko (2016)

2.3.4 Polyurethanes

The addition polymerization of di- or polyisocyanates with polyols results in the synthesis of polyurethanes (PU), the reaction of both monomers produce carbamate (urethane) linkages. The PU coatings, spandex fibers, elastomeric wheel, and tires, basic polyurethanes foam products, adhesives, and carpet underlay are some examples of the most known applications of this type of polymer. The applications depend on the monomers used, and the type of reaction applied, which will provide different physical and chemical properties to the final product. Some examples of diisocyanates used are toluene 2,4-diisocyanate, 4,4′-methylene diphenyl diisocyanate (Sáenz-Pérez et al. 2016; Gang et al. 2017), hexamethylene diisocyanate (Karak 2016), and others. Diisocyanates are highly reactive, which is the main advantage of this reaction. However, it is a toxic compound that reacts readily with human cells causing health problems (Hu et al. 2014; Upton and Kasko 2016). By 2021, it is projected to have a market demand of about 6.3 million metric tons of polyurethane for use only in wood and furniture, and the overall global demand is expected to increase even higher than 19 million metric tons (Statista 2019). A report showed that in 2017, the total income generated by the polyurethane industry was USD 37.9 billion (Kevin et al. 2018).

As it was discussed in the previous sections, lignin contains different functional groups and reactive sites in its structure that can be used for the synthesis of polyurethanes. A vast amount of studies have reported the use of renewable materials as a building block to replace petroleum chemicals, and lignin has been proved as a potential material for the polyurethane industry. Recently, biodegradable RPUF was synthesized, lignin with polyol obtained from soy oil, and a polymeric methyl diphenyl diisocyanate were processed through one-pot and self-rising method, resulting in a 3D macromolecular structure with thermal and mechanical properties improved; the biodegradable rigid foams obtained are shown in Fig. 8. The concentrations of lignin powders used on each foam from a to f were 0, 5, 10,

Fig. 8 Biobased polyurethane foams obtained using lignin and soy oil-derived polyols at different concentrations of lignin. Reproduced from Ref. Luo et al. (2018b) with permission from Elsevier

15, 20, and 25% respectively; the foam color shows a variation from light to dark brown color with the increase of lignin percentage (Luo et al. 2018b).

Pan and Sadler evaluated the use of lignin as polyol to develop RPUF. Two types of lignin were evaluated: hardwood ethanol organosolv lignin and hardwood kraft lignin. The majority of lignin introduced in the foam obtained was chemically cross-linked, the density decreased by 30% with the addition of 50% lignin in the formulation which is attributed to the production of larger cells and less uniform cellular structure. As well as in the other type of polymers, the poor accessibility of hydroxyl groups in lignin affects the final properties. However, the use of butanediol, a chain extender, significantly improved the access of hydroxyl groups in lignin, increasing the compressive strength in the foam (Pan and Saddler 2013). In another study, different amounts of industrial soda lignin were mixed with xylaric acid using methylene diphenyl diisocyanate as a coupling agent resulting in a bio-based polyurethane with high Young's modulus values and thermal stability up to 550 °C (Pohjanlehto et al. 2014).

Saito (Saito et al. 2013) reported that coupling of high molecular weight lignin with rubbery diisocyanate could result in tough thermoplastic polyurethanes. More network formation was observed with a concentration of 65 wt% lignin, which resulted in higher modulus in the rubbery plateau region. A different approach was used for the synthesis of sustainable polyurethane, where carbonated soybean oils were treated with kraft lignin by a three-step reaction; the percentage of these biosources was as high as 85%; lignin contributed to the rigidity of the final product (Lee and Deng 2015). Hayati et al. reported the dispersion of an industrial kraft lignin in a polyether polyol using glycerol as a compatibilizer and cross-linker (Hayati et al. 2018).

3 Lignin Depolymerization

As it was mentioned at the beginning of this chapter, lignin is a complex molecule composed mainly of three-unit blocks (*p*-coumaryl, coniferyl, and sinapyl alcohol) which make lignin the only bio-resource able to provide an extensive amount of biobased chemical compounds mostly aromatics (Jiang and Loos 2016). For the production of biochemicals from lignin, it is necessary to fragment the molecule into small compounds; this process is called depolymerization; some of the methods can be classified as chemicals, thermochemical, and biological depolymerizations (Fig. 9) (Jiang and Loos 2016). However, the depolymerization process is still challenging and requires more improvement to achieve high yields of conversion; some of the compounds that already have been obtained from lignin are phenols, hydrocarbons, and others which are represented in Fig. 10 (Pandey and Kim 2011).

Various parameters affect the yield of the lignin depolymerization such as temperature, pH, pressure, oxidants, the type of catalyst used, lignin sources, and lignin recovery procedure (Wendisch et al. 2018). Chemical depolymerization

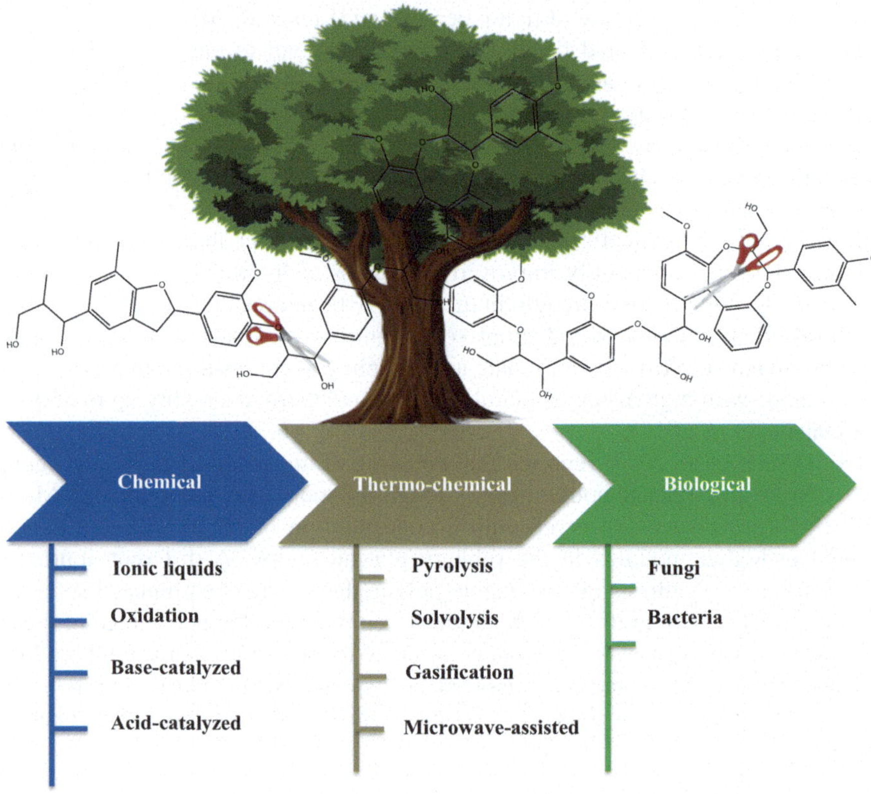

Fig. 9 A summary of procedures for lignin depolymerization

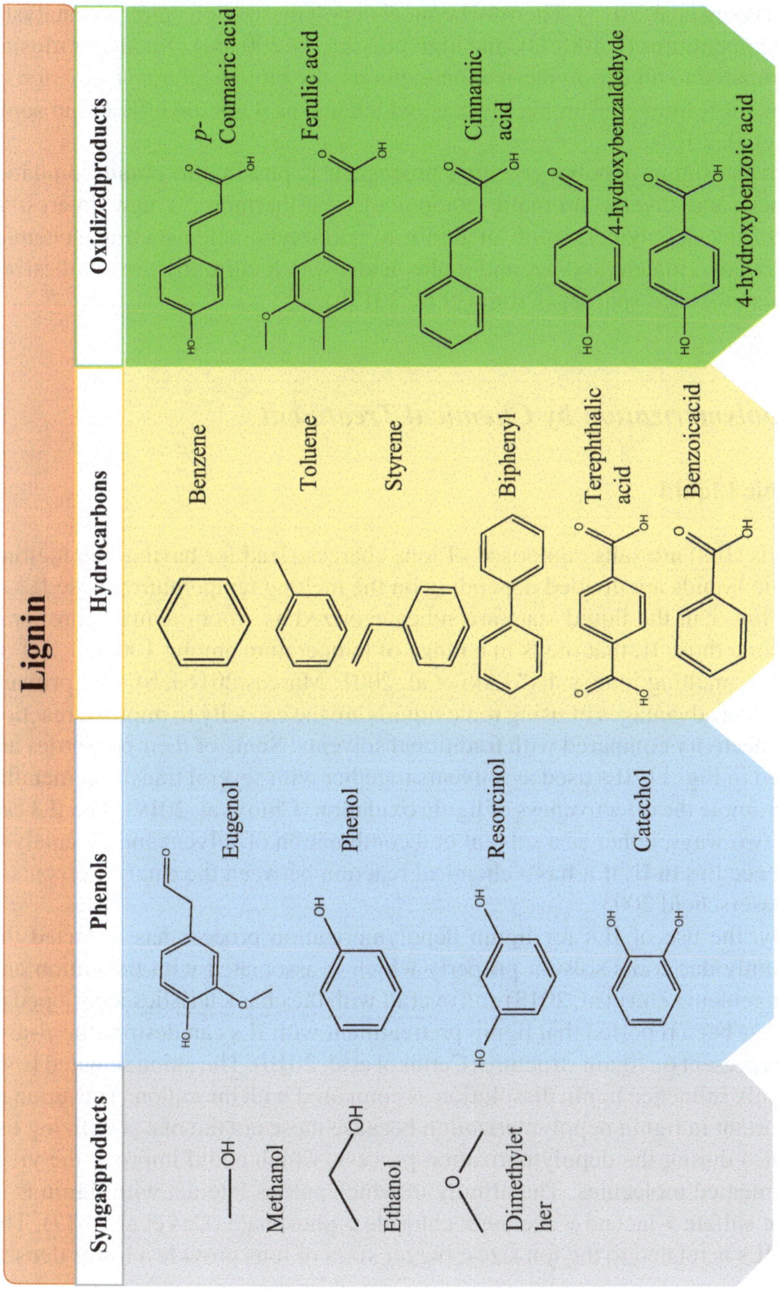

Fig. 10 Main chemicals compound obtained from lignin depolymerization

technique has shown some advantages such as reaction control and high selectivity providing excellent potential for the lignin conversion into renewable fuels and chemicals (Wang et al. 2013). Thermochemical depolymerization employs catalysts, moderate temperatures (~ 400 °C), and high pressure (> 200 bar) (Xu and Ferdosian 2017). Compared to all depolymerization methods, the biological transformation of lignin is the most time-consuming process, which makes it not too efficient in some cases (Xu et al. 2018).

Through the lignin depolymerization process, it is possible to obtain aliphatic, cycloalkanes, and diverse aromatic compounds. Furthermore, a new interest is emerging in the depolymerization of lignin to carboxylic acids such as succinic, fumaric, malonic, maleic, oxalic, and malic acids, which currently are synthesized from petroleum-based sources (Cronin et al. 2017b).

3.1 Depolymerization by Chemical Treatment

3.1.1 Ionic Liquid

Ionic liquids (ILs) are salts composed of ions characterized for having low melting points. Ionic liquids are divided depending on the melting temperature; those ILs at 100 °C or lower in the liquid state are subcategorized as "room temperature ionic liquids," while those IL that melts in a range of temperature among 100 and 250 °C is called "low-melting ionic salts" (Aki et al. 2001; Marcus 2016a, b). The primary importance and advantages of using ionic liquids are the capacity to improve reaction rates and selectivity compared with traditional solvents. Some of their properties are summarized in Fig. 11. ILs used as solvents together with several transition metallic catalysts promote the effectiveness of lignin oxidation (Chio et al. 2019). The ILs can be used in two ways, either as a solvent or a combination of solvent and co-catalyst. The difference lies in IL if it has a chemical reaction between the catalyst precursor or not (Wasserscheid 2003).

Recently, the use of ILs for lignin depolymerization process has attracted the interest mainly due to its solvent property which is associated with the cation and anion arrangement (Zhu et al. 2018) and overall with the characteristics mentioned in Fig. 11. It has been reported that lignin pretreatment with ILs can destroy the β-aryl ether bonds present on lignin structure (Çetinkol et al. 2010). The anion selected from ILs can highly influence lignin dissolution as compared with the cation. The anion in ILs is important in lignin depolymerization because these are the one stabilizing the intermediates during the depolymerization process, which could improve the yield of the fragmented molecules. The affinity in which anions interact with lignin is in the order of sulfate > lactate > acetate > chloride > phosphate (Das et al. 2017). The density of ILs is related to the ion sizes; bigger sizes of ions provide a lower density value.

There are more than 100 types of ionic liquids commercially produced which are sold as chemicals rather than solvents. The weaknesses of the market of IL

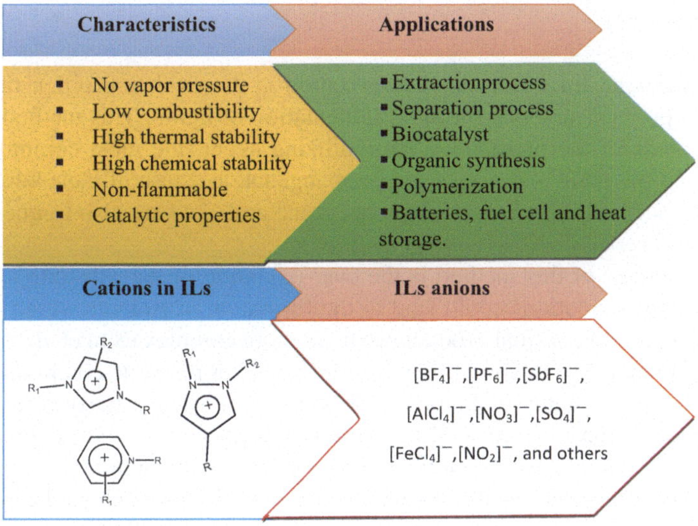

Fig. 11 Some general properties and structures of ionic liquids (Wasserscheid 2003; Martínez-Palou 2010)

is the relatively high costs, which make them less suitable for its application on industries and also have lower specified purity (Wasserscheid 2003; Rogers and Voth 2007). They have been called "green solvents" mainly due to the insignificant vapor pressure produced compared with organic solvents (Wasserscheid 2003; Martínez-Palou 2010).

ILs have shown excellent potential for its use with microwave-assisted degradation. It allows the increase of temperature in a short time, indicating that ILs are good absorbers. Pan and coworkers used lignin model compounds for evaluating the microwave-assisted depolymerization of lignin using 29 different kinds of imidazolium-based ILs. The result showed that 1-butyl-3-methylimidazolium hydrogen sulfate ([BMIM]HSO_4) was the most effective. The acidity of ILs plays a role in favor of catalytic activity of the benzyl phenyl ether and guaiacol (the lignin model compounds).

Furthermore, it was proved that ([BMIM]HSO_4) can be reused five times without any loss of catalytic activity or change on its chemical structure (Pan et al. 2014). Besides, the ILs using microwave technique was evaluated by Thierry et al., which tested a methylimidazolium-based ionic liquid for the cleavage of alkyl-aryl ether bonds in lignin. The obtained results proved that the IL is a useful solvent/reagent for the depolymerization of lignin using microwave-assisted technique (Thierry et al. 2018). It was reported that the aerobic oxidation of linkages present on lignin could be improved by using 1-octyl-3-methylimidazolium acetate [OMIm][OAc] under metal-free environment. The acetate anion of the ILs attacked the hydrogen from the β-carbon, which induced the cleavage of C–O bond of the aromatic ether (Yang et al. 2017).

3.1.2 Oxidation

Another technique for lignin depolymerization is the oxidation (OL); this is one of the most popular methods for the fragmentation of lignin. This method involves the conversions in the presence of an oxidizing agent; the most commonly used are hydrogen peroxide and potassium permanganate because of their low cost and commercial accessibility; however, nitrobenzene and CuO are also frequently used in OL (Schutyser et al. 2018; Chio et al. 2019).

A disadvantage of this method is the over-oxidation of the substrate to gaseous products, as the oxidations could lead to the formation of isomers which make the process of separation of final product mixtures more complex (Sun et al. 2018). The products obtained by oxidation process could be either phenolic compounds by the cleavage of the side chains or aliphatic carboxylic acids by cleavage of aromatic rings which depend on the reaction conditions used, changes on O_2 pressure, pH, reaction time, and other parameters. Phenolic compounds are the main products obtained by oxidation under alkaline conditions with oxygen or air, although acidic conditions and ionic liquids have been studied as well (Schutyser et al. 2018).

Lignin oxidation under alkaline conditions has received more attention than acidic for many years because it allows the selective production of the phenolic aldehyde vanillin. This reaction has been studied extensively but still is the subject of much debate (Sun et al. 2018; Schutyser et al. 2018). Vanillin is the only molecular phenolic compounds obtained on a large scale from a biosource, this compound has gained importance due to its number of advantages such as safe compound, aromaticity, and it allows two reactive sites which can be modified chemically. Aromaticity and bifunctionality which allows the use of vanillin as a monomeric compound for the synthesis of some polymeric materials. The oxidation of lignin in alkaline medium using nitrobenzene is the technique used since the 1940s to produce lignin fragmentation into vanillin and other phenolic compounds (Fache et al. 2016).

In alkaline oxidation, oxygen is used as the primary oxidant in a concentrated aqueous solution of NaOH with a concentration between 0.5 and 4 M, and also KOH has been used for increasing the pH of the solution. For this type of reaction, high pH is needed to ease the oxidation by ionizing the free phenolic hydroxyl groups found in lignin and also retard the degradation of aromatic aldehydes (Schutyser et al. 2018). A reduction of the pH from 10.8 to 10.3 would result in complete inhibition of vanillin formation because of acidic dissociation reactions of the intermediates, and this shows that the oxidation is influenced by an acidic environment (Tarabanko et al. 2004).

Phenolic compounds can also be obtained under acidic and neutral pH conditions. For oxidations in an acidic environment, the reactions are developed in aerobic conditions with diluted organic acids or concentrated acetic acid, while for neutral pH, solvents like methanol and ionic liquids are used. The latter produces aromatic aldehydes, acids, and unsubstituted methoxyphenols. The other type of compounds that can be obtained by oxidation is carboxylic and dicarboxylic acids by the oxidative cleavage of phenolic compounds (Schutyser et al. 2018).

Dicarboxylic acids (DCAs) are organic compounds characterized by the presence of two carboxyl functional groups (–COOH). These are mostly used in the preparation of copolymers, such as polyesters and polyamides (Chadwick 1988). Nowadays, the production of DCAs is mostly from the petroleum-based feedstock, which is broken down by catalytic oxidation. However, the interest of obtaining DCAs from biomass, such as lignin, has gained more attention; a report of the US Department of Energy presented DCAs of four-carbons as the first place in a list of 12 chemical compounds produced from biomass. These components have a potential application as an alternative to current commercial insecticides, disinfectants, and antifungal fertilizers (Yin et al. 2015; Cronin et al. 2017a). Ma et al. (2014) produced DCAs by catalyzed oxidation of diluted acid maize stover lignin and spruce lignin. Hydrogen peroxide and chalcopyrite ($CuFeS_2$) were both used in the reaction. As a result, a narrow range of DCAs was obtained such as malonic, succinic, maleic, and malic acids after 5 h reaction providing highest mass conversion yield and selectivity with 14 and 91%, respectively, from diluted acid corn stover lignin. Z. Bi et al. (2018) studied the effect of particle size OL, as a way to increase the contact surface area between the metal ions and H_2O_2 in deionized water whereby this research was done using nanoparticles of the chalcopyrite catalyst. The optimum results used a catalyst concentration of 10 wt% with a reaction temperature of 60 °C for 5 h; the yield of DCAs was over 85% based on the $CuFeS_2/H_2O_2$ reaction system and high selectivity of up to 30% mainly giving oxalic acid (Bi et al. 2018).

3.1.3 Base-Catalyzed

In base-catalyzed depolymerization (BCD) of lignin, mainly ether linkages (α- and β-aryl ether), are cleaved; α-aryl ether linkages produce quinone methide intermediate from phenol units which are transformed into guaiacol and coniferyl alcohol. On the other hand, β- aryl ether (β-O-4) linkages are the most abundant aryl-alkyl bond in lignin on cleavage. This chemical reaction provides an oxirane ring that reacts with hydroxyl ions to produce a glycol group (Wang et al. 2013; Agarwal et al. 2018).

The main catalysts used in BCD are usually inorganic bases like hydroxides of alkali metals which often results in the formation of simple aromatic compounds. Moreover, this is one of the advantages of BCD; the catalysts with basic properties are cheap and commercially available (Thring 1994). Yield of depolymerization depends on the type of catalyst selected, with strong bases such as NaOH, KOH, and CsOH; higher conversions of lignin into ether soluble materials can be obtained, contrary to weak bases like $Ca(OH)_2$, LiOH, and Na_2CO_3 which lead to the formation of char and detain the reactivity of the phenolic compounds (Miller et al. 1999).

An inconvenience of this reaction is the high temperatures used which are mostly above 260 °C in the presence of nitrogen, while showed low selectivity of final products, and the reactors tend to suffer corrosion. Therefore, a neutralization step is required. In addition to that, one of the significant issues is the repolymerization, which gives an increase of solid residues. Also, the formation of carboxylic acids during the depolymerization process could increase acidity and consequently reduce efficiency (Agarwal et al. 2018; Chio et al. 2019). Toledano and coworkers compared the efficiency of two capping agents, boric acid and phenol as suppressors of lignin repolymerization, the results showed a decrease in char formation when phenol is applied onto the reaction (Toledano et al. 2014). The combination of water as a solvent with low concentrations of base catalyst gives low yields of monomeric compounds from lignin. Thus, in BCD, organic solvents are most commonly used such as ethanol, methanol, polyethylene glycol, and isopropanol (Miller et al. 1999; Chio et al. 2019).

3.1.4 Acid-Catalyzed

Similar to base-catalyzed, in an acidic medium occurs hydrolysis of α- and β-aryl ether linkages (Agarwal et al. 2018). Most of the reactions of acid-catalyzed depolymerizations (ACD) are performed at high temperatures (above 250 °C); however, the temperatures between 140 and 180 °C are also used. ACD requires Lewis acids (e.g., chlorides of boron, aluminum, iron, copper, nickel, and zinc and others), minerals acids, soluble Brønsted acids, and solid Brønsted acids for carrying out the depolymerization process. Recently, ionic liquids and zeolites have been tested as well, while the solvents used were either water, an organic solvent or a mixture of solvents (Sun et al. 2018; Schutyser et al. 2018; Agarwal et al. 2018; Chio et al. 2019).

However, this strategy for converting lignin into value-added chemicals has significant disadvantages as severe reaction conditions such as higher temperature, high pressure required, long reaction times, and harsh chemicals are needed. These conditions create environmental pollutants and a problem for waste management after the reactions. Besides that, repolymerization is also a drawback in acid-catalyzed reactions as well because the generated intermediates could condensate and form oligomers reducing the yield of the reactions (Chio et al. 2019).

Guvenatam and collaborators evaluated the lignin depolymerization using different Lewis acid salts; higher yield of monomeric products was obtained using ethanol as the solvent since it induces the alkylation of the aromatic compounds avoiding the polymerization to oligomers. The ideal Lewis acid catalyst was Fe, Cu, Ni, and Al chlorides (Güvenatam 2015). Furthermore, the authors explored the use of metal chlorides (Fe^{+2}, Co^{+2}, Cu^{+2}, Ni^{+2}, and Al^{+3}), acetates (Fe^{+2}, Co^{+2}, Cu^{+2}, Ni^{+2}), and triflates(Cu^{+2}, Ni^{+2}, Al^{+3}, and Sc^{+3}), and the latter one provided higher yields (Güvenatam et al. 2016).

3.2 Depolymerization by the Thermochemical Process

3.2.1 Pyrolysis

Pyrolysis is characterized as a thermal degradation in the absence of oxygen; this technique has been evaluated for different authors showing it as a promising method for lignin depolymerization. Pyrolysis classification depends on the temperature used and processing time; some authors refer to two categories: (1) primary pyrolysis which ranges between 200 and 400 °C and (2) secondary pyrolysis with temperatures higher than 400 °C. Besides, pyrolysis can be classified into three types depending on the heating rate: slow, flash, and fast pyrolysis; the parameters of each one is described in Fig. 12. After pyrolysis, the final products obtained highly depend on the parameters applied, but these fall into three groups: (1) gases (e.g., H_2, CO, CH_4, CO_2, and others) and (2) liquid product or also known as bio-oil and tar; in this fraction, most of the low molecular weight compounds are present such as phenols, ketones, furans, esters, aldehydes, alcohols, carbohydrates, and others, and (3) an insoluble residue or solid phase, commonly called biochar (Wang et al. 2012; Zhou et al. 2013; Mašek 2016; Kawamoto 2017; Ranzi et al. 2017; Fahmy et al. 2020; Agarwal et al. 2018).

Pyrolysis can use different sources of heating; however, at this time, microwave irradiation is the one that has attracted the most (Fahmy et al. 2020). Diverse studies have focused on using a microwave to increase the selectivity and yields by using lower processing time and temperature. It has been applied on depolymerization of lignin from various sources (Bu et al. 2014; Fu et al. 2014; Duan et al. 2018a, b; Rajasekhar Reddy and Vinu 2018; Yerrayya et al. 2018). A more in-depth explanation about microwave pyrolysis is explained in the following section of this chapter.

Different research papers have been published in the past years about optimization of the pyrolysis parameters for obtaining higher yields of monomeric compounds from lignin. Zhou and collaborators reported the pyrolysis of organosolv lignin in a temperature range from 360 to 700 °C (Zhou et al. 2013). Kawamoto et al. studied the primary pyrolysis of some model dimers that content part of the lignin linkages to evaluate lignin depolymerization under the absence of oxygen at 400 °C (Kawamoto et al. 2007).

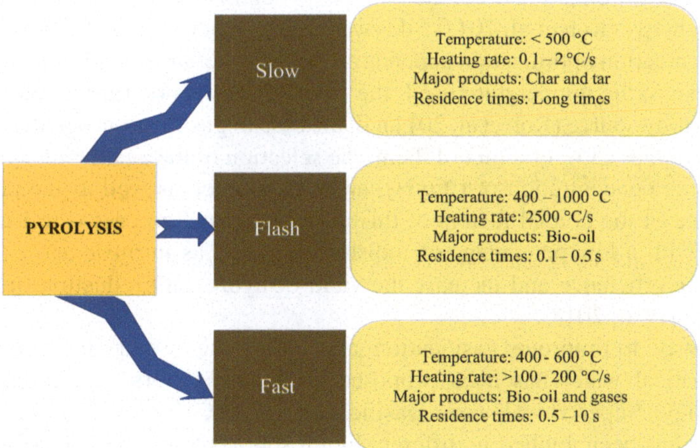

Fig. 12 Classification of pyrolysis based on reaction times (Mašek 2016; Fahmy et al. 2020)

Furthermore, recent publications have presented some improvement on pyrolysis by the addition of catalyst such as transition, noble, and inorganic metals and zeolites (Kim et al. 2015). Regarding zeolites, HZSM-5 has been considered as one of the most effective in lignin pyrolysis; it was used on the fast pyrolysis of lignin from black liquor, resulting in a yield increase of low molecular compounds such as phenol, toluene, guaiacol, and others (Shen et al. 2015). Mullen and Boateng evaluated an acidic zeolite (HZSM-5) and a combination of two metal oxide catalysts (CoO/MoO_3) for the pyrolysis of four sources of lignin at 650 °C. The addition of one of the catalyst yielded various aromatic compounds being HZSM-5 more productive, while the absence of catalyst resulted in a small number of hydrocarbons (Mullen and Boateng 2010). Zhen et al. proved that the dimensions of the catalyst and raw material could affect the lignin fragmentation as well as other biomass compounds; the crystal catalyst ZSM-5 size of 200 nm showed higher aromatic yield due to an increase of the micropore superficial area (Zheng et al. 2014). Likewise, Yu et al. reported the importance of the shape selectivity by testing different types of zeolite catalyst; also, ZSM-5 proved to have higher surface area compared to the others, and this characteristic enhanced the efficacy of the thermal process which resulted in higher fragmentation of lignin structure into low molecular chemical compounds (Yu et al. 2012).

3.2.2 Gasification

Another process to fragment lignin into small molecules is gasification; this technique consists on the application of high temperature, high pressure, and steam to transform an organic raw material into gases which are mostly the constituents of syngas; at the same time, there is a formation of a solid phase (char) and a liquid of high viscosity (tar or bio-oil). Different raw materials are used for gasification, including fossil fuels, but when biomass is used, then the gases obtained can turn into a sustainable source of energy (Koh et al. 2011; Sikarwar et al. 2016; Ciuta et al. 2018).

The gasification optimization depends on diverse operating conditions such as the reactor design, the temperature used, the environment inside (gasification agents), and the heating source (Koh et al. 2011). In the case of gasification, agent used can be air, O_2, steam, or CO_2, or a mix of them; the selection of the agent will significantly influence the concentration of CO, H_2, and N_2 obtained as well as the production of tars. One of the crucial factors of this type of thermal process is to use the raw materials with a low percentage of moisture as much as intrinsic and extrinsic to increase the efficiency and increase the yield of hydrogen production (moisture $<$ 10%) (Ciuta et al. 2018).

As a way to improve gasification and avoid the pre-drying step, the use of supercritical water has been proposed by many authors as a more efficient methodology. Supercritical water gasification (SCWG) ($T = 374$ °C and $P =$ 22.1 MPa) has been applied in different lignin sources to produce hydrogen which makes it more profitable than conventional gasification (Kang et al. 2015; Sikarwar et al. 2016; Rodriguez Correa and Kruse 2018). Furthermore, the addition of catalyst has also an effect on the gases compound yields, and according to this, the SCWG can be divided into two groups depending if it contains catalyst or not; they are called low-temperature (300–400 °C) and high-temperature (>600 °C) gasification, respectively (Cao et al. 2017). Cao et al. showed that higher temperature resulted in an increase of H_2 production from lignin contained in black liquor (Cao et al. 2017). Osada et al. reported a yield of 30% of gases; mostly, CH_4 was obtained from SCWG of organosolv lignin at 400 °C using ruthenium/titanium dioxide as a catalyst; there was no char formation in the final product (Osada et al. 2004). Watanabe and collaborators studied the partial oxidation of organosolv lignin using supercritical water with two catalysts, namely NaOH and ZrO_2, which improved the production of CO and H_2; also this process avoided the formation of a solid phase (Watanabe et al. 2003). Likewise, other types of catalyst have been evaluated in SCWG such as nickel (Lu et al. 2019), nickel with magnesium oxide (Sato et al. 2006), ruthenium salts (Osada et al. 2006; Yamaguchi et al. 2008), platinum and palladium (Yamaguchi et al. 2012), and others showing higher gasification efficiency

3.2.3 Solvolysis

Solvolysis is another method used to depolymerize lignin into aromatic monomers; this reaction takes place in the presence of a solvent with temperatures between 105 and 400 °C and pressure between 2 and 20 MPa (Fang and Smith 2016). This process is commonly preferred over pyrolysis because the reaction parameters are relatively milder such as lower temperature; less time is required for the reaction process, and the drying step is not necessary (Rachel-Tang et al. 2017). The principal component of solvolytic reaction is the solvents, such as water (Miller et al. 2002; Toledano et al. 2012; Onwudili and Williams 2014), organic solvents(Rezzoug and Capart 2002; Xu et al. 2012; Shi et al. 2016; Duan et al. 2018c) or a mixture of both or different organic solvents (Mishra and Saka 2013; Singh and Ekhe 2014; Xie et al. 2015; Goldmann et al. 2019); the difference among using water or organic solvents relies on the vapor pressure generated inside the vessel. This reaction usually is carry on in conventional heating; however, recent technology has been applied to enhance the yield and selectivity of depolymerization process, for example, using microwave-assisted, which is explained further in Sect. 2.2.4.

Similarly, to gasification technique, solvolysis has been developed with supercritical solvents to increase the conversion of lignin into low molecular compounds (Belkheiri et al. 2018; Funkenbusch et al. 2019). Cardoso et al. obtained high yields of mainly methoxyphenols from depolymerization process of kraft lignin using as solvent supercritical water together with Ni-$CeCO_2$ to catalyze the reaction (Cardoso et al. 2018). However, the use of supercritical water could reduce the solubility of salts becoming a problem in reaction with alkali salts used as a catalyst such as KOH, NaOH, K_2CO_3, and others. To overcome the insolubility problem, subcritical water ($T = 100 - 374.1$ °C and $P = 1$–221 bar) (Todd and Baroutian 2017) have been studied as a more suitable solvent in an alkali environment. Recently, Belkheiri et al. reported the depolymerization of softwood kraft lignin using subcritical water with three different catalysts ZrO, K_2CO_3, and KOH (Belkheiri et al.

2018). Likewise, Singh and collaborators used methanol in subcritical conditions with a different catalyst; the results exhibited that methanol played an essential role in the cleavage of ether linkages, and the catalyst HZSM-5 (solid zeolite) resulted in higher yields (Singh et al. 2016).

3.2.4 Microwave-Assisted Conversion

The main focus in the last years has been shifted in developing the depolymerization reactions by using microwaves as the source of heat because of its several advantages over conventional heating. Microwaves are defined as high-frequency electromagnetic radiation with frequencies between 0.3 and 300 GHz; in the electromagnetic, spectrum microwaves are located between infrared and radiofrequency region (Zhen Fang et al. 2015). Using microwaves as a heating technique improve the efficiency as the conventional heating sources need reaction times of about hours today, while microwaves could carry out the same experiments in a couple of minutes (Xiao et al. 2011; Zhen Fang et al. 2015; Chio et al. 2019).

The organic molecules penetrated by microwave irradiation produce their oscillation with extremely high frequency, thus producing a vast amount of heat on the sample by the friction caused between the molecules. Some of the advantages and disadvantages of using microwave-assisted conversion are summarized in Fig. 13. When comparing this technique with conventional heating, some distinct advantages are observed, such as high efficiency, selectivity process in which only polar compounds are directly heated. Microwaves can instantly be turned on or off, and the devices are not expensive and can efficiently operate (Wang et al. 2012; Yunpu et al. 2016).

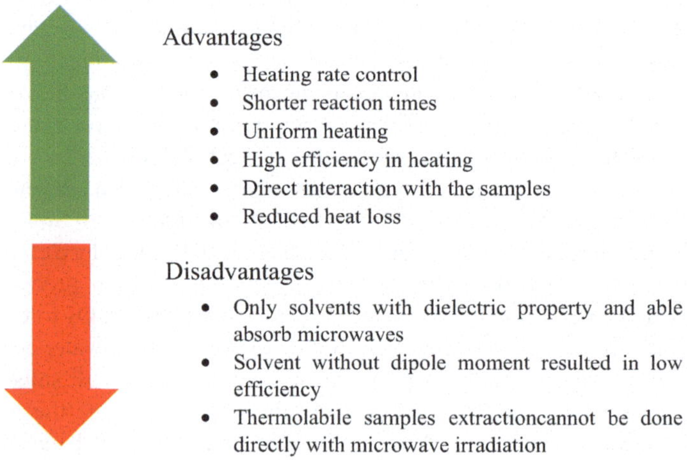

Fig. 13 Advantages and disadvantages of the microwave-assisted technique (Mandal et al. 2015; Dhar and Vinu 2017)

Microwaves have been used for the lignin depolymerization. Its application for the conversion of biosources can be divided into two main groups. Microwave-assisted pyrolysis which is based on the conversion of lignin in an atmosphere without oxygen, under high temperatures and with or without the use of catalysts; the second group is the microwave-assisted solvolysis which is under mild conditions either in the presence or absence of catalysts (Yunpu et al. 2016).

Microwave heating pyrolysis has been applied to different lignocellulosic materials such as rice straw (Huang et al. 2008), waste office paper (Zhang et al. 2015b), rice husk (Zhang et al. 2015a), corn stover(Lei et al. 2009), pinewood sawdust (Chen et al. 2008), and others. The resulting products of microwave pyrolysis are permanent gases such as H_2, CH_4, CO, and CO_2. After the simple treatment by microwave heating, it turns into a dark liquid called bio-oil (Huang et al. 2016). Fu and collaborators extracted guaiacol and 4-methyl guaiacol from the bio-oil obtained by using switchable hydrophilicity solvents to depolymerize lignin under microwave pyrolysis (Fu et al. 2014). In conventional pyrolysis of lignin, the main product obtained is guaiacols, while by catalytic microwave pyrolysis chemical compounds such as phenols, guaiacols, hydrocarbons, and esters are obtained with a yield up to 70% of the bio-oil (Bu et al. 2014).

Microwave solvolysis is becoming one of the most promising techniques for lignin fragmentation due to the moderate reaction conditions needed, efficient precipitation of specific products, and affordable costs (Dong et al. 2014). Liu and collaborators evaluated the lignin depolymerization in isopropanol using microwave heating, with a reaction at 120 °C for 30 min; the highest yield of aromatic compounds was up to 45%; some of the products were 1-(4-hydroxy-3-methoxyphenyl) and ethanone, 1-(4-hydroxy-3, 5- dimethoxy phenyl) (Liu et al. 2017).

The catalytic effect of different acids on organosolv lignin fragmentation using microwave irradiation was evaluated; the best catalytic effects were obtained by using sulfuric acid showing fast thermal decomposition and smaller fragments than the others (Tayier et al. 2018). Black liquor lignin was depolymerized using the microwave as a heating source by acid-catalyzed with formic acid obtaining a maximum yield of 64.08% (Dong et al. 2014). The organic solvents: ethylene alcohol, methyl sulfoxide, and dimethylformamide were evaluated for the depolymerization of lignin using microwave technique, which could be an effective method for obtaining low molecular chemical compounds. Phenolic compounds such as acetosyringone, guaiacol, syringaldehyde, anisole, and lignin dimers were obtained with a yield of 20 wt% in the presence of dimethylformamide and dimethyl sulfoxide which cleaved etheric C_β–O and C_α–C_β bonds (Dhar and Vinu 2017). Comparison between traditional and microwave heating methods was carried out for the liquefaction of corn stover, showing that microwave radically improved the liquefaction extent and rate (Xiao et al. 2011).

3.3 Depolymerization by Biological Transformations

3.3.1 Fungi

For the lignin depolymerization by biological transformations, fungi are the most studied. Nowadays, there are millions of species of fungi that have been discovered. However, only a few can degrade wood; those are classified into three groups: soft-rot, brown-rot, and white-rot; the latter one has shown the great ability for fragmentation of lignin. This fungus can degrade and mineralize all the major wood components like cellulose, hemicellulose, and lignin; some white-rot fungi preferentially attack lignin. It has a moist, soft, and spongy consistency of rotten wood. Some examples of white-root fungi are *Coriolus versi-color, Dichomitus squalens, Lentinusedodes, Phlebiaradiata, Panustigrinus, Pleurotus ostreatus,* and *Pycnoporus cinnabarinus* and a few species of *Asco-mycota* (Sigoillot et al. 2012; Abdel-Hamid et al. 2013; Chio et al. 2019).

Dias et al. evaluated the effect of pretreatment of wheat straw with basidiomycetous fungi, Euc-1 and *Irpexlateus.* Both showed highly lignolytic activity, and high selectivity delignification was observed. Both strains could be used to promote the lignocellulosic bioconversion to value-added products (Dias et al. 2010). The white-root fungus can be used for the treatment of biomass under room temperature without the production of any harmful residues (Zhi and Wang 2014). *Coloriolus Versicolor* and *Trametes gallica* are a type of white-rot fungi that have been applied in paper mill sludge and lignin degradation with positive results (Hong et al. 2015). Also, the white-rot fungal has shown its capability to improve fermentative biohydrogen yield from lignocellulosic biomass (Zhi and Wang 2014). The brown-rot fungi are less capable than white-rot fungi in the fragmentation of lignin because it also degrades wood polysaccharides; some examples are *Gloeophyllum trabeum, Laetiposus portentosus, and Fomitopsisli lacinogilva* (Abdel-Hamid et al. 2013).

3.3.2 Bacteria

Bacteria have not been studied extensively as fungi on lignin fragmentation until a few years ago that has been observed, and also some bacterial showed high activity on metabolizing lignin (Brown and Chang 2014; Lambertz et al. 2016). Both fungi and bacteria have some similar enzymatic systems such as Mn-peroxidase, lignin peroxidases, and dye-decoloring peroxidases. These enzymes are capable of initiating C_{α}–C_{β} cleavage, alkyl-aryl cleavage, cross-linking, demethoxylation, and ring cleavage of aromatic compounds. However, bacteria have some advantages over fungi that make it more attractive as bacteria adapt to diverse environments, easy to handle, can be subject to gene manipulation, and have a wider distribution of activities, pH, and thermal dependability as well as highly tolerant to anaerobic conditions (Wu and He 2013; Salvachúa et al. 2015; Xu et al. 2018; Agarwal et al. 2018).

Similarly, fungi and bacteria on lignin degradation produce similar aromatic compounds, but the use of bacterial metabolic pathway resulted in additional phenolic compounds, for example 2-methoxy-4-vinyl phenol and 4-hydroxy-3-methoxy acetophenone which have been produced by *Pseudomonas* and *Rhodococcus* species as a degraded product (Suman et al. 2016).

All these advantages of bacteria over fungi have allowed the increase of interest to study more on utilization of bacteria for the depolymerization of lignin. Due to bacteria capability of adapting at different environmental conditions, a wide range of them is studied nowadays such as bacteria from the soil, rotten wood, activated sludge, compost, animal gut, and wastewater treatment plant (Xu et al. 2018). Figure 14 shows the cycle phylogram of some bacteria that can fragment the molecule of lignin. Some bacterial strains used on lignin depolymerization have been studied such as *Streptomyces viridosporus*, Rhodococci (e.g., *Rhodococcus jostii* RHA1 and *Rhodococcus opacus*) certain pseudomonads (e.g., *Pseudomonas putida, Pseudomonas fluorescent*), and the aerobic bacterium *Enterobacter alginolyticus* (Sainsbury et al. 2013; Beckham et al. 2016). Bacteria should allow its genetic

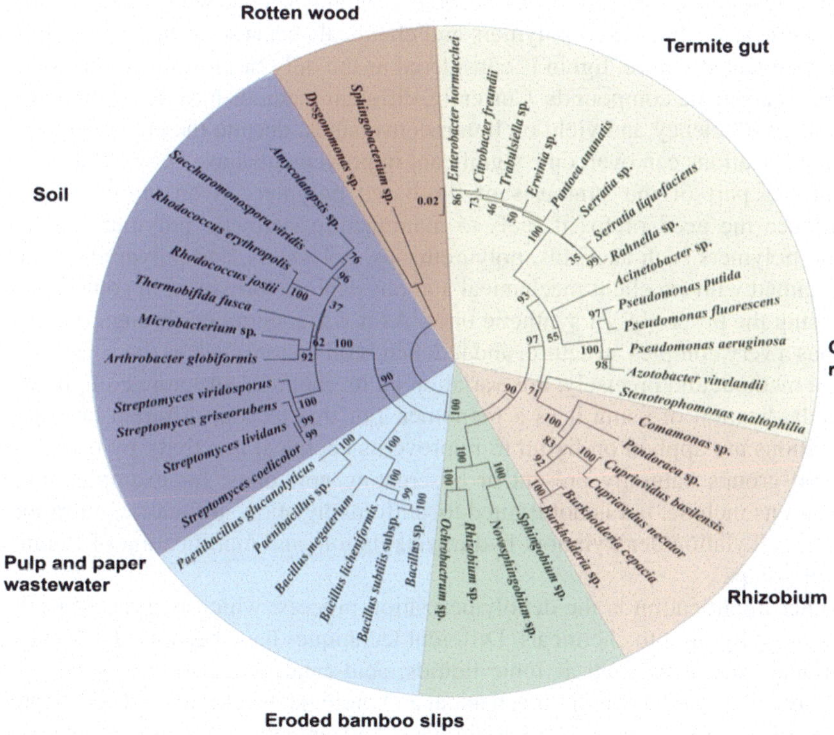

Fig. 14 Bacteria from different habitats with the ability to degrade lignin. Reproduced from Xu et al. (2018) with permission from Elsevier

modification, tolerate stress, and the most importantly easy to domesticate to escalate the lignin depolymerization process to industrial bioreactors (Beckham et al. 2016).

The complex lignin structure can be degraded by strains like *Sphingomonas, Bacillus,* and *Pandoraea*. Besides them, laccases are also capable of lignin degradation; these are a class of bacterial enzymes and part of the "denominated blue oxidases." Some examples are *Azospirillum lipoferum, B. lichiniformis, B. subtilis, E. coli, Nitrosomonas europea, S. griseu, and Thermus thermophiles* (Xu et al. 2018).

4 Conclusions

In this chapter, some of the lignin modifications and products that can be obtained were summarized to give an overview of applications of lignin and its potential as bio-based raw material. In the past, lignin was only considered as the by-product of paper manufacture and ethanol production from cellulose and incinerated to convert it to energy. On the other hand, in recent times, there has been a shift to use more renewable resources, and lignin has become a natural source with high potential for the synthesis of bio-based polymers and chemicals because of the nature of its unique chemical structure; lignin is considered as the only bio-resource constituted by different aromatic compounds. Currently, different methods have been studied to achieve high efficiency and yield on lignin conversions, despite the challenges that lignin modifications can overcome significant improvements have been obtained.

Lignin as part of the synthesis of bio-based polymers is an attractive way of to reduce the need of fossil fuels to manufacture synthetic polymers; at this moment, polymers such as resins, polyurethanes, polyesters, and polyamides have been obtained with excellent mechanical and physical performance; in some cases, overpassing the properties of synthetic ones. As it was mentioned through the text, lignin has a very complex structure, and it depends on factors such as the source and isolation method; and precisely, this variation on its chemical structure complicates the standardization of using it as a monomer. For this reason, different chemical modifications are applied on lignin to improve its chemical reactivity by using the functional groups naturally present or by creating new ones; for example, some modifications include, but is not limited to, sulfomethylation/sulfonation, nitration, amination, alkylation/dealkylation, hydroxyalkylation, and modification of lignin's hydroxyl groups.

Another modification is the depolymerization process, which is directed to the conversion of lignin into chemicals. Different techniques have been used as the use of chemical treatments such as ionic liquids, acid-catalyzed, base-catalyzed, and oxidations and as well as thermomechanical and biological techniques. This chapter gave a brief review of some methodologies used for fragmenting the lignin molecule, mentioning the advantages and disadvantages; for instance, ionic liquids give a high yield of aromatic compounds, but the price is costly compared with other techniques, which could make it unsuitable for its use a big scale. Acid-catalyzed and basic-catalyzed produce specific chemical compounds, but high reaction temperatures are required, which means higher costs of productions. Conversely, pyrolysis is an effective method to convert lignin into aromatics; however, it has low product selectivity. In general, to select a depolymerization method will depend on what are the final requirements that want to be obtained and decide which can be adjusted the best.

Lignin, as a macromonomer or as the source of biobased chemicals, shows a great potential worthy of exploring more and improving it in a way to increase its market that can compete with commercial polymers and chemicals. Besides that, the possibility of scale the production also allows producing more sustainable products and does not depend only on fossil fuels.

Acknowledgements This book chapter was possible thanks to the financial support received from the National Council of Science and Technology (CONACyT, Mexico).

References

Abas N, Kalair A, Khan N (2015) Review of fossil fuels and future energy technologies. Futures 69:31–49. https://doi.org/10.1016/J.FUTURES.2015.03.003

Abdel-Hamid AM, Solbiati JO, Cann IKO (2013) Insights into lignin degradation and its potential industrial applications. Adv Appl Microbiol 82:1–28. https://doi.org/10.1016/B978-0-12-407679-2.00001-6

Agarwal A, Rana M, Park JH (2018) Advancement in technologies for the depolymerization of lignin. Fuel Process Technol 181:115–132. https://doi.org/10.1016/j.fuproc.2018.09.017

Aki SNVK, Brennecke JF, Samanta A (2001) How polar are room-temperature ionic liquids? Chem Commun 413–414. https://doi.org/10.1039/b008039j

Aro T, Fatehi P (2017) Production and application of lignosulfonates and sulfonated lignin. ChemSuschem 10:1861–1877

Asada C, Basnet S, Otsuka M et al (2015) Epoxy resin synthesis using low molecular weight lignin separated from various lignocellulosic materials. Int J Biol Macromol 74:413–419. https://doi.org/10.1016/j.ijbiomac.2014.12.039

Beckham GT, Johnson CW, Karp EM et al (2016) Opportunities and challenges in biological lignin valorization. Curr Opin Biotechnol 42:40–53. https://doi.org/10.1016/j.copbio.2016.02.030

Belkheiri T, Andersson SI, Mattsson C et al (2018) Hydrothermal liquefaction of Kraft lignin in subcritical water: influence of phenol as capping agent. Energy Fuels 32:5923–5932. https://doi.org/10.1021/acs.energyfuels.8b00068

Bernardini J, Cinelli P, Anguillesi I et al (2015) Flexible polyurethane foams green production employing lignin or oxypropylated lignin. Eur Polym J 64:147–156. https://doi.org/10.1016/j.eurpolymj.2014.11.039

Bi Z, Li Z, Yan L (2018) Catalytic oxidation of lignin to dicarboxylic acid over the $CuFeS_2$ nanoparticle catalyst. Green Process Synth 7:306–315. https://doi.org/10.1515/gps-2017-0056

Brown ME, Chang MCY (2014) Exploring bacterial lignin degradation. Curr Opin Chem Biol 19:1–7. https://doi.org/10.1016/j.cbpa.2013.11.015

Bu Q, Lei H, Wang L et al (2014) Bio-based phenols and fuel production from catalytic microwave pyrolysis of lignin by activated carbons. Bioresour Technol. 162:142–147. https://doi.org/10.1016/j.biortech.2014.03.103

Calvo-Flores FG, Dobado JA (2010) Lignin as renewable raw material. ChemSuschem 3:1227–1235. https://doi.org/10.1002/cssc.201000157

Cao C, Xu L, He Y et al (2017) High-efficiency gasification of wheat straw black liquor in supercritical water at high temperatures for hydrogen production. Energy Fuels 31:3970–3978. https://doi.org/10.1021/acs.energyfuels.6b03002

Capellán-Pérez I, Mediavilla M, de Castro C et al (2014) Fossil fuel depletion and socio-economic scenarios: an integrated approach. Energy 77:641–666. https://doi.org/10.1016/J.ENERGY.2014.09.063

Cardoso A, Ramirez Reina T, Suelves I et al (2018) Effect of carbon-based materials and CeO 2 on Ni catalysts for kraft lignin liquefaction in supercritical water. Green Chem 20:4308–4318. https://doi.org/10.1039/c8gc02210k

Cateto CA, Barreiro MF, Rodrigues AE, Belgacem MN (2009) Optimization study of lignin oxypropylation in view of the preparation of polyurethane rigid foams. Ind Eng Chem Res 48:2583–2589. https://doi.org/10.1021/ie801251r

Cateto CA, Barreiro MF, Ottati C et al (2014) Lignin-based rigid polyurethane foams with improved biodegradation. J Cell Plast 50:81–95. https://doi.org/10.1177/0021955X13504774

Çetinkol ÖP, Dibble DC, Cheng G et al (2010) Understanding the impact of ionic liquid pretreatment on eucalyptus. Biofuels 1:33–46. https://doi.org/10.4155/bfs.09.5

Chadwick SS (1988) Ullmann's encyclopedia of industrial chemistry. Wiley

Chen MQ, Wang J, Zhang MX et al (2008) Catalytic effects of eight inorganic additives on pyrolysis of pine wood sawdust by microwave heating. J Anal Appl Pyrolysis 82:145–150. https://doi.org/10.1016/j.jaap.2008.03.001

Cheng S, Yuan Z, Anderson M et al (2012) Synthesis of biobased phenolic resins/adhesives with methylolated wood-derived bio-oil. J Appl Polym Sci 126:E431–E441. https://doi.org/10.1002/app.35655

Chio C, Sain M, Qin W (2019) Lignin utilization: a review of lignin depolymerization from various aspects. Renew Sustain Energy Rev 107:232–249. https://doi.org/10.1016/j.rser.2019.03.008

Ciuta S, Tsiamis D, Castaldi MJ (2018) Fundamentals of gasification and pyrolysis. In: Ciuta S, Tsiamis D, Castaldi MJ (eds) Gasification of waste materials. Academic Press, pp 13–36

Cotana F, Cavalaglio G, Nicolini A et al (2014) Lignin as co-product of second generation bioethanol production from ligno-cellulosic biomass. Energy Procedia 45:52–60. https://doi.org/10.1016/j.egypro.2014.01.007

Cronin DJ, Dunn K, Zhang X, Doherty WOS (2017a) Relating dicarboxylic acid yield to residual lignin structural features. ACS Sustain Chem Eng 5:11695–11705. https://doi.org/10.1021/acssuschemeng.7b03164

Cronin DJ, Zhang X, Bartley J, Doherty WOS (2017b) Lignin depolymerization to dicarboxylic acids with sodium percarbonate. ACS Sustain Chem Eng 5:6253–6260. https://doi.org/10.1021/acssuschemeng.7b01208

D'Souza J, George B, Camargo R, Yan N (2015) Synthesis and characterization of bio-polyols through the oxypropylation of bark and alkaline extracts of bark. Ind Crops Prod 76:1–11. https://doi.org/10.1016/j.indcrop.2015.06.037

Das L, Xu S, Shi J (2017) Catalytic oxidation and depolymerization of lignin in aqueous ionic liquid. Front Energy Res 5:1–12. https://doi.org/10.3389/fenrg.2017.00021

Dehne L, Vila Babarro C, Saake B, Schwarz KU (2016) Influence of lignin source and esterification on properties of lignin-polyethylene blends. Ind Crops Prod 86:320–328. https://doi.org/10.1016/j.indcrop.2016.04.005

Dhar P, Vinu R (2017) Understanding lignin depolymerization to phenols via microwave-assisted solvolysis process. J Environ Chem Eng 5:4759–4768. https://doi.org/10.1016/j.jece.2017.08.031

Dias AA, Freitas GS, Marques GSM et al (2010) Enzymatic saccharification of biologically pre-treated wheat straw with white-rot fungi. Bioresour Technol 101:6045–6050. https://doi.org/10.1016/j.biortech.2010.02.110

Ding N, Wang X, Tian Y, et al (2014) A renewable agricultural waste material for the syntehsis of the novel thermal stability epoxy resins. Polym Eng Sci 2777–2784. https://doi.org/10.1002/pen.23838

Dong C, Feng C, Liu Q et al (2014) Mechanism on microwave-assisted acidic solvolysis of black-liquor lignin. Bioresour Technol 162:136–141. https://doi.org/10.1016/j.biortech.2014.03.060
Du X, Li J, Lindström ME (2014) Modification of industrial softwood kraft lignin using Mannich reaction with and without phenolation pretreatment. Ind Crops Prod 52:729–735. https://doi.org/10.1016/j.indcrop.2013.11.035
Duan D, Ruan R, Lei H et al (2018a) Microwave-assisted co-pyrolysis of pretreated lignin and soapstock for upgrading liquid oil: effect of pretreatment parameters on pyrolysis behavior. Bioresour Technol 258:98–104. https://doi.org/10.1016/j.biortech.2018.02.119
Duan D, Ruan R, Wang Y et al (2018b) Microwave-assisted acid pretreatment of alkali lignin: effect on characteristics and pyrolysis behavior. Bioresour Technol 251:57–62. https://doi.org/10.1016/j.biortech.2017.12.022
Duan D, Wang Y, Ruan R et al (2018c) Comparative study on various alcohols solvolysis of organosolv lignin using microwave energy: physicochemical and morphological properties. Chem Eng Process Process Intensif 126:38–44. https://doi.org/10.1016/j.cep.2017.10.023
Duval A, Lawoko M (2014) A review on lignin-based polymeric, micro- and nano-structured materials. React Funct Polym 85:78–96. https://doi.org/10.1016/j.reactfunctpolym.2014.09.017
El Mansouri N-E, Pizzi A, Salvado J (2007) Lignin-based polycondensation resins for wood adhesives. J Appl Polym Sci 103:1690–1699. https://doi.org/10.1002/app.25098
El Mansouri N-E, Yuan Q, Huang F (2011a) Study of chemical modification of alkaline lignin by the glyoxalation reaction. BioResources 6:4523–4536
El Mansouri N-E, Yuan Q, Huang F (2011b) Synthesis and characterization of kraft lignin-based epoxy resins. BioResources 6:2492–2503
Espinoza-Acosta JL, Torres-Chávez PI, Olmedo-Martínez JL et al (2018) Lignin in storage and renewable energy applications: a review. J Energy Chem 27:1422–1438. https://doi.org/10.1016/J.JECHEM.2018.02.015
Fache M, Auvergne R, Boutevin B, Caillol S (2015) New vanillin-derived diepoxy monomers for the synthesis of biobased thermosets. Eur Polym J 67:527–538. https://doi.org/10.1016/j.eurpolymj.2014.10.011
Fache M, Boutevin B, Caillol S (2016) vanillin production from lignin and its use as a renewable chemical. ACS Sustain Chem Eng 4:35–46. https://doi.org/10.1021/acssuschemeng.5b01344
Fahmy TYA, Fahmy Y, Mobarak F, et al (2020) Biomass pyrolysis: past, present, and future. Environ Dev Sustain 22:17–32. https://doi.org/10.1007/s10668-018-0200-5
Fang Z, Smith RL (eds) (2016) Production of Biofuels and Chemicals from Lignin. Springer Singapore, Singapore
Fang Z, Smith RL Jr, Xinhua Qi (2015) Production of biofuels and chemicals with microwave. Springer, Dordrecht
Fernández-Rodríguez J, Erdocia X, Sánchez C et al (2017) Lignin depolymerization for phenolic monomers production by sustainable processes. J Energy Chem 26:622–631. https://doi.org/10.1016/j.jechem.2017.02.007
Figueiredo P, Lintinen K, Hirvonen JT et al (2018) Properties and chemical modifications of lignin: towards lignin-based nanomaterials for biomedical applications. Prog Mater Sci 93:233–269. https://doi.org/10.1016/j.pmatsci.2017.12.001
Foyer G, Chanfi BH, Boutevin B et al (2016) New method for the synthesis of formaldehyde-free phenolic resins from lignin-based aldehyde precursors. Eur Polym J 74:296–309. https://doi.org/10.1016/j.eurpolymj.2015.11.036
Fu D, Farag S, Chaouki J, Jessop PG (2014) Extraction of phenols from lignin microwave-pyrolysis oil using a switchable hydrophilicity solvent. Bioresour Technol 154:101–108. https://doi.org/10.1016/j.biortech.2013.11.091
Funkenbusch LLT, Mullins ME, Vamling L et al (2019) Technoeconomic assessment of hydrothermal liquefaction oil from lignin with catalytic upgrading for renewable fuel and chemical production. WIREs Energy Environ 8:e319 8. https://doi.org/10.1002/wene.319

Gang H, Lee D, Choi K-Y et al (2017) Development of High performance polyurethane elastomers using vanillin-based green polyol chain extender originating from lignocellulosic biomass. ACS Sustain Chem Eng 5:4582–4588. https://doi.org/10.1021/acssuschemeng.6b02960

Goldmann WM, Anthonykutty JM, Ahola J et al (2019) Effect of process variables on the solvolysis depolymerization of pine kraft lignin. Waste Biomass Valorization. https://doi.org/10.1007/s12649-019-00701-1

Gordobil O, Herrera R, Llano-Ponte R, Labidi J (2017) Esterified organosolv lignin as hydrophobic agent for use on wood products. Prog Org Coat 103:143–151. https://doi.org/10.1016/J.PORGCOAT.2016.10.030

Güvenatam B (2015) Catalytic pathways for lignin depolymerization. Technische Universiteit Eindhoven, Eindhoven

Güvenatam B, Heeres EHJ, Pidko EA, Hensen EJM (2016) Lewis-acid catalyzed depolymerization of protobind lignin in supercritical water and ethanol. Catal Today 259:460–466. https://doi.org/10.1016/J.CATTOD.2015.03.041

Hayati AN, Evans DAC, Laycock B et al (2018) A simple methodology for improving the performance and sustainability of rigid polyurethane foam by incorporating industrial lignin. Ind Crops Prod 117:149–158. https://doi.org/10.1016/j.indcrop.2018.03.006

Hong Y, Dashtban M (2015) Lignin in paper mill sludge is degraded by white-rot fungi in submerged fermentation. J Microb Biochem Technol 07:04

Höök M, Tang X (2013) Depletion of fossil fuels and anthropogenic climate change—a review. Energy Policy 52:797–809. https://doi.org/10.1016/J.ENPOL.2012.10.046

Hu L, Pan H, Zhou Y, Zhang M (2011) Methods to improve lignin's reactivity as a phenol substitute and as replacement for other phenolic compounds: a brief review. BioResources 6:3515–3525. https://doi.org/10.15376/biores.6.3.3515-3525

Hu S, Luo X, Li Y (2014) Polyols and polyurethanes from the liquefaction of lignocellulosic biomass. ChemSuschem 7:66–72. https://doi.org/10.1002/cssc.201300760

Huang YF, Kuan WH, Lo SL, Lin CF (2008) Total recovery of resources and energy from rice straw using microwave-induced pyrolysis. Bioresour Technol 99:8252–8258. https://doi.org/10.1016/J.BIORTECH.2008.03.026

Huang Y-F, Chiueh P-T, Lo S-L (2016) A review on microwave pyrolysis of lignocellulosic biomass. Sustain Environ Res 26:103–109. https://doi.org/10.1016/J.SERJ.2016.04.012

Jablonskis A, Arshanitsa A, Arnautov A et al (2018) Evaluation of Ligno Boost™ softwood kraft lignin epoxidation as an approach for its application in cured epoxy resins. Ind Crops Prod 112:225–235. https://doi.org/10.1016/j.indcrop.2017.12.003

Jiang Y, Loos K (2016) Enzymatic synthesis of biobased polyesters and polyamides. Polymers 8:243. https://doi.org/10.3390/polym8070243

Jiang X, Liu J, Du X et al (2018) Phenolation to improve lignin reactivity toward thermosets application. ACS Sustain Chem Eng 6:5504–5512. https://doi.org/10.1021/acssuschemeng.8b00369

Jiao GJ, Peng P, Sun SL et al (2019) Amination of biorefinery technical lignin by Mannich reaction for preparing highly efficient nitrogen fertilizer. Int J Biol Macromol 127:544–554. https://doi.org/10.1016/j.ijbiomac.2019.01.076

Kai D, Zhang K, Jiang L et al (2017) Sustainable and antioxidant lignin-polyester copolymers and nanofibers for potential healthcare applications. ACS Sustain Chem Eng 5:6016–6025. https://doi.org/10.1021/acssuschemeng.7b00850

Kalami S, Arefmanesh M, Master E, Nejad M (2017) Replacing 100% of phenol in phenolic adhesive formulations with lignin. J Appl Polym Sci 134:45124. https://doi.org/10.1002/app.45124

Kalami S, Chen N, Borazjani H, Nejad M (2018) Comparative analysis of different lignins as phenol replacement in phenolic adhesive formulations. Ind Crops Prod 125:520–528. https://doi.org/10.1016/j.indcrop.2018.09.037

Kang K, Azargohar R, Dalai AK, Wang H (2015) Noncatalytic gasification of lignin in supercritical water using a batch reactor for hydrogen production: an experimental and modeling study. Energy Fuels 29:1776–1784. https://doi.org/10.1021/ef5027345

Karak N (2016) Biopolymers for paints and surface coatings. In: Pacheco-Torgal F, Ivanov V, Niranjan K, Jonkers H (eds) Biopolymers and biotech admixtures for eco-efficient construction materials. Woodhead Publishing, pp 333–368

Kawamoto H (2017) Lignin pyrolysis reactions. J Wood Sci 63:117–132. https://doi.org/10.1007/s10086-016-1606-z

Kawamoto H, Horigoshi S, Saka S (2007) Pyrolysis reactions of various lignin model dimers. J Wood Sci 53:168–174. https://doi.org/10.1007/s10086-006-0834-z

Kevin S, Martha M, Emily S, Heather R-G (2018) The economic benefits of the U.S. Polyurethane Industry

Khabarov YG, Kuzyakov NY, Veshnyakov VA et al (2016) Nitration of sulfate lignin under homogeneous conditions studied by electron spectroscopy. Russ Chem Bull 65:2925–2931. https://doi.org/10.1007/s11172-016-1679-2

Kim JY, Lee JH, Park J et al (2015) Catalytic pyrolysis of lignin over HZSM-5 catalysts: effect of various parameters on the production of aromatic hydrocarbon. J Anal Appl Pyrolysis 114:273–280. https://doi.org/10.1016/j.jaap.2015.06.007

Kleinert M, Barth T (2008) Phenols from lignin. Chem Eng Technol 31:736–745. https://doi.org/10.1002/ceat.200800073

Koh SCL, Maguire S, Koh SCL, Maguire S (2011) WTE-Technology. In: Rogoff MJ, Screve F (eds) Waste-to-energy: technologies and project implementation. Second, IGI Global, pp 266–284

Lambertz C, Ece S, Fischer R, Commandeur U (2016) Progress and obstacles in the production and application of recombinant lignin-degrading peroxidases. Bioengineered 7:145–154. https://doi.org/10.1080/21655979.2016.1191705

Laurichesse S, Avérous L (2014) chemical modification of lignins: towards biobased polymers. Prog Polym Sci 39:1266–1290. https://doi.org/10.1016/j.progpolymsci.2013.11.004

Lee A, Deng Y (2015) Green polyurethane from lignin and soybean oil through non-isocyanate reactions. Eur Polym J 63:67–73. https://doi.org/10.1016/j.eurpolymj.2014.11.023

Lee Y, Park CH, Lee EY (2017) Chemical modification of methanol-insoluble kraft lignin using oxypropylation under mild conditions for the preparation of bio-polyester. J Wood Chem Technol 37:334–342. https://doi.org/10.1080/02773813.2017.1303512

Lei H, Ren S, Julson J (2009) The effects of reaction temperature and time and particle size of corn stover on microwave pyrolysis. Energy Fuels 23:3254–3261. https://doi.org/10.1021/ef9000264

Liu Y, Li K (2006) Preparation and characterization of demethylated lignin-polyethylenimine adhesives. J Adhes 82:593–605. https://doi.org/10.1080/00218460600766632

Liu Q, Li P, Liu N, Shen D (2017) Lignin depolymerization to aromatic monomers and oligomers in isopropanol assisted by microwave heating. Polym Degrad Stab 135:54–60. https://doi.org/10.1016/j.polymdegradstab.2016.11.016

Lu Y, Jin H, Zhang R (2019) Evaluation of stability and catalytic activity of Ni catalysts for hydrogen production by biomass gasification in supercritical water. Carbon Resour Convers 2:95–101. https://doi.org/10.1016/j.crcon.2019.03.001

Luo S, Cao J, McDonald AG (2017) Esterification of industrial lignin and its effect on the resulting poly(3-hydroxybutyrate-co-3-hydroxyvalerate) or polypropylene blends. Ind Crops Prod 97:281–291. https://doi.org/10.1016/j.indcrop.2016.12.024

Luo S, Cao J, McDonald AG (2018a) Cross-linking of technical lignin via esterification and thermally initiated free radical reaction. Ind Crops Prod 121:169–179. https://doi.org/10.1016/J.INDCROP.2018.05.007

Luo X, Xiao Y, Wu Q, Zeng J (2018b) Development of high-performance biodegradable rigid polyurethane foams using all bioresource-based polyols: lignin and soy oil-derived polyols. Int J Biol Macromol 115:786–791. https://doi.org/10.1016/j.ijbiomac.2018.04.126

Luong ND, Binh NTT, Duong LD et al (2012) An eco-friendly and efficient route of lignin extraction from black liquor and a lignin-based copolyester synthesis. Polym Bull 68:879–890. https://doi.org/10.1007/s00289-011-0658-x

Ma R, Guo M, Zhang X (2014) Selective conversion of biorefinery lignin into dicarboxylic acids. Chemsuschem 7:412–415. https://doi.org/10.1002/cssc.201300964

Mai C, Elder T (2016) Wood: chemically modified. Ref Modul Mater Sci Mater Eng 1–6. https://doi.org/10.1016/B978-0-12-803581-8.03537-2
Mandal SC, Mandal V, Das AK (2015) Classification of extraction methods. In: Mandal SC, Mandal V, Das AK (eds) Essentials of botanical extraction. Academic Press, pp 83–136
Marcus Y (2016a) Low-melting ionic salts. Ionic liquid properties. Springer, Cham, Switzerland, pp 109–122
Marcus Y (2016b) Room temperature ionic liquids. Ionic liquid properties. Springer, Cham, Switzerland, pp 123–220
Martínez-Palou R (2010) Microwave-assisted synthesis using ionic liquids. Mol Divers 14:3–25. https://doi.org/10.1007/s11030-009-9159-3
Mašek O (2016) Biochar in thermal and thermochemical biorefineries-production of biochar as a coproduct. In: Luque R, Lin CSK, Wilson K, Clark JH (eds) Handbook of biofuels production: processes and technologies, 2nd edn. Woodhead Publishing, pp 655–671
Matsushita Y (2015) Conversion of technical lignins to functional materials with retained polymeric properties. J Wood Sci 61:230–250. https://doi.org/10.1007/s10086-015-1470-2
Matsushita Y, Yasuda S (2003) Reactivity of a condensed-type lignin model compound in the Mannich reaction and preparation of cationic surfactant from sulfuric acid lignin. J Wood Sci 49:166–171. https://doi.org/10.1007/s100860300026
Hong Y, Dashtban M, Chen S, Song R, Qin W (2015) Lignin in paper mill sludge is degraded by white-rot fungi in submerged fermentation. J Microb Biochem Technol 7:177–181. https://doi.org/10.4172/1948-5948.1000201
Meister JJ (2002) MODIFICATION OF LIGNIN*. J Macromol Sci Part C Polym Rev 42:235–289. https://doi.org/10.1081/MC-120004764
Miller JE, Evans L, Littlewolf A, Trudell DE (1999) Batch microreactor studies of lignin and lignin model compound depolymerization by bases in alcohol solvents. Fuel 78:1363–1366. https://doi.org/10.1016/S0016-2361(99)00072-1
Miller J, Evans L, Mudd JE, Brown KA (2002) Batch microreactor studies of lignin depolymerization by bases. 2. Aqueous Solvents. Sandia Natl Rep 1–52. https://doi.org/10.2172/800964
Mishra G, Saka S (2013) Effects of water in water/phenol mixtures on liquefaction of Japanese beech as treated under subcritical conditions. Holzforschung 67:241–247. https://doi.org/10.1515/hf-2012-0050
Mohanty AK, Misra M, Drzal LT (2005) Natural fibers, biopolymers, and biocomposites. Taylor & Francis
Mullen CA, Boateng AA (2010) Catalytic pyrolysis-GC/MS of lignin from several sources. Fuel Process Technol 91:1446–1458. https://doi.org/10.1016/j.fuproc.2010.05.022
Nikafshar S, Zabihi O, Moradi Y et al (2017) Catalyzed synthesis and characterization of a novel lignin-based curing agent for the curing of high-performance epoxy resin. Polymers (Basel) 9:1–16. https://doi.org/10.3390/polym9070266
Onwudili JA, Williams PT (2014) Catalytic depolymerization of alkali lignin in subcritical water: influence of formic acid and Pd/C catalyst on the yields of liquid monomeric aromatic products. Green Chem 16:4740–4748. https://doi.org/10.1039/C4GC00854E
Osada M, Sato T, Watanabe M et al (2004) Low-temperature catalytic gasification of lignin and cellulose with a ruthenium catalyst in supercritical water. Energy Fuels 18:327–333. https://doi.org/10.1021/ef034026y
Osada M, Sato O, Watanabe M et al (2006) Water density effect on lignin gasification over supported noble metal catalysts in supercritical water. Energy Fuels 20:930–935. https://doi.org/10.1021/ef050398q
Ouyang X, Ke L, Qiu X et al (2009) Sulfonation of alkali lignin and its potential use in dispersant for cement. J Dispers Sci Technol 30:1–6. https://doi.org/10.1080/01932690802473560
Oveissi F, Fatehi P (2015) Characterization of four different lignins as a first step toward the identification of suitable end-use applications. J Appl Polym Sci 132:42336. https://doi.org/10.1002/app.42336

Pan X, Saddler JN (2013) Effect of replacing polyol by organosolv and kraft lignin on the property and structure of rigid polyurethane foam. Biotechnol Biofuels 6:12. https://doi.org/10.1186/1754-6834-6-12

Pan J, Fu J, Deng S, Lu X (2014) microwave-assisted degradation of lignin model compounds in imidazolium-based ionic liquids. Energy Fuels 28:1380–1386. https://doi.org/10.1021/ef402062w

Pandey MP, Kim CS (2011) Lignin depolymerization and conversion: a review of thermochemical methods. Chem Eng Technol 34:29–41. https://doi.org/10.1002/ceat.201000270

Park J, Riaz A, Insyani R, Kim J (2018) Understanding the relationship between the structure and depolymerization behavior of lignin. Fuel 217:202–210. https://doi.org/10.1016/j.fuel.2017.12.079

Park I-K, Sun H, Kim S-H et al (2019) Solvent-free bulk polymerization of lignin-polycaprolactone (PCL) copolymer and its thermoplastic characteristics. Sci Rep 9:7033. https://doi.org/10.1038/s41598-019-43296-2

Podschun J, Saake B, Lehnen R (2015) Reactivity enhancement of organosolv lignin by phenolation for improved bio-based thermosets. Eur Polym J 67:1–11. https://doi.org/10.1016/j.eurpolymj.2015.03.029

Pohjanlehto H, Setälä HM, Kiely DE, McDonald AG (2014) Lignin-xylaric acid-polyurethane-based polymer network systems: preparation and characterization. J Appl Polym Sci 131:39714. https://doi.org/10.1002/app.39714

Rachel-Tang DY, Islam A, Taufiq-Yap YH (2017) Bio-oil production via catalytic solvolysis of biomass. RSC Adv 7:7820–7830. https://doi.org/10.1039/c6ra27824h

Rajasekhar Reddy B, Vinu R (2018) Microwave-assisted co-pyrolysis of high ash Indian coal and rice husk: product characterization and evidence of interactions. Fuel Process Technol 178:41–52. https://doi.org/10.1016/j.fuproc.2018.04.018

Ranzi E, Debiagi PEA, Frassoldati A (2017) Mathematical modeling of fast biomass pyrolysis and bio-oil formation. Note I: kinetic mechanism of biomass pyrolysis. ACS Sustain Chem Eng 5:2867–2881. https://doi.org/10.1021/acssuschemeng.6b03096

Rezzoug S-A, Capart R (2002) Liquefaction of wood in two successive steps: solvolysis in ethylene-glycol and catalytic hydrotreatment. Appl Energy 72:631–644. https://doi.org/10.1016/S0306-2619(02)00054-5

Cetin NS, Özmen N (2002) Use of organosolv lignin in phenol-formaldehyde resins for particleboard production: II. Particleboard production and properties. Int J Adhes Adhes 22:481–486. https://doi.org/10.1016/S0143-7496(02)00059-3

Rodriguez Correa C, Kruse A (2018) Supercritical water gasification of biomass for hydrogen production—review. J Supercrit Fluids 133:573–590

Rogers RD, Voth GA (2007) Ionic liquids. Accounsts Chem Res 40:1077–1078

Sáenz-Pérez M, Lizundia E, Laza JM et al (2016) Methylene diphenyl diisocyanate (MDI) and toluene diisocyanate (TDI) based polyurethanes: thermal, shape-memory and mechanical behavior. RSC Adv 6:69094–69102. https://doi.org/10.1039/C6RA13492K

Sainsbury PD, Hardiman EM, Ahmad M et al (2013) Breaking down lignin to high-value chemicals: the conversion of lignocellulose to vanillin in a gene deletion mutant of *rhodococcus jostii* RHA1. ACS Chem Biol 8:2151–2156. https://doi.org/10.1021/cb400505a

Saito T, Perkins JH, Jackson DC et al (2013) Development of lignin-based polyurethane thermoplastics. RSC Adv 3:21832–21840. https://doi.org/10.1039/c3ra44794d

Salvachúa D, Karp EM, Nimlos CT et al (2015) Towards lignin consolidated bioprocessing: simultaneous lignin depolymerization and product generation by bacteria. Green Chem 17:4951–4967. https://doi.org/10.1039/c5gc01165e

Sanchez-Vazquez SA, Hailes HC, Evans JRG (2013) Hydrophobic polymers from food waste: resources and synthesis. Polym Rev 53:627–694. https://doi.org/10.1080/15583724.2013.834933

Santos JI, Martín-sampedro R, Fillat U et al (2015) Evaluating lignin-rich residues from biochemical ethanol production of wheat straw and olive tree pruning by FTIR and 2D-NMR. Int J Polym Sci 2015:1–11. https://doi.org/10.1155/2015/314891

Sato T, Furusawa T, Ishiyama Y et al (2006) Effect of water density on the gasification of lignin with magnesium oxide supported nickel catalysts in supercritical water. Ind Eng Chem Res 45:615–622. https://doi.org/10.1021/ie0510270

Schutyser W, Renders T, Van den Bosch S et al (2018) Chemicals from lignin: an interplay of lignocellulose fractionation, depolymerisation, and upgrading. Chem Soc Rev 47:852–908. https://doi.org/10.1039/C7CS00566K

Shen D, Zhao J, Xiao R, Gu S (2015) Production of aromatic monomers from catalytic pyrolysis of black-liquor lignin. J Anal Appl Pyrolysis 111:47–54. https://doi.org/10.1016/j.jaap.2014.12.013

Shi Y, Xia X, Li J et al (2016) Solvolysis kinetics of three components of biomass using polyhydric alcohols as solvents. Bioresour Technol 221:102–110. https://doi.org/10.1016/J.BIORTECH.2016.09.008

Sigoillot JC, Berrin JG, Bey M, et al (2012) Fungal strategies for lignin degradation. In: Advances in botanical research. Academic Press, pp 263–308

Sikarwar VS, Zhao M, Clough P et al (2016) An overview of advances in biomass gasification. Energy Environ Sci 9:2939–2977

Singh SK, Ekhe JD (2014) Solvent effect on HZSM-5 catalyzed solvolytic depolymerization of industrial waste lignin to phenols: superiority of the water–methanol system over methanol. RSC Adv 4:53220–53228. https://doi.org/10.1039/C4RA10240A

Singh SK, Nandeshwar K, Ekhe JD (2016) Thermochemical lignin depolymerization and conversion to aromatics in subcritical methanol: effects of catalytic conditions. New J Chem 40:3677–3685. https://doi.org/10.1039/c5nj02916c

Speight JG (2017) Industrial organic chemistry. Environ Org Chem Eng 87–151. https://doi.org/10.1016/B978-0-12-804492-6.00003-4

Statista (2019) Polyurethane global demand 2022 | Statistic. In: Statista. https://www.statista.com/statistics/747004/polyurethane-demand-worldwide/. Accessed 5 Feb 2020

Suman SK, Dhawaria M, Tripathi D et al (2016) Investigation of lignin biodegradation by Trabulsiella sp. isolated from termite gut. Int Biodeterior Biodegrad 112:12–17. https://doi.org/10.1016/j.ibiod.2016.04.036

Sun Z, Fridrich B, De Santi A et al (2018) Bright side of lignin depolymerization: toward new platform chemicals. Chem Rev 118:614–678. https://doi.org/10.1021/acs.chemrev.7b00588

Tarabanko VE, Petukhov DV, Selyutin GE (2004) New mechanism for the catalytic oxidation of lignin to vanillin. Kinet Catal 45:569–577. https://doi.org/10.1023/B:KICA.0000038087.95130.a5

Tayier M, Duan D, Zhao Y et al (2018) Catalytic effects of various acids on microwave-assisted depolymerization of organosolv lignin. BioResources 13:412–424. https://doi.org/10.15376/biores.13.1.412-424

Thanh Binh NT, Luong ND, Kim DO et al (2009) Synthesis of lignin-based thermoplastic copolyester using kraft lignin as a macromonomer. Compos Interfaces 16:923–935. https://doi.org/10.1163/092764409X12477479344485

Thierry M, Majira A, Pégot B et al (2018) Imidazolium-based ionic liquids as efficient reagents for the C−O bond cleavage of lignin. Chemsuschem 11:439–448. https://doi.org/10.1002/cssc.201701668

Thring RW (1994) Alkaline degradation of Alcell lignin. Biomass Bioenerg 7:125–130. https://doi.org/10.1016/0961-9534(94)00051-T

Todd R, Baroutian S (2017) A techno-economic comparison of subcritical water, supercritical CO_2 and organic solvent extraction of bioactives from grape marc. J Clean Prod 158:349–358. https://doi.org/10.1016/J.JCLEPRO.2017.05.043

Toledano A, Serrano L, Labidi J (2012) Organosolv lignin depolymerization with different base catalysts. J Chem Technol Biotechnol 87:1593–1599. https://doi.org/10.1002/jctb.3799

Toledano A, Serrano L, Labidi J (2014) Improving base catalyzed lignin depolymerization by avoiding lignin repolymerization. Fuel 116:617–624. https://doi.org/10.1016/j.fuel.2013.08.071

Upton BM, Kasko AM (2016) Strategies for the conversion of lignin to high-value polymeric materials: review and perspective. Chem Rev 116:2275–2306

Vázquez G, González J, Freire S, Antorrena G (1997) Effect of chemical modification of lignin on the gluebond performance of lignin-phenolic resins. Bioresour Technol 60:191–198. https://doi.org/10.1016/S0960-8524(97)00030-8
Venkatesagowda B (2019) Enzymatic demethylation of lignin for potential biobased polymer applications. Fungal Biol Rev. 33:190–224. https://doi.org/10.1016/j.fbr.2019.06.002
Venkatesagowda B, Dekker RFH (2019) A rapid method to detect and estimate the activity of the enzyme, alcohol oxidase by the use of two chemical complexes—acetylacetone (3,5-diacetyl-1,4-dihydrolutidine) and acetylacetanilide (3,5-di-N-phenylacetyl-1,4-dihydrolutidine). J Microbiol Methods 158:71–79. https://doi.org/10.1016/j.mimet.2019.01.021
Vithanage AE, Chowdhury E, Alejo LD et al (2017) Renewably sourced phenolic resins from lignin bio-oil. J Appl Polym Sci 134:44827. https://doi.org/10.1002/app.44827
Wang H, Sun X, Seib P (2001) Strengthening blends of poly (lactic acid) and starch with methylenediphenyl diisocyanate. J Appl Polym Sci 82:1761–1767. https://doi.org/10.1002/app.2018
Wang L, Lei H, Ren S et al (2012) Aromatics and phenols from catalytic pyrolysis of Douglas fir pellets in microwave with ZSM-5 as a catalyst. J Anal Appl Pyrolysis 98:194–200. https://doi.org/10.1016/j.jaap.2012.08.002
Wang H, Tucker M, Ji Y (2013) Recent development in chemical depolymerization of lignin: a review. J Appl Chem 2013:1–9. https://doi.org/10.1155/2013/838645
Wang B, Chen TY, Wang HM et al (2018) Amination of biorefinery technical lignins using Mannich reaction synergy with subcritical ethanol depolymerization. Int J Biol Macromol 107:426–435. https://doi.org/10.1016/j.ijbiomac.2017.09.012
Wasserscheid P (2003) Potential to apply ionic liquids in industry. In: Rogers RD, Seddon KR, Volkov S (eds) Green industrial applications of ionic liquids. Springer, Dordrecht, Dordrecht, pp 29–47
Watanabe M, Inomata H, Osada M et al (2003) Catalytic effects of NaOH and ZrO_2 for partial oxidative gasification of n-hexadecane and lignin in supercritical water. Fuel 82:545–552. https://doi.org/10.1016/S0016-2361(02)00320-4
Wendisch VF, Kim Y, Lee J-H (2018) Chemicals from lignin: recent depolymerization techniques and upgrading extended pathways. Curr Opin Green Sustain Chem.14:33–39. https://doi.org/10.1016/j.cogsc.2018.05.006
Wikberg H, Ohra-aho T, Leppävuori J, Liitiä T, Kanerva H (2018) Method for producing reactive lignin. WO 2018/115592 A1, Teknologian Tutkimuskeskus Vtt OY
Wu Y-R, He J (2013) Characterization of anaerobic consortia coupled lignin depolymerization with biomethane generation. Bioresour Technol 139:5–12. https://doi.org/10.1016/J.BIORTECH.2013.03.103
Xiao W, Han L, Zhao Y (2011) Comparative study of conventional and microwave-assisted liquefaction of corn stover in ethylene glycol. Ind Crops Prod 34:1602–1606. https://doi.org/10.1016/j.indcrop.2011.05.024
Xie J, Qi J, Hse C, Shupe TF (2015) Optimization for microwave-assisted direct liquefaction of bamboo residue in glycerol/methanol mixtures. J For Res 26:261–265. https://doi.org/10.1007/s11676-015-0032-1
Xu C, Ferdosian F (2017) Conversion of lignin into bio-based chemicals and materials. Springer, Germany
Xu J, Jiang J, Hse C, Shupe TF (2012) Renewable chemical feedstocks from integrated liquefaction processing of lignocellulosic materials using microwave energy. Green Chem 14:2821–2830. https://doi.org/10.1039/c2gc35805k
Xu R, Zhang K, Liu P et al (2018) Lignin depolymerization and utilization by bacteria. Bioresour Technol 269:557–566. https://doi.org/10.1016/j.biortech.2018.08.118
Yamaguchi A, Hiyoshi N, Sato O et al (2008) Lignin gasification over supported ruthenium trivalent salts in supercritical water. Energy Fuels 22:1485–1492. https://doi.org/10.1021/ef8001263

Yamaguchi A, Hiyoshi N, Sato O, Shirai M (2012) Gasification of organosolv-lignin over charcoal supported noble metal salt catalysts in supercritical water. Top Catal 55:889–896. https://doi.org/10.1007/s11244-012-9857-4

Yang Y, Fan H, Meng Q et al (2017) Ionic liquid [OMIm][OAc] directly inducing oxidation cleavage of the β-O-4 bond of lignin model compounds. Chem Commun 53:8850–8853. https://doi.org/10.1039/c7cc04209d

Yerrayya A, Suriapparao DV, Natarajan U, Vinu R (2018) Selective production of phenols from lignin via microwave pyrolysis using different carbonaceous susceptors. Bioresour Technol 270:519–528. https://doi.org/10.1016/j.biortech.2018.09.051

Yin G, Jin F, Yao G, Jing Z (2015) Hydrothermal conversion of catechol into four-carbon dicarboxylic acids. Ind Eng Chem Res 54:68–75. https://doi.org/10.1021/ie5036447

Yu Y, Li X, Su L et al (2012) The role of shape selectivity in catalytic fast pyrolysis of lignin with zeolite catalysts. Appl Catal A Gen 447–448:115–123. https://doi.org/10.1016/j.apcata.2012.09.012

Yue X, Chen F, Zhou X (2011) Improved interfacial bonding of pvc/wood-flour composites by lignin amine modification. BioResources 6:2022–2034. https://doi.org/10.15376/biores.6.2.2022-2044

Yunpu W, Leilei D, Liangliang F et al (2016) Review of microwave-assisted lignin conversion for renewable fuels and chemicals. J Anal Appl Pyrolysis 119:104–113. https://doi.org/10.1016/j.jaap.2016.03.011

Zhang L, Huang J (2001) Effects of nitrolignin on mechanical properties of polyurethane-nitrolignin films. J Appl Polym Sci 80:1213–1219. https://doi.org/10.1002/app.1206

Zhang S, Dong Q, Zhang L, Xiong Y (2015a) High quality syngas production from microwave pyrolysis of rice husk with char-supported metallic catalysts. Bioresour Technol 191:17–23. https://doi.org/10.1016/J.BIORTECH.2015.04.114

Zhang Z, Macquarrie DJ, De Bruyn M et al (2015b) Low-temperature microwave-assisted pyrolysis of waste office paper and the application of bio-oil as an Al adhesive. Green Chem 17:260–270. https://doi.org/10.1039/c4gc00768a

Zheng A, Zhao Z, Chang S et al (2014) Effect of crystal size of ZSM-5 on the aromatic yield and selectivity from catalytic fast pyrolysis of biomass. J Mol Catal A: Chem 383–384:23–30. https://doi.org/10.1016/j.molcata.2013.11.005

Zhi Z, Wang H (2014) White-rot fungal pretreatment of wheat straw with Phanerochaete chrysosporium for biohydrogen production: Simultaneous saccharification and fermentation. Bioprocess Biosyst Eng 37:1447–1458. https://doi.org/10.1007/s00449-013-1117-x

Zhou S, Garcia-Perez M, Pecha B et al (2013) Effect of the fast pyrolysis temperature on the primary and secondary products of lignin. Energy Fuels 27:5867–5877. https://doi.org/10.1021/ef4001677

Zhu X, Peng C, Chen H et al (2018) Opportunities of ionic liquids for lignin utilization from biorefinery. ChemistrySelect 3:7945–7962. https://doi.org/10.1002/slct.201801393

Zou L, Ross BM, Hutchison LJ et al (2015) Fungal demethylation of kraft lignin. Enzyme Microb Technol 73–74:44–50. https://doi.org/10.1016/J.ENZMICTEC.2015.04.001

Lignin and Its Composites

Muhammad Ghozali, Evi Triwulandari, Witta Kartika Restu, Sri Fahmiati and Yenny Meliana

Abstract Lignin as a novel biomaterial has functional properties for the development of new biocomposite materials, and it would be an appropriate candidate for modification and chemical reactions. This chapter deals with the capability of lignin for reinforcing other materials in polymer biocomposites, especially incorporation of lignin in thermoplastic and thermoset composites. Lignin can be composited with thermoplastic polymers (polyethylene, polypropylene and polylactic acid) and also thermoset materials (polyurethane and epoxy). The addition of lignin to non-biodegradable materials improved the biodegradability of resulted composites. The degradation of resulted composites related to the existence of lignin which can be seen through the hydrophilic functional groups of lignin. This encourage the biodegradation of the obtained composites.

Keywords Composites · Lignin · Polypropylene · Polyethylene · Polylactic acid · Polyurethane · Epoxy

1 Introduction

Study on the application and development of natural-based polymeric materials for the polymer industry has been attracting the interest of several researchers since the last few decades. Basically, the development of natural-based polymeric materials comes from the waste problem and the abundance of natural resources which can replace petroleum-based materials.

Several biomasses which can be used as natural-based polymeric materials are starch, lignin, cellulose, protein, chitosan, and vegetable oils. Among those raw materials, lignocellulose is considered as one of the most potent materials. Because it not edible material and does not crash with corps. Commonly, lignocellulose

M. Ghozali (✉) · E. Triwulandari · W. K. Restu · S. Fahmiati · Y. Meliana
Polymer Chemistry Group, Research Center for Chemistry, Indonesian Institute of Sciences (LIPI), 452 Building, Kawasan Puspiptek Serpong, Tangerang Selatan, Banten 15314, Indonesia
e-mail: muhammad.ghozali@lipi.go.id

S. Sharma and A. Kumar (eds.), *Lignin*, Springer Series on Polymer and Composite Materials, https://doi.org/10.1007/978-3-030-40663-9_6

is derived from wood which contains cellulose, hemicellulose, and lignin. The utilization of lignin as novel biomaterial would be an appropriate candidate for modification and chemical reactions due to its functional properties. Yet, the unclear structure and its diversity such as origins, separations, and fragmentation processes became its major challenge on its applications. Chemical-modified lignin is categorized into three types such as fragmented lignin into phenol or aromatic compounds, active sites, and functionalization process for hydroxyl groups (Laurichesse and Avérous 2014).

The availability of lignin has reached almost 30% of nonfossil organic carbon and is commonly produced by paper industrial in a significant amount. The number of hydroxyl groups in lignin gives a relatively polar characteristic which provides great affinity to the polar polymer. However, it also causes a problem when it was composited with nonpolar material which is not compatible with stress transfer in terms of mechanical properties. Therefore, it needs compatibilizer to improve interphase analysis. Moreover, it also can be improved by reducing hydroxyl group content (Duval and Lawoko 2014). There are studies which concern on the application of lignin as well as focused on the production of lignin-based thermoplastic and thermoset materials. In these studies, lignin was mixed with thermoplastic and thermoset material. Its result was a composite that has good biodegradability properties.

2 Lignin in Thermoplastic Composites

2.1 Polyethylene/Lignin Composites

This research is to investigate the effects of lignin addition morphology, biodegradability, and mechanical. Then, the thermal properties of low linear density polyethylene (LLDPE)/lignin biocomposites were reported (Ghozali et al. 2017). LLDPE, lignin, and PE-g-MA are mixed in the mixer. Five to twenty per hundred resin (phr) of lignin was added with 5 phr of compatibilizer. It was carried out in the mixer at 200 °C and 70 rpm for 30 min. The pieces produced from the mixture are then made into a sheet form using a hot-press at a temperature of 210 °C with a pressure of 6 bar for 20 min.

The lignin composition added to LLDPE affects the color of LLDPE/lignin biocomposite. The effect of lignin composition on the color of biocomposite films is shown in Fig. 1. The more lignin was added to the biocomposite, the darker LLDPE/lignin biocomposite film produced.

LLDPE/lignin biocomposites were easier to degrade at a temperature below 450–460 °C compared with LLDPE itself. However, it is more resistant to thermal degradation instead of lignin. By contrast, at a temperature over 460 °C, LLDPE/lignin biocomposites were more resistant with thermal degradation than LLDPE. Yet, it was easier to degrade instead of lignin. The declination of thermal

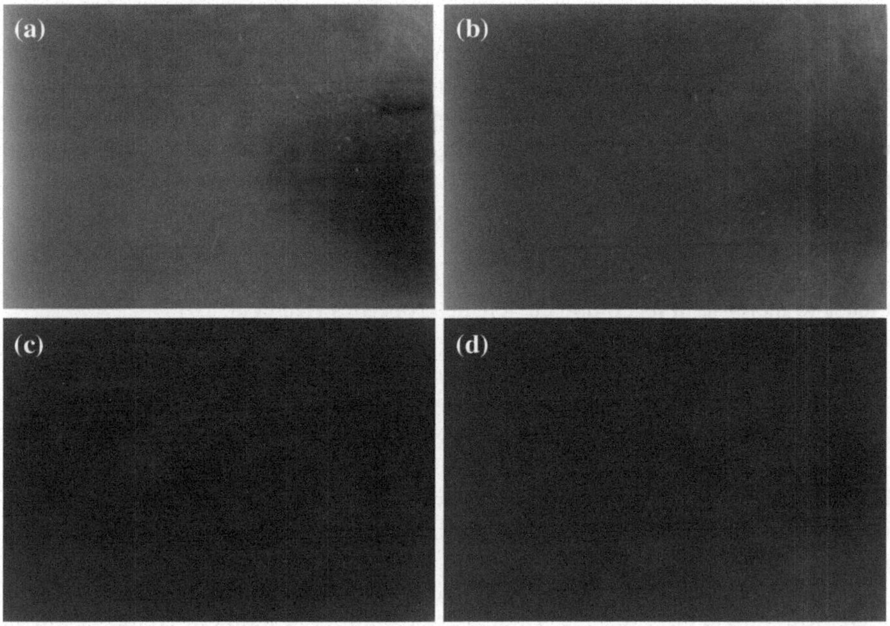

Fig. 1 LLDPE/lignin biocomposite films: (**a**) LLDPE/lignin 5 phr; (**b**) LLDPE/lignin 10 phr;(**c**) LLDPE/lignin 15 phr and (**d**) LLDPE/lignin 20 phr. Reprinted with permission from Ghozali et al. (2017)

stability of LLDPE/lignin was occurred due to the vacant space in the LLDPE/lignin biocomposite itself which is filled by trapped oxygen or air. After that, it initiated the degradation process which results in lower thermal stability (Dikobe and Luyt 2010). The thermal resistance of lignin was better than LLDPE at a temperature over 450 °C which then affects the thermal resistance of LLDPE/lignin biocomposites. It is caused by the existence of aromatic phenyl groups. In addition, the existence of hydroxyl groups contributes to thermal stability as well and also improves the stability of the aromatic structure in lignin (Gordobil et al. 2015). High residual mass was caused by the chemical structure of lignin. Due to the high charring ability and low heat release when it burned, the flammability of materials gave less contribution (Detyothin et al. 2013; Yang et al. 2015). It also shows that high lignin composition in LLDPE/lignin biocomposites was increased the char content (Sahoo et al. 2011; Sailaja and Deepthi 2010; Spiridon et al. 2015).

On the other hand, tensile properties of all LLDPE/lignin biocomposites at several compositions were lower, compared to LLDPE (Cazacu et al. 2004; Alexy et al. 2000). Lower tensile strength and elongation at break were considered as a result of nonuniform lignin distribution on LLDPE/lignin biocomposite (Anwer et al. 2015). The elongation at break of LLDPE/lignin biocomposites was dramatically decreased, compared to LLDPE. It was probably caused by the existence of lignin on LLDPE/lignin biocomposites. It also results in fragility, due to the lack of interface adhesion. This lack of interface adhesion can be attributed to reduce the flexibility of LLDPE/lignin biocomposites. Moreover, the existence of lignin also increased the brittleness and affected the viscoelastic property of LLDPE/lignin biocomposites. The hydrophilic characteristic properties of lignin produce a mismatch of lignin with the hydrophobic characteristics of LLDPE and the tendency to form aggregates. It is also the cause of lignin which is not resistant to moisture which leads to high water absorption. Therefore, it caused a poor tensile property of LLDPE/lignin biocomposite. In this case, chemical modification is needed to increase the hydrophobicity of lignin and reduce water absorption; then, it leads to increase in its composite tensile properties (Ku et al. 2011).

The effects during the degradation process are highly dependent on hydrophilicity or hydrophobicity and particle dispersion. In addition, the delay in biodegradability is caused by an increase in the properties of materials which can prevent water diffusion through its bulk (Lee et al. 2002). This degradation of LLDPE/lignin biocomposite is related to the existence of lignin which has many hydrophilic functional groups that encourage it to have biodegradable properties. When the amount of lignin was increased, biocomposite becomes more hydrophilic. Therefore, the material becomes more biodegradable

The effect of lignin on the morphological, mechanical, and thermal properties of HDPE/lignin composites was studied. The addition of lignin to HDPE results in poor mechanical properties due to its low compatibility (Hu et al. 2015). Sailaja and Deepthi (2010) studied the mechanical and thermal properties of the esterified polyethylene and lignin composites. Maleic grafted LDPE anhydride has been used to increase interface adhesion. The results of the study showed that the addition of a compatibilizer improved thermal stability and mechanical properties of polyethylene/lignin composites (Sailaja and Deepthi 2010).

2.2 Polypropylene/Lignin Composites

Polypropylene (PP) is a thermoplastic polymer which has some different arrangements of methyl group, i.e., isotactic, syndiotactic, and atactic (Baker 2018). Polypropylene is also known to have good mechanical properties and relatively low price. These features make polypropylene which is applied to wide applications such as plastics and fiber. One of the disadvantages of PP is instability to high temperature and UV light (Dopico-Garcia et al. 2007). To overcome thermal oxidative degradation, the antioxidant is usually added to the polymer (Gao et al. 2008). Irganox 1010 is the primary antioxidant for polypropylene (Gadioli et al. 2016). However, this antioxidant gives after disposal effect which decreases the decomposition process of polypropylene (Morandim-Giannetti et al. 2012). Compared to commercial antioxidants, lignin is a natural alternative which is biodegradable, abundant biopolymer and has good stability and good mechanical properties (Triwulandari et al. 2019).

A limitation of using lignin to be blended with nonpolar polymers like polyethylene and polypropylene (PP) is the poor compatibility which can cause a poor dispersion of lignin in these polymers (Chen et al. 2016). To overcome this, lignin was modified chemically or physically and combined with various coupling agents.

Triwulandari et al. (2019) studied the effect of various concentrations of lignin that incorporated into neat PP, neat PLA, and PP/PLA blends using the extrusion method. The compatibilizer, PP-g-Maleic anhydride, was used. Lignin addition was found to reduce the tensile strength of PP/Lignin composite compared to neat PP. Lignin addition as 5 phr increased the elongation at break of PP/lignin composite, but as the concentration of lignin was higher (10, 15 and 20 phr), it tends to decrease as compared to neat PP. This is supposedly attributed to the weak interfacial adhesion between PP and lignin.

Canetti et al. (2004) studied the influence effect of lignin on the thermal properties and kinetic crystallization of isotactic polypropylene. Thermal degradation temperature increased with the increasing of lignin content in PP–lignin blend. This study suggested that PP crystallized in two crystalline forms when lignin was incorporated.

Blanco et al. (2017) studied a mixture of polypropylene–lignin. Lignin was obtained from liquid wood, a mixture of cellulose, hemp, fax, and lignin with Maleic Anhydride–grafted Polypropylene (MAPP) was employed as a coupling agent with various contents of PP/MAPP and lignin. The blending was proceeded by mechanical mixing followed by thermal extrusion. The DSC showed that the melting temperature was increased as lignin content increased; however, interestingly, they found the glass transition shortly before 60 °C. This phenomenon was thought to be the effect of the presence of PLA in liquid wood and enhanced successful blending between PP, liquid wood, and MAPP.

Chen et al. (2016) studied the physical properties of PP blend with modified alkylated lignin. They found that the incorporation of 5 wt% of alkylated lignin elevated the thermal degradation temperature of neat PP about 27 °C. For physical properties, the impact strength of the PP/lignin composites was increased with a maximum of 35% of 10 wt% alkylated lignin. SEM image showed that alkylated lignin can be acted as a compatibilizing agent for the composite.

Another chemical modification of lignin to enhance its compatibility with PP was done by Ye et al. (2016). They proceeded butylation reaction to modify kraft lignin by using n-butyric anhydride, considering the fact that solubility of butylated lignin is near close to polypropylene. The thermal oxidation stability of polypropylene was increased by the addition of butylated lignin. The blending of butylated lignin with PP showed better compatibility and mixing compared to unmodified kraft lignin. Thermal oxidation stability of butylated lignin/PP was increased as 29–66%. The addition of butylated lignin also improved the onset temperature and $T_5\%$ of PP by about 10 °C with only 1 wt% of it. It was found that there was no significant effect of butylated lignin on mechanical properties (e.g., tensile strength and elongation) of PP.

Besides the chemical modification, physical modification of lignin was also used to modify the compatibility between lignin and PP. Chen et al. (2016) used a freeze-drying method of alkali lignin to obtain a sheet-like morphology and then incorporated it into polypropylene (PP) by melt compounding. Figure 2 shows the different morphology of freeze-dried lignin in the PP matrix. TEM images in Fig. 3 exhibited the relatively good dispersion of the freeze-dried lignin in PP as compared with neat lignin which was exhibited in TEM images. Tensile modulus of the composite was increased by the addition of lignin; in contrary, yield stress and the strain were decreased slightly.

Another physical modification of lignin was also done by surface modification of lignin by introducing the hydrophobic molecule into lignin surface, in order to reduce surface tension. Atifi et al. (2017) used cetyltrimethylammonium bromide (CTAB), poly(ethylene oxide), polyethylene-block-poly(ethylene glycol), dodecenyl succinic anhydride, and alkyl ketene dimer (AKD) to modify the surface of lignin.

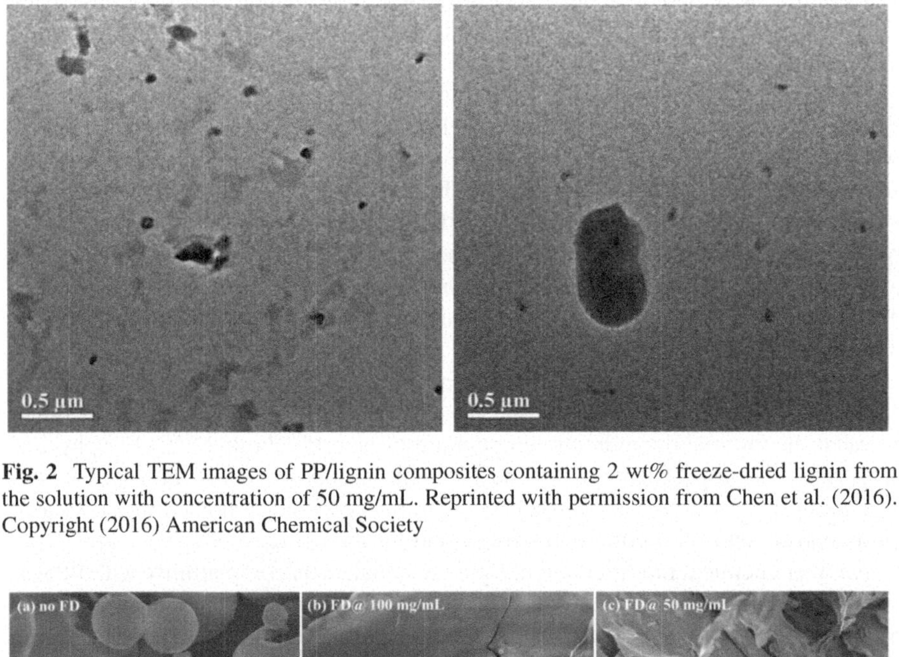

Fig. 2 Typical TEM images of PP/lignin composites containing 2 wt% freeze-dried lignin from the solution with concentration of 50 mg/mL. Reprinted with permission from Chen et al. (2016). Copyright (2016) American Chemical Society

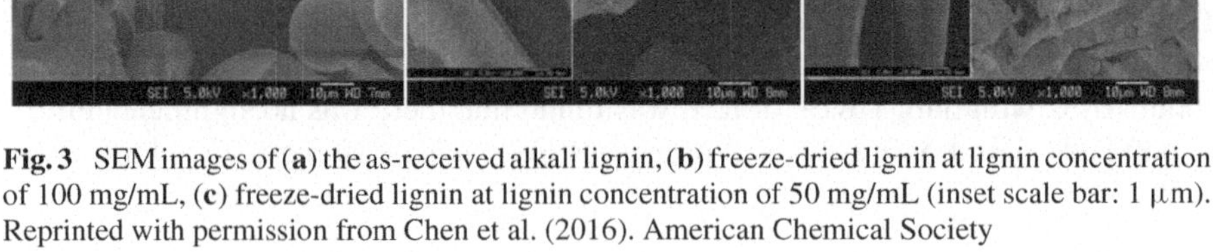

Fig. 3 SEM images of (**a**) the as-received alkali lignin, (**b**) freeze-dried lignin at lignin concentration of 100 mg/mL, (**c**) freeze-dried lignin at lignin concentration of 50 mg/mL (inset scale bar: 1 μm). Reprinted with permission from Chen et al. (2016). American Chemical Society

The composite of PP–alkyl ketene dimer-treated lignin showed better mechanical properties. The yield stress was maintained (relative to neat PP), while it was higher than that for untreated lignin. However, Young's modulus and stress at yield for both composites decreased when lignin was treated with poly(ethylene oxide) or dodecenyl succinic anhydride.

To increase mechanical properties and components sterilization, polypropylene (PP) was irradiated by electron beam irradiation (Sugano-Segura et al. 2017). Effects of irradiation dose on the mechanical and thermal properties of a PP matrix reinforced with technical kraft lignin were investigated. This study found that lignin reduced the effects of radiation on yield stress and the strain at break. The lignin content of 10 wt% was found as the most efficient condition for maintaining mechanical and thermal properties and preventing the degradation of PP.

2.3 Polylactic Acid/Lignin Composites

Poly(lactic acid) (PLA) is a kind of biodegradable polymer that extracted from crops such as wheat, sugar beets, and corn (Gordobil et al. 2014; Lim et al. 2008). The aim of exploring PLA is to replace commodity polymers. This polymer is one of the most promising polymers that environmentally friendly for the sustainable development (Courgneau et al. 2012; Ouyang et al. 2012; Pillai et al. 2013). Like other thermoplastic polymers, PLA shows the melting temperature (Tm) and the glass transition temperature (Tg). The Tg of PLA was influenced by the molecular weight and purity, while the physical and mechanical properties depend on the morphology and crystallinity. The tendency in thermal degradation is one of the shortcomings in synthesizing PLA. This is correlated with the process temperature and time length in the process (Vouyiouka et al. 2013). Therefore, PLA composite in which reinforced by lignin is one of the promising approaches to solve the shortcoming of PLA. Like the abundant natural resources, lignin is composed of functional groups such as carbonyl, phenolic or aliphatic hydroxyls, and carboxyl (Brebu and Vasile 2010). Here, several PLA/lignin composite will be discussed in terms of their improvement in the physicochemical, mechanical, and morphological properties.

Yang et al. (2015) conducted two different methods in producing the composites, melt extrusion and solvent casting, in which each method had a different percentage of the amount of lignin added. In thermal properties, the glass transition (Tg) was influenced by the method, in which the cast system had lower Tg than the extruded samples. This was due to the residual solvent remaining in the system. While for the second heating, this did not happen. The addition of 1 wt% lignin decreases glass transition temperature due to the plasticizing effect of low molecular weight fraction of lignin (Domenek et al. 2013). In terms of Cold Crystallization Temperature (Tcc), the increasing amount of lignin added to PLA for extruded sample shifted to the lower temperature. While for the second heating, similar curves were observed for a sample from extruded and solvent cast method. This indicated the same behaviors and phases for their crystallization. The thermal characteristic was also investigated by utilizing analysis in thermogravimetric (TGA) which measured the loss of weight from the composites. During the heating process (80–150 °C), the PLA films with a solvent cast method decreased about 10% from their initial weight due to the removal of solvent (He et al. 2008).

Different from Yang group, Spiridon et al. (2015) evaluated the properties of PLA/lignin due to the acceleration of weathering by utilizing lignin from softwood and hardwood for PLA composites. Accelerated weathering was done by exposing samples to the artificial light such as from a mercury lamp. Mercury lamps as the resource was utilized at a humidity of 60% and a temperature of 30 °C while the samples were exposed up to 600 h. The melt temperature of the blends increased due to the hydrogen bonding from the groups of carbonyl from PLA and the functional groups from lignin. There was no important difference in Tg values in the first and second heating. The weathering resulted in a decrease of the transition temperature, namely PLA/hardwood, PLA/softwood, and then PLA. The degradation for PLA and PLA/softwood composite was completed at the same temperature. There was no dramatic change in loss of weight from the composite materials when compared with the matrix of PLA.

Gordobil et al. (2015) worked on utilizing kraft lignin derived from kraft black liquor which had a high level of purity than ashes and sugars at low concentration. The kraft lignin (KL) was modified into acetylated kraft lignin (AKL) due to higher molecular weight and less polydisperse. In the obtained curves of TGA and DTG, the addition of KL and AKL increased the thermal stability of PLA. All composites showed an increase in the temperature compared to PLA. The thermal stability obtained from lignin became high due to the complex units of phenylpropanoid, which is composed of aromatic groups. Another study of Gordobil et al. (2014) showed the effects of acetylated lignin on the properties of composites or blends in this study. The lignin that was derived from almond shells and alkaline lignin was used as fillers for PLA to achieve the compatibility from the mixture of two components. The addition of different lignin increased the stability in thermal properties of all blends that explained in the previous study. In addition, some of the hydroxyl groups in lignin that were added to the blend affected the stability (Gordobil et al. 2015). Generally, char residue increased as lignin was added into the composites, and this was reported in the previous studies (Sahoo et al. 2011).

While Wang et al. (2014) investigated the PLA/lignin composite-containing cellulose nanofibrils (LCNF). The composites were produced by the solution casting method. This work showed that lignin composed with cellulose in nanofibrils form could be disseminated in PLA with the enhancement in thermal, mechanical, and crystallization properties (Mu et al. 2014). The thermal properties were affected by the increase of lignin's composition. Lignin has a better effect on the solubility between PLA and LCNFs, and the increase in lignin content also influences the resistance in thermal for LCNFs. Regarding the lignin added into composite which consists of two polymers, Triwulandari et al. had explored the significance of lignin addition toward PLA and polypropylene (PP) composite (Triwulandari et al. 2019). As a result, the glass temperature transition, crystallization process, and melting temperature from the obtained PLA/lignin and PP/PLA/lignin composites films decreased after the addition of lignin (Triwulandari et al. 2019).

In Yang et al. (2015), in the melt extrusion method, the introduction of lignin of 1 wt% resulted in the increasing value for each parameter and the opposite for the solvent casting method, while the increase of lignin content in composite resulted in decreasing of tensile strength, modulus, and elongation at break for both methods. This is because of the existence of lignin particles. It was also observed that the number of lignin particles could influence the flexibility of the structure due to the linkage of benzene, and ether rings created internal plasticizers (Mu et al. 2014; Wang et al. 2012). However, the speeding up weathering resulted in no important effect toward the elasticity characteristic of the PLA/lignin composites. The tensile and impact strength decreased slowly. The tensile properties were affected by the intermolecular interaction of lignin–lignin, PLA–PLA and also PLA–lignin. It was found that the decrease in tensile strength of the blends was caused by the lower value of tensile strength from lignin if compared to PLA, while the rigid characteristic of lignin yielded in a decrease of elongation at break for the blends. These results proved that the composites were composed of lignin that was derived from hardwood, showed the highest value in mechanical characteristics for before and after 600 h weathering (Spiridon et al. 2015).

In terms of modified lignin (Gordobil et al. 2015), the maximum strength decreased greatly, particularly in materials with KL addition of 10 and 20%. The low concentration (0.5–5%) of KL and AKL did not affect the strength at break. While at higher concentration (greater than 5%), the strength decreases along with the increase in lignin content. This was particularly for composite from PLA and non-modified lignin. A significant difference was revealed between PLA composite with the addition of KL and AKL. This might be caused by the best interaction from PLA with AKL compare to PLA/non-modified lignin. The same went to other sources of lignin in which the optimum strain did not influence by the lower percentages (0.5–1%) of non-modified lignin. However, at higher percentages of more than 5%, the strain decreased along with the increase in lignin composition, particularly in the blends of PLA with commercial lignin.

The properties of PLA that were blended with acetylated lignin maintained a similar characteristic of neat PLA. Generally, the value of elongation at break increased. However, the case was different in lignin-containing cellulose (Wang et al. 2014). There was a decrease in tensile strength and modulus for PLA and PLA/CNFs composite from 35.7 MPa and 1.8 GPa to 31.6 MPa and 1.5 GPa, respectively. Compared with pure PLA, the addition of lignin–cellulose until 9% improved the tensile strength and modulus because of the better compatibility of the LCNFs with PLA. The increasing lignin–cellulose more than 9% was slightly lower due to the interaction from the LCNFs and PLA which enhances the increasing lignin concentration in the LCNFs; the mechanical characteristics of the LCNFs are also influenced by the amount of lignin. This showed that the flexibility characteristic of PLA increases due to the blending process with LCNFs (Wang et al. 2014). Lignin content also affected the composite which is composed of two polymers. The presence of lignin in a composite of PP/lignin, PLA/lignin, and PP/PLA/lignin resulted in a reduction of tensile properties such as tensile strength and elongation at break. The higher the lignin content, the lower the tensile properties for these composites (Triwulandari et al. 2019).

In terms of surface structure, the melt extrusion and solvent casting method resulted in a different way (Yang et al. 2015). Morphology with a smoother surface could be observed in the extruded film sample because of an even distribution of lignin in the PLA matrix. However, the sample from the solvent casting method showed many micro-holes and less homogeneous surface. These phenomena were in correlation with the evaporated chloroform solvent and separation of PLA and lignin. In term of morphology, the solubility value could be adjusted to reckon the mixing ability of the blends. The solubility value of PLA and lignin are 20.2 and 19.02 MPa, respectively (Karst and Yang 2005), which indicated better miscibility between the polymer and the lignin. The surface of aged PLA and PLA-based composites displayed matrix cracks and milky patches. It is hard to search for lignin particles in the spherical form within nonaged PLA matrix samples. This result showed that the two kinds of lignin could be dispersed in the PLA matrix while in the melting process. There is also a good adhesion of two lignins toward PLA. Different from the previous research group, in Gordobil et al. (2015), all blends with non-modified lignin showed aggregation because of the incompatibility of PLA and lignin. In their process, the blends displayed a homogeneous appearance after acetylation completed. However, at a microscopic view, it revealed the presence of small size of lignin particles. In this scale, the particles size was much smaller than the composites from non-modified lignin. Therefore, acetylation process of lignin resulted in a mixture with higher compatibility and greater dispersion of lignin particles in the mixture. When PLA was mixed with LCNFs, lignin could act as a compatibilizer in order to form a mixed system uniformly. When LCNFs with various lignin concentrations were added to the matrix of PLA, the fracture surfaces of the composites showed a different microstructure. The results showed that the increase of lignin content from the LCNFs promotes ductile rupture of PLA (Wang et al. 2014). Therefore, lignin and its amount have significant roles in enhancing the capability of PLA. Besides, lignin also affected the composites of two polymers like PP/PLA. From the morphological analysis, the interfaced adhesion of lignin in PP/PLA composite could be seen, in which it was demonstrated by microvoid structures in the surface of composites (Triwulandari et al. 2019).

3 Lignin in Thermoset Composites

3.1 Polyurethane/Lignin Composites

Polyurethanes (PUs) are a unique group of polymeric materials which contain urethane group (–NH–CO–O–) with high potential to be used in various applications, particularly due to their structure-property relationships. PUs with their special character are essentially different from most other plastics types. PUs contain two types of monomers, isocyanates and polyols; both polymerize one after the other and also can be group as alternating copolymers. This polyurethane group can be incorporated into many different items with broad applications, such as coatings, elastomers, insulators, elastic fibers, foams, and integral skins (Akindoyo et al. 2016).

Historically, urethane was created in 1849 by Wurtzring aliphatic monofunctional isocyanates and alcohols. Then in 1937, Otto Bayer processed PUs by reacted a diisocyanate and a polyester diol. Polyurethanes typically synthesized through the reaction between the hydroxyl (OH) groups of a polyol with the isocyanate functional group (NCO) groups of an isocyanate. The name of PU was in correlation with the resulting urethane linkage (Zhang et al. 2015).

Currently, the polyols that are used in the process of PUFs are derived from petroleum-based resources. However, the increasing concern regarding the environmental issue and the scarcity for the future of petroleum encouraged the development of PUFs from bio and renewable raw materials (Scheme 1) (Gama et al. 2018). The growing interest in using renewable materials led to an increase in developing materials based on renewable resources such as natural oils, starch, cellulose, and lignin. Some studies investigated the composite of this polyurethane with the lignin (Zhang et al. 2015). Lignin, in the common term, is a large group of aromatic polymers resulting from the oxidative combinatorial coupling of 4-hydroxyphenylpropanoids (Vanholme et al. 2010). Lignin has compatibility with several basic chemicals. Therefore, lignin has potential in polyurethane applications by using lignin as a low-cost bio-renewable resource in the polyurethane industry.

Luo et al. (2013) studied lignin as a filler and polyurethane based on soy oil to be used as water-blown rigid biofoam composites. The results exhibit that the densities of composites were increased along with increasing the lignin content. In terms of mechanical properties, the samples were enhanced by increasing the lignin content. The addition of 10 wt% lignin resulted in the best mechanical properties. A similar result was applied in glass transition temperature (Tg) and storage modulus, about and after Tg, which rose more than the biofoam when it contained no lignin. The analysis for dynamic mechanical (DMA) was in line with the enhancement of mechanical characteristics, and it resulted in the better stability in thermal for the composites compared to the pure biofoam, and the analysis of thermogravimetric also resulted in an increase in thermal stability. Lignin as filler modified with urethane in polyurethane composites was studied by Zhang et al. (2015). Here, when the lignin was modified with octadecyl isocyanate (lignin urethane), it was found that lignin was

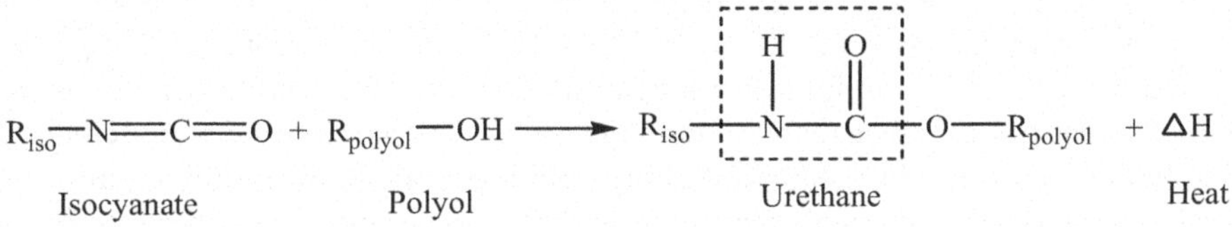

Scheme 1 Reaction scheme of urethane production

more compatible with the polyurethane matrix compared to the lignin that modified with butyric anhydride. With the high lignin content (up to 30 wt%), the Young modulus value increased and the starting of thermal degradation become lowered. The incorporation of lignin urethane did not give much differences in change of the glass transition temperature of the polyurethane composites, but it increased the dielectric constant compared to the pure polyurethane.

Lignin also can composite with polyurethane films (Ciobanu et al. 2004). The range of thermal degradation of polyurethane and blends is not much different; also, it showed that the presence of lignin speeded up the decomposition at lower temperatures. When the composite contained lignin about 4.2 wt%, the loss of the weight of the blends reached almost one-half more until the temperature reached 300 °C and then almost two times when lignin content at 9.3 wt% compared to neat polyurethane. The lignin content of 9.3 wt% lowered the intensity of the main degradation that correlates to the soft segment of polyurethane which around 120 °C (from 360 to 235–245 °C). The important factors in degradation depend on the lignin concentration and the polyurethane soft segment properties. It was shown that for the blends, the tensile strength increased up to 370%, toughness up to 470%, and elongation at break up to 160%, if compared to the pure polyurethane film. In general, the characteristic of blends enhanced due to the increase of the strain rate from 4.6 to 120 mm/mind, whereas the polyurethane elastomer was influenced. When lignin content about 10 wt%, it was sufficient enough to increase the polyurethane strength in term of elasticity. However, when the lignin decreased to 4.2 wt%, the better thermo-mechanical characteristics were achieved.

Functionalized lignin for rigid polyurethane foam (PUF) that acted as halogen-free flame retardant was studied by Xing et al. (2013). This functionalized lignin provided higher thermal stability, Flame Retardant (FR) properties, and mechanical characteristics of the foams. The modified foam with lignin and Phenolic Encapsulated Ammonium Polyphosphate (PFAPP) exhibited much higher char formation about 42.7% char residue compared with pure foam that has only 0.2%; both of them were at 700 °C. The FR characteristics of the modified foams were conducted by the means of cone calorimeter and Limiting Oxygen Index (LOI). This modified lignin as acted in the foam system decreased the release rate of the heat and exhibited the process of combustion. There was a significant increase for the compressive strength due to the addition of 15% PFAPP and 10% modified lignin to the rigid polyurethane foam (RPUF) compared to neat RPUF.

PUF composites with alkali lignin/renewable PUF were studied by Li et al. (2011). The renewable PUF influenced the thermal stability of the composite of polyurethane foam. The concentration of renewable PUF affected the compressive strength ($\sigma 10$) in which the value was from 203 to 125 kPa. Polyurethane–nitrolignin composite films show with 2.8% nitrolignin amount, tensile strength, and elongation at the break that was two twice higher than polyurethane film. The densities of polyurethane–nitrolignin crosslink films enhanced along with the increase of nitrolignin amount until around 3%. The thermal stability of polyurethane–nitrolignin composite films was higher than polyurethane. The covalent bonds that occurred from a polyurethane prepolymer and the nitrolignin in the polyurethane–nitrolignin films formed crosslink networks which resulted in the improvement of mechanical characteristic and the stability in thermal properties (Zhang and Huang 2001).

3.2 Epoxy/Lignin Composites

Lignin has a chance for the use of synthesis and modification of polymer composite due to owing various functional groups like aromatic, phenolic, methoxy, aliphatic hydroxyl, and carbonyl so has moderate reactivity with different chemicals (Simionescu et al. 1993; Yin et al. 2012). Epoxy is one of the interest thermosetting resin to be composited with lignin because it has wide applications such as adhesive, composite materials, and coating (Behin et al. 2018; Engelmann and Ganster 2015; Xu and Ferdosian 2017). The preparation method of the lignin–epoxy composite can be conducted by mixing the epoxy resin and lignin directly to create the structure of interpenetrating polymer network to improve its polarity, compatibility with other material and mainly improve the reactivity (Engelmann and Ganster 2015; Polyurethane 2019; Simionescu et al. 1993).

The blending between lignin, epoxy resin, and polyamide (curing agent) is a simple method to prepare composite materials (Xu and Ferdosian 2017; Yin and Di 2012). In the blending method, the lignin content is influencing the mechanical properties of a composite. The increased lignin content decreases the bending strength of the composite. From Fig. 4, the bending strength decreases when the lignin content is 35 to 70% in quantity of the composite (lignin, epoxy, and polyamide), while the impact strength of composite first increased at lignin content 45% and then decreased

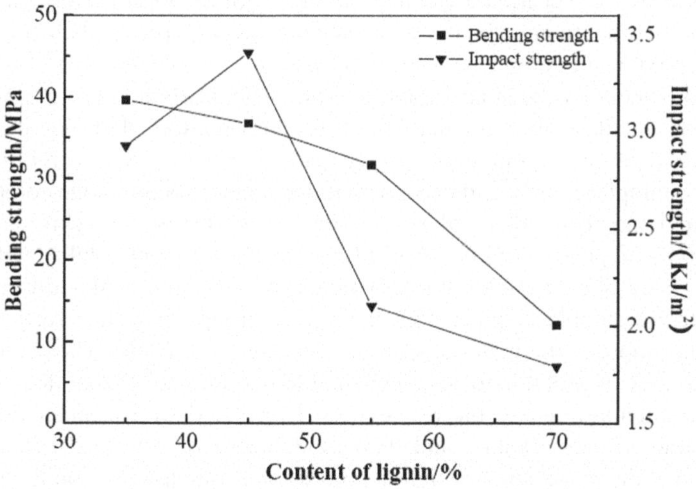

Fig. 4 Effect of the lignin content on bending strength and impact strength of the composite. Reprinted with permission from Yin and Di (2012)

with the increase of lignin content above 45% (Yin and Di 2012). However, the limitation of this method is that the lignin can replace only a little amount (<20–45 wt%) of epoxy resin (Xu and Ferdosian 2017; Yin and Di 2012).

Enzymatic hydrolysis lignin (EHL) is a form of lignin that in recent times received much attention. In a previous study, EHL has the probability to be used in the manufacture of the lignin–epoxy composite up to 60% (Engelmann and Ganster 2015; Wang et al. 2017; Yin et al. 2012). EHL has high pureness and reactivity, so the chemical activity and application is greater than lignosulfonate or kraft lignin. EHL can be isolated from residues of the biomass via enzymatic hydrolysis and carried out under relatively lightweight condition. Therefore, the many functional groups, i.e., alcoholic hydroxyl and phenolic hydroxyl, were well maintained. Due to its reason, the preparation of EHL–epoxy composite can be conducted by directly blending of the polymers to enhance the polymer material's properties (Wang et al. 2017; Yin et al. 2012).

Compared with the synthesis or modification of new polymer materials, the blending process of two or more polymers is the most prospect methods to get novel material with good performance and can reduce also save time. In the blending process, molding temperature and pressure influence the mechanical and thermal properties of the composite. The mechanical properties (bending strength, impact strength) and thermal stability of lignin–epoxy resin composite showed an increase when the molding temperature was up to 130 °C but become decreased when the molding temperature was above 130 °C. An increase of the molding pressure caused to increase the bending strength, glass transition temperature, and thermal stability of the composite. Different from it, the impact strength increased initially and then decreased (Yin et al. 2012).

EHL–epoxy composite has good thermostability, but the mechanical strength and other properties are inadequate when compared with the high-performance native epoxy resin. In order to more increase the mechanical properties and other properties of lignin–epoxy resin composite, especially based on enzymatic hydrolysis lignin, fiber reinforcement has been attempted to composite. Carbon fibers, aramid fibers, basalt fibers, and glass fibers are main fibers used for reinforced the composite. But, the production process of the lignin–epoxy composite using fibers as reinforcement is relatively complex, costly and needed coupling agent to enhance the compatibility of fibers and the resin. Plant fibers can be an alternative for reinforcement of lignin–epoxy composite because plant fibers have good compatibility with epoxy and lignin. Plant fibers have similar structure unit with lignin and hydroxyl groups on its structure so able to react with the epoxy groups to form strong chemical bonding and improve the material's properties (He et al. 2016). Therefore, in line with the increase of attention to environmental issues, the use of plant fibers received widespread attention among the researchers. Correlated with it, plant fibers have relatively low cost, renewable, and biodegradable (Faruk et al. 2014). He et al. (2016) studied the preparation of lignin–epoxy composite using wood flour as plant fiber reinforcement. This study showed that the presence of wood flour significantly improves the mechanical characteristic of the lignin–epoxy composite in line with the content of wood flour. The bending strength and impact strength of composite approximately increase 33% and 57%, respectively, when the lignin was replaced with 15% of the wood fiber. When the content of wood flour increases up to 60%, the bending strength reaches approximately 144% and the impact strength 468%. The mechanical characteristic of the composite is also affected by the particle size of the wood fiber. The bending and impact strength of composite increase when the particle size of wood fiber is 16–60 mesh and then decrease when the particle size was 60–120 mesh (Fig. 5). The possible reason that the largest particle size decreases the bending and impact strength is that it may easily form holes in the matrix. The holes then become stress concentration points under an applied load (He et al. 2016).

Due to the presence of the reactive functional group and polyphenolic on the lignin structure, it can be applied as filler or modifier and as the curing agent of epoxy (Kun and Pukánszky 2017; Wang et al. 2017). When the blending process for producing lignin–epoxy composite is conducted without an added curing agent, the lignin can act in the curing reaction (Polyurethane 2019). While in the presence of curing agent, lignin influences the curing reaction of the curing agent, epoxy resin, and the product performance (Wang et al. 2017). Comparison between lignin–epoxy and lignin–epoxy–polyamine composite was studied. According to the DSC and FTIR analysis, lignin can cure the epoxy. From FTIR analysis, data showed that in the lignin–epoxy resin structure, there was no stretching vibration absorption from the epoxy group at 914 cm^{-1}. The data is similar to the lignin–epoxy resin–polyamine composite, where the absorption peak of epoxy at 914 cm^{-1} is also disappeared (Fig. 6). This indicated that the epoxy group reacted and make chemical bonding with the active group of lignin (carboxyl group or hydroxyl group) and/or polyamine during the curing process.

The DSC data showed that lignin–epoxy resin and lignin–epoxy resin–polyamine had only one exothermic peak (Fig. 7). The peak temperatures of the lignin–epoxy resin and lignin–epoxy resin–polyamine are 174.4 cm^{-1} and 109.7 cm^{-1}, respectively. Both of the composite showed only one exothermic peak; it is indicated that lignin can act as a curing agent of epoxy which assumed was catalyzed by the

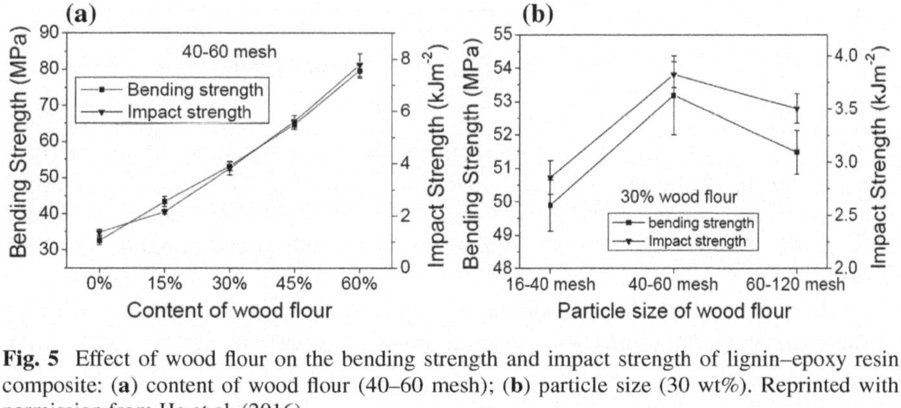

Fig. 5 Effect of wood flour on the bending strength and impact strength of lignin–epoxy resin composite: (**a**) content of wood flour (40–60 mesh); (**b**) particle size (30 wt%). Reprinted with permission from He et al. (2016)

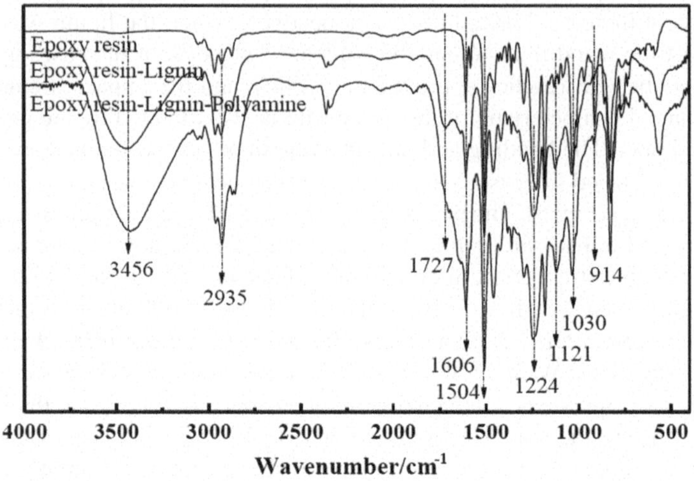

Fig. 6 FTIR spectra for epoxy resin, epoxy resin–lignin blends, and epoxy–lignin–polyamine blends. Reprinted with permission from Yin et al. (2012)

Fig. 7 DSC curves for two blends at a heating rate of 5 °C/min. Reprinted with permission from Yin et al. (2012)

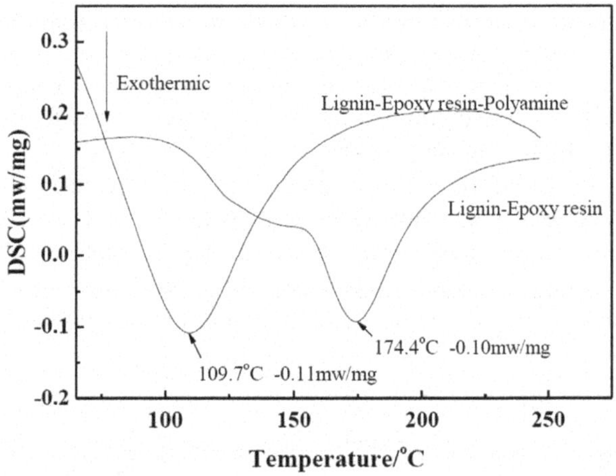

polyphenol group of lignin. Furthermore, the addition of polyamine did not interfere with the curing of the epoxy resin. However, it lowering the curing temperature of lignin–epoxy–polyamine which is showed from the peak temperature. The peak temperature of lignin–epoxy–polyamine is lower than lignin–epoxy resin (Yin et al. 2012).

Khalil et al. studied about producing the epoxy–lignin green composite without curing agent and used lignin (from oil palm biomass empty fruit bunches) as a curing agent. As a comparison, composite also was cured using isophorone diamine curing agent. The lignin content variations were 15, 20, 25, and 30%, and it influences the mechanical, thermal, and morphology properties. The mechanical properties

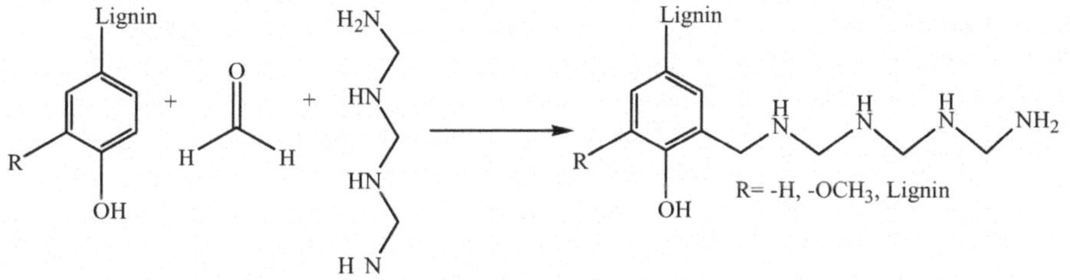

Fig. 8 Scheme of Mannich-functionalized lignin molecule reaction

of epoxy composite cured with lignin were lower than amine-cured composite. Nevertheless, the mechanical properties of the epoxy composite lignin cured have a tendency to increase in line with the increase of lignin content at 15–25% but decrease when the lignin content 30%. Therefore, the optimum properties of epoxy composite that lignin-cured based is shown at lignin content of 25%. This is supported by mechanical, morphological, and thermal properties (Abdul Khalil et al. 2011).

Actually, the application of lignin as curing agent can be an alternative for green curing agent of epoxy. The curing agent based on aromatic and aliphatic amine petroleum is generally dangerous and probably causes some problems for the health and environmental issues. However, the limitation of lignin is possessed a small number of amine groups, so the properties of lignin–epoxy are lower than lignin–epoxy–polyamine/amide. Research about modified lignin to increase the characteristic has been studied. One of the lignin modifications is introduced of primary amine groups into the structure of lignin (Nikafshar et al. 2017). Several methods to introduce primary amine groups into the structure of lignin can be conducted by using hydration, Mannich functionalization, and synthesized using nanocatalyst of cobalt/copper supported on nano-alumina to increase the hydrogen amine (Figs. 8, 9) (Mendis et al. 2014; Nikafshar et al. 2017). The optimum Izod impact strength and tensile strength of epoxy lignin cured showed at the presence of aminated lignin 12.9%, where Izod impact strength and tensile strengths are 35.73 (kJ/m^2) and 86.91 $\pm$ 4.3 MPa, respectively, while the optimum tensile modulus and elongation at break showed at the presence of aminated lignin 9.9% with the value 4104.71 $\pm$ 38.5 MPa and 4.46 $\pm$ 0.37, respectively (Nikafshar et al. 2017).

4 Conclusions

In this chapter, authors have summarized the potential material which called lignin for the development of new biocomposite materials in order to replace petroleum-based material, and it would solve the environmental problem which is caused by the generated waste. Lignin has been incorporated in thermoplastic (polyethylene, polypropylene, and polylactic acid) and thermosetting (polyurethane and epoxy) resin. Basically, the purpose of the addition of lignin into thermoplastic and

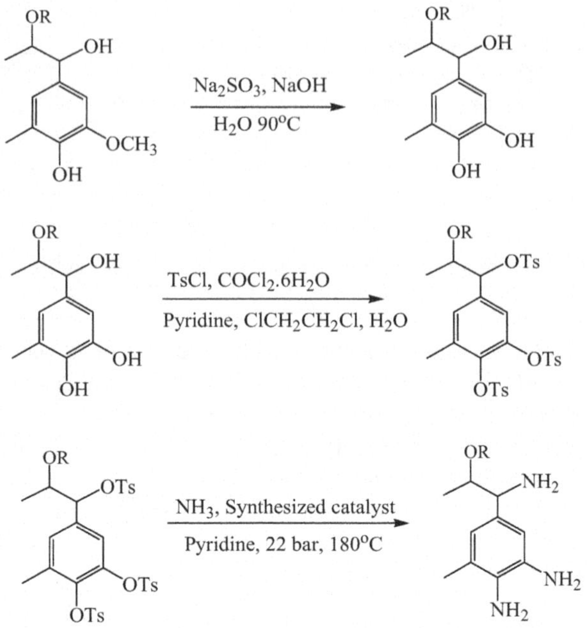

Fig. 9 Reaction step introduced of primary amine groups into the structure of lignin using nanocatalyst of cobalt/copper supported on nano-alumina

thermosetting resins is to enhance some properties, especially the biodegradability of the resulted composites. However, it causes problems when it is composed of nonpolar materials because it is not compatible with each other. Therefore, compatibilizer is needed to improve its interphases adhesion so that the material becomes more compatible.

Generally, the addition of lignin into thermoplastic and thermosetting resin will lead to a decrease in mechanical properties. The existence of lignin has increased the brittleness and affected its viscoelastic character which reduces the flexibility of resulted composites. However, thermal stability and residual mass are increased. It was caused by the existence of aromatic phenyl groups and hydroxyl groups which contributes to the thermal stability.

Moreover, the high residual mass was caused by lignin's chemical structure. High residual mass was caused by the chemical structure of lignin. Due to the high charring ability and low heat release when it burned, the flammability of materials was given less contribution. This degradation of LLDPE/lignin biocomposite is related to the existence of lignin which has many hydrophilic functional groups that encourage it to have biodegradable properties. When the amount of lignin was increased, biocomposite becomes more hydrophilic. Therefore, the material becomes more biodegradable.

References

Abdul Khalil HPS, Marliana MM, Issam AM, Bakare IO (2011) Exploring isolated lignin material from oil palm biomass waste in green composites. Mater Des 32(5):2604–2610. https://doi.org/10.1016/j.matdes.2011.01.035

Akindoyo JO, Beg MDH, Ghazali S, Islam MR (2016) RSC Advances 6:114453–114482. https://doi.org/10.1039/C6RA14525F

Alexy P, Košíková B, Podstránska G (2000) The effect of blending lignin with polyethylene and polypropylene on physical properties. Polymer 41:4901–4908. https://doi.org/10.1016/S0032-3861(99)00714-4

Anwer MAS, Naguib HE, Celzard A, Fierro V (2015) Comparison of the thermal, dynamic mechanical and morphological properties of PLA-Lignin & PLA-Tannin particulate green composites. Compos B Eng 82:92–99. https://doi.org/10.1016/j.compositesb.2015.08.028

Atifi S, Miao C, Hamad WY (2017) Surface modification of lignin for applications in polypropylene blends. J Appl Polym Sci 134(29):1–10. https://doi.org/10.1002/app.45103

Baker I (2018) Fifty materials that make the world. pp 1–271. https://doi.org/10.1007/978-3-319-78766-4

Behin J, Rajabi L, Etesami H, Nikafshar S (2018) Enhancing mechanical properties of epoxy resin using waste lignin and salicylate alumoxane nanoparticles. Korean J Chem Eng 35(2):602–612. https://doi.org/10.1007/s11814-017-0301-0

Blanco I, Cicala G, Latteri A, Saccullo G, El-Sabbagh AMM, Ziegmann G (2017) Thermal characterization of a series of lignin-based polypropylene blends. J Therm Anal Calorim 127(1):147–153. https://doi.org/10.1007/s10973-016-5596-2

Brebu M, Vasile C (2010) Thermal degradation of lignin—a review. Cellul Chem Technol 44(9):353–363

Canetti M, De Chirico A, Audisio G (2004) Morphology, crystallization and melting properties of isotactic polypropylene blended with lignin. J Appl Polym Sci 91(3):1435–1442. https://doi.org/10.1002/app.13311

Cazacu G, Pascu MC, Profire L, Kowarski AI, Mihaes M, Vasile C (2004) Lignin role in a complex polyolefin blend. Ind Crops Prod 20(2):261–273. https://doi.org/10.1016/j.indcrop.2004.04.030

Chen F, Liu W, Seyed Shahabadi SI, Xu J, Lu X (2016) Sheet-like lignin particles as multifunctional fillers in Polypropylene. ACS Sustain Chem Eng 4(9):4997–5004. https://doi.org/10.1021/acssuschemeng.6b01369

Ciobanu C, Ungureanu M, Ignat L, Ungureanu D, Popa VI (2004) Properties of lignin—polyurethane films prepared by casting method. Ind Crops Prod 20:231–241. https://doi.org/10.1016/j.indcrop.2004.04.024

Courgneau C, Domenek S, Leboss R, Guinault A, Averous L Ducruet, Ducruet V (2012) Effect of crystallization on barrier properties of formulated polylactide. Polym Int 61:180–189. https://doi.org/10.1002/pi.3167

Detyothin S, Selke SEM, Narayan R, Rubino M, Auras R (2013) Reactive functionalization of poly(lactic acid), PLA: effects of the reactive modifier, initiator and processing conditions on the final grafted maleic anhydride content and molecular weight of PLA. Polym Degrad Stab 98(12):2697–2708. https://doi.org/10.1016/j.polymdegradstab.2013.10.001

Dikobe DG, Luyt AS (2010) Comparative study of the morphology and properties of PP/LLDPE/wood powder and MAPP/LLDPE/wood powder polymer blend composites. Express Polym Lett 4(11):729–741. https://doi.org/10.3144/expresspolymlett.2010.88

Domenek S, Louaifi A, Guinault A, Baumberger S (2013) Potential of lignins as antioxidant additive in active biodegradable packaging materials. J Polym Environ 21(3):692–701. https://doi.org/10.1007/s10924-013-0570-6

Dopico-Garcia MS, Lopez-Vilarino JM, Gonzales-Rodriguez MV (2007) Antioxidant Content of and Migration from Commercial Polyethylene, Polypropylene, and Polyvinyl Chloride Packages. J Agr Food Chem 55:3225–3231. https://doi.org/10.1021/jf070102+

Duval A, Lawoko M (2014) A review on lignin-based polymeric, micro- and nano-structured materials. React Funct Polym 85:78–96. https://doi.org/10.1016/j.reactfunctpolym.2014.09.017
Engelmann G, Ganster J (2016) Lignin reinforcement in thermosets composites. Lignin in Polym Compos, pp.119-151. https://doi.org/10.1016/B978-0-323-35565-0.00007-2
Faruk O, Bledzki AK, Fink HP, Sain M (2014) Progress report on natural fiber reinforced composites. Macromol Mater Eng 299(1):9–26. https://doi.org/10.1002/mame.201300008
Gadioli R, Waldman WR, De Paoli MA (2016) Lignin as a green primary antioxidant for polypropylene. J Appl Polym Sci 133(45):1–7. https://doi.org/10.1002/app.43558
Gama NV, Ferreira A, Barros-timmons A (2018) Polyurethane foams: past, present, and future. Materials 11:1841. https://doi.org/10.3390/ma11101841
Gao X, Meng X, Wang H, Wen B, Ding Y, Zhang S, Yang M (2008) Antioxidant behaviour of a nanosilica-immobilized antioxidant in polypropylene. Polym Degrad Stab 93(8):1467–1471. https://doi.org/10.1016/j.polymdegradstab.2008.05.009
Ghozali M, Triwulandari E, Haryono A, Yuanita E (2017) Effect of lignin on morphology, biodegradability, mechanical and thermal properties of low linear density polyethylene/lignin biocomposites. IOP Conf Ser Mater Sci Eng 223(1):1–13. https://doi.org/10.1088/1757-899X/223/1/012022
Gordobil O, Delucis R, Egüés I, Labidi J (2015) Kraft lignin as filler in PLA to improve ductility and thermal properties. Ind Crops Prod 72:46–53. https://doi.org/10.1016/j.indcrop.2015.01.055
Gordobil O, Egüés I, Llano-Ponte R, Labidi J (2014) Physicochemical properties of PLA lignin blends. Polym Degrad Stab 108:330–338. https://doi.org/10.1016/j.polymdegradstab.2014.01.002
He S, Liu T, Di M (2016) Preparation and properties of wood flour reinforced lignin-Epoxy Resin composite. BioResources 11(1):2319–2333. https://doi.org/10.15376/biores.11.1.2319-2333
He Y, Wu T, Wei J, Fan Z, Li S (2008) Morphological investigation on melt crystallized Polylactide Homo- and Stereocopolymers by Enzymatic degradation with Proteinase K. J Polym Sci Part B Polym Phys 46:959–970. https://doi.org/10.1002/polb.21430
Hu L, Stevanovic T, Rodrigue D (2015) Unmodified and esterified Kraft lignin-filled polyethylene composites: Compatibilization by free-radical grafting. J Appl Polym Sci 132(7):1–8. https://doi.org/10.1002/app.41484
Karst D, Yang Y (2005) Using the solubility parameter to explain disperse dye sorption on polylactide. J Appl Polym Sci 96(2):416–422. https://doi.org/10.1002/app.21456
Ku H, Wang H, Pattarachaiyakoop N, Trada M (2011) A review on the tensile properties of natural fiber reinforced polymer composites. Compos B Eng 42(4):856–873. https://doi.org/10.1016/j.compositesb.2011.01.010
Kun D, Pukánszky B (2017) Polymer/lignin blends: interactions, properties, applications. European Polym J 93:618–641. https://doi.org/10.1016/j.eurpolymj.2017.04.035
Laurichesse S, Avérous L (2014) Chemical modification of lignins: towards biobased polymers. Prog Polym Sci 39(7):1266–1290. https://doi.org/10.1016/j.progpolymsci.2013.11.004
Lee WK, Nowak RW, Gardella JA (2002) Hydrolytic degradation of polyester blend monolayers at the air/water interface: effects of a slowly degrading component. Langmuir 18(6):2309–2312. https://doi.org/10.1021/la011663c
Li C, Liu L, Zhu C (2011) Characterization of renewable PUF and preparation of Polyurethanefoam composites with Alkali Lignin/ renewable PUF. Open Mater Sci J 5:130–133. https://doi.org/10.2174/1874088X01105010130
Lim LT, Auras R, Rubino M (2008) Processing technologies for poly(lactic acid). Prog Polym Sci 33(8):820–852. https://doi.org/10.1016/j.progpolymsci.2008.05.004 (Oxford)
Luo X, Mohanty A, Misra M (2013) Lignin as a reactive reinforcing filler for water-blown rigid biofoam composites from soy oil-based polyurethane. Ind Crops Prod 47:13–19. https://doi.org/10.1016/j.indcrop.2013.01.040
Mendis GP, Hua I, Youngblood JP, Howarter JA (2014) Enhanced dispersion of lignin in epoxy composites through hydration and mannich functionalization. J Appl Polym Sci 132(1):1–8. https://doi.org/10.1002/app.41263

Morandim-Giannetti AA, Agnelli JAM, Lanças BZ, Magnabosco R, Casarin SA, Bettini SHP (2012) Lignin as additive in polypropylene/coir composites: thermal, mechanical and morphological properties. Carbohyd Polym 87(4):2563–2568. https://doi.org/10.1016/j.carbpol.2011.11.041
Mu C, Xue L, Zhu J, Jiang M, Zhou Z (2014) Mechanical and thermal properties of toughened Poly(L-lactic) acid and Lignin blends. BioResources 9(3):5557–5566. https://doi.org/10.15376/biores.9.3.5557-5566
Nikafshar S, Zabihi O, Moradi Y, Ahmadi M, Amiri S, Naebe M (2017) Catalyzed synthesis and characterization of a novel lignin-based curing agent for the curing of high-performance epoxy resin. Polymers 9(7):1–16. https://doi.org/10.3390/polym9070266
Ouyang W, Huang Y, Luo H, Wang D (2012) Poly(Lactic Acid) blended with cellulolytic Enzyme Lignin: mechanical and thermal properties and morphology evaluation. J Polym Environ 20(1):1–9. https://doi.org/10.1007/s10924-011-0359-4
Pillai SK, Ray SS, Scriba M, Ojijo V, Hato MJ (2013) Morphological and thermal properties of photodegradable biocomposite films. J Appl Polym Sci 129:362–370. https://doi.org/10.1002/app.38763
Huang J, Gan L, Wang Y (2019) Lignin-modified thermosetting materials. Lignin Chem Appl, pp. 163–180. https://doi.org/10.1016/b978-0-12-813941-7.00006-0
Sahoo S, Misra M, Mohanty AK (2011) Enhanced properties of lignin-based biodegradable polymer composites using injection moulding process. Compos A Appl Sci Manuf 42(11):1710–1718. https://doi.org/10.1016/j.compositesa.2011.07.025
Sailaja RRN, Deepthi MV (2010) Mechanical and thermal properties of compatibilized composites of polyethylene and esterified lignin. Mater Des 31(9):4369–4379. https://doi.org/10.1016/j.matdes.2010.03.046
Simionescu CI, Rusan V, Macoveanu MM, Cazacu G, Lipsa R, Vasile C, Stoleriu A, Ioanid A (1993) Lignin/epoxy composites. Compos Sci Technol 48(1–4):317–323. https://doi.org/10.1016/0266-3538(93)90149-B
Spiridon I, Leluk K, Resmerita AM, Darie RN (2015) Evaluation of PLA-lignin bioplastics properties before and after accelerated weathering. Compos B Eng 69:342–349. https://doi.org/10.1016/j.compositesb.2014.10.006
Sugano-Segura ATR, Tavares LB, Rizzi JGF, Rosa DS, Salvadori MC, dos Santos DJ (2017) Mechanical and thermal properties of electron beam-irradiated polypropylene reinforced with Kraft lignin. Radiat Phys Chem 139:5–10. https://doi.org/10.1016/j.radphyschem.2017.05.016
Triwulandari E, Ghozali M, Sondari D, Septiyanti M, Sampora Y, Meliana Y, Fahmiati S, Restu WK, Haryono A (2019) Effect of lignin on mechanical, biodegradability, morphology, and thermal properties of polypropylene/polylactic acid/lignin biocomposite. Plast Rubber Compos 48(2):82–92. https://doi.org/10.1080/14658011.2018.1562746
Vanholme R, Demedts B, Morreel K, Ralph J, Boerjan W (2010) Lignin biosynthesis and structure. Plant Physiol 153:895–905. https://doi.org/10.1104/pp.110.155119
Vouyiouka S, Theodoulou P, Symeonidou A, Papaspyrides CD, Pfaendner R (2013) Solid state polymerization of poly(lactic acid): some fundamental parameters. Polym Degrad Stab 98(12):2473–2481. https://doi.org/10.1016/j.polymdegradstab.2013.06.012
Wang K, Bauer S, Sun RC (2012) Structural transformation of miscanthus × giganteus lignin fractionated under mild formosolv, basic organosolv, and cellulolytic enzyme conditions. J Agric Food Chem 60(1):144–152. https://doi.org/10.1021/jf2037399
Wang S, Lai Y, Yu Y, Di M, Shi J (2017) Effect of enzymatically hydrolyzed lignin on the curing characteristics of epoxy resin/polyamine blends. BioResources 12(4):7793–7806. https://doi.org/10.15376/biores.12.4.7793-7806
Wang X, Sun H, Bai H, Zhang L (2014) Thermal, mechanical, and degradation properties of nanocomposites prepared using Lignin-Cellulose Nanofibers and Poly(Lactic Acid). BioResources 9:3211–3224. https://doi.org/10.15376/biores.9.2.3211-3224
Xing W, Yuan H, Zhang P, Yang H, Song L, Hu Y (2013) Functionalized Lignin for Halogen free flame retardant rigid Polyurethane foam preparation, thermal stability, fire performance and mechanical properties. J Polym Res 20:234. https://doi.org/10.1007/s10965-013-0234-1

Xu C, Ferdosian F (2017) Lignin-based epoxy resin. Conversion of lignin into bio-based chemicals and materials, pp. 111-131. https://doi.org/10.1007/978-3-662-54959-9_7

Yang W, Fortunati E, Dominici F, Kenny JM, Puglia D (2015) Effect of processing conditions and lignin content on thermal, mechanical and degradative behavior of lignin nanoparticles/polylactic (acid) bionanocomposites prepared by melt extrusion and solvent casting. Eur Polymer J 71:126–139. https://doi.org/10.1016/j.eurpolymj.2015.07.051

Ye D, Li S, Lu X, Zhang X, Rojas OJ (2016) Antioxidant and thermal stabilization of Polypropylene by addition of Butylated Lignin at low loadings. ACS Sustain Chem Eng 4(10):5248–5257. https://doi.org/10.1021/acssuschemeng.6b01241

Yin QF, Di MW (2012) Preparation and mechanical properties of Lignin/Epoxy resin composites. Adv Mater Res 482–484: 1959–1962. https://doi.org/10.4028/www.scientific.net/amr.482-484.1959

Yin Q, Yang W, Sun C, Di M (2012) Preparation and properties of lignin-epoxy resin composite. BioResources 7(4):5737–5748. https://doi.org/10.15376/biores.7.4.5737-5748

Zhang C, Wu H, Kessler MR (2015) High bio-content polyurethane composites with urethane modified lignin as filler. Polymer 69:52–57. https://doi.org/10.1016/j.polymer.2015.05.046

Zhang L, Huang JIN (2001) Effects of Nitrolignin on mechanical properties of Polyurethane—Nitrolignin Films. J Appl Polym Sci 80:1213–1219. https://doi.org/10.1002/app.1206

Advances in Nanotechnology based Strategies for Synthesis of Nanoparticles of Lignin

Raj Kumar, Abhishek Gupta, Mohit Chawla, Keshaw Ram Aadil, Sunil Dutt, Vijay Bhooshan Kumar and Abhishek Chaudhary

Abstract In recent years, environmentally friendly polymers are gaining considerable interest. Biorenewable polymers are promising materials which fall in this category. Among the biorenewable polymers known, Lignin is one of the most vastly used polymers. It is the most abundant natural polymer after cellulose. Lignin can be an excellent material having a potential of replacing other synthetic polymers in drug delivery and biomedical applications due to its excellent biocompatibility and nontoxic nature. Hence, lignin micro- and nanoparticles are gaining high interest

R. Kumar (✉)
Faculty of Engineering and Bar Ilan Institute of Nanotechnology and Advanced Materials (BINA), Bar Ilan University, Ramat Gan 5290002, Israel
e-mail: rk7410@gmail.com

A. Gupta
Department of Desalination and Water Treatment, Zuckerberg Institute for Water Research, The Jacob Blaustein Institutes for Desert Research, Ben-Gurion University of the Negev, Sede-Boqer Campus, Be'er Sheva 84990, Israel
e-mail: abhigupta.iitmandi@gmail.com

M. Chawla
Gyan Sankul, Gyan Sankul Complex, Bhutti Colony, Kullu, Himachal Pradesh 175126, India
e-mail: mohitchawla_msit@yahoo.com

K. R. Aadil
Center for Basic Sciences, Pt. Ravishanker Shukla University, Raipur, Chhattisgarh 492010, India
e-mail: kaadil7@gmail.com

S. Dutt
Department of Chemistry, Government Degree College Daulatpur Chowk, Ghanari, Una, Himachal Pradesh 177204, India
e-mail: sunildutt.iitmandi@gmail.com

V. B. Kumar
Department of Chemistry and Bar Ilan Institute of Nanotechnology and Advanced Materials (BINA), Bar Ilan University, Ramat Gan 5290002, Israel
e-mail: vijaybhushan86@gmail.com

A. Chaudhary
Department of Biotechnology and Bioinformatics, Jaypee University of Information Technology, Waknaghat, Solan, Himachal Pradesh 173234, India
e-mail: abhishekcbt@gmail.com

S. Sharma and A. Kumar (eds.), *Lignin*, Springer Series on Polymer and Composite Materials, https://doi.org/10.1007/978-3-030-40663-9_7

in the research fraternity. Syntheses of a wide range of polymer nanoparticles have been reported. However, the usage of the reports to the synthesis of lignin micro- and nanoparticles is limited due to the physicochemical properties of lignin. The present chapter discusses the various approaches suitable for preparation of lignin micro- or nanoparticles. The effect of different experimental parameters and their physicochemical properties at nanoscale are also thoroughly explored. A brief discussion on the applications of lignin micro- or nanoparticles in drug delivery, biomedical science, nanomedicine, and therapeutics is also explored.

Keywords Biorenewable · Polymer · Lignin · Nanoparticles · Preparation · Biomedical applications

1 Introduction

Polymeric nanomaterials are significantly contributing and influencing modern human life as there is a wide range of polymers available in daily life from household to space technology (Suyama and Shirai 2009). Polymer material such as elastomers, plastic, and artificial fibers are being widely used in almost all walks of life (Huang et al. 2000). This has led to an exponential increase in the number of polymer industries. Polymers have been replacing and alternative materials for metals and glass-based materials due to simple processing, low cost, and ease of availability (Iden et al. 2003). Since the last two decades or so, polymer-based research is enormously expanding its role in a vast range of areas such as electronic (Jung et al. 2006), photonics (Dalton 2002), sensors (Song et al. 2017), drug delivery (Kumar et al. 2016), medicine (Zhao et al. 2019), biotechnology (del Agua et al. 2018), environmental science (Ghosh and Khastgir 2018), toxicology (Kabanov 2006), and conducting materials (Zhang et al. 2017). This is also evident from the increase in number of publications and citations in the last 10 years. The range of applications of bulk polymer materials is limited to their usage in advanced technological applications due to physicochemical properties. However, the introduction of nanotechnology has completely changed the scenario of polymer-based research and has led to the applications of polymers (Paul and Robeson 2008) in several innovative and advanced technological applications. The polymers display excellent physicochemical properties at the nanoscale due to the reduction in particle size and enormous increase in surface to volume ratio (Paul and Robeson 2008).

Based on the synthesis process, polymers can be classified into two basic types, viz. natural polymers and synthetic polymers. It is widely reported that the natural polymers have many advantages over synthetic polymers such as toxicity, biodegradability, and biocompatibility (Mogoşanu and Grumezescu 2014). However, synthetic polymers have their own advantages and are suitable for specific applications. Moreover, high usage of polymers results in environmental issues, depletion of petroleum natural resources, and health-related concern. Thus, the focus of the research has now moved from the development of polymeric materials to renewable materials (Thakur et al. 2014). The main renewable sources are biopolymers such as polysaccharides (starch, chitin/chitosan, cellulose, alginate, and carbohydrates) and animal protein-based biopolymers (gelatin, wool, silk, collagen) (Cui et al. 2013; Thunga et al. 2014). Presently, biorenewable precursors from natural resources are being used in biomedical and automotive applications. The excellent properties of biorenewable precursors such as renewable nature, biodegradability, recyclability, high toughness, good specific strength, facile separation, good thermal properties, nontoxic nature, lower density, and improved recovery of energy have made them priority materials (Anderson et al. 1972, 2004; Francis Suh and Matthew 2000; Huebsch and Mooney 2009; Lopes et al. 2012; Reddy et al. 2015). Due to their availability in large quantities at lower costs, these materials also meet the demand of industries (Kogan et al. 2006). In the last two to three decades, the use of biorenewable polymer-based nanomaterials is steadily gaining interest. (Stewart 2008).

Among the large known natural renewable biopolymers, to name a few are lignin, cellulose, chitosan, collagen, hyaluronic acid (Dang and Leong 2006), cellulose is the most abundant natural polymer on the earth (Thakur and Thakur 2014). Cellulose and cellulose-related polymers are well explored, studied, and their use in a range of applications is widely reported (Edgar et al. 2001). Lignin is the second most abundant natural biopolymer (Davin and Lewis 2005). However, lignin is one of the natural renewable biopolymer that has not been thoroughly studied and explored at nanoscale level yet to the best of our knowledge. Lignin is a biopolymer and renewable natural source. One-fourth part of the wood contains lignin along with cellulose-related components in trees and plants (Uggla et al. 1998). It is a major by-product in paper-manufacturing industries. The major use of lignin so far is for energy production (Reddy and Yang 2005). The basic physicochemical properties of lignin are listed in Table 1. During the last decade, the use of lignin has been explored significantly. It is a cheap source of phenolic biopolymer. The availability of phenolic group makes it useful in acting as filler and additive which induces oxidative and biostability (Sahoo et al. 2013). Lignin can be an excellent source material for the production of valuable products such as chemical and biofuels (Naik et al. 2010; Fitz Patrick et al. 2010). Hence, lignin is gaining popularity in the research community. In addition, due to its excellent biocompatibility and biodegradability, it can be an ideal material for drug delivery and biomedical applications (Figueiredo et al. 2017, 2018). Figure 1 presents the overview of the most important applications of lignin micro- and nanostructures. Lignin is excellent antioxidant and photoprotective agent. Lignin antioxidant activity is well known as it comes across the mechano-bio-chemical stress of plants. Due to free radical scavenging capacity, it is widely used in pharmaceutical, food, and cosmetic industries (Aadil et al. 2014). Glasser et al. investigated the blend system (Atkins et al. 1993; De Oliveira and Glasser 1994) and Sarkanen et al. developed several lignin-based compounds (Li and Sarkanen ; Sarkanen et al. 2016). Kosikova et al. studied the mechanical behavior of lignin and lignin-based compounds (Nitz et al. 2001; Košíková et al. 2007). All these promising results were encouraging for further investigation of lignin at nanoscale level.

The properties of lignin can be further enhanced by synthesizing lignin at the micro- or nanoscales. Hence, the researchers are showing significant interest in the

Table 1 Basic physicochemical properties of lignin

Serial number	Physicochemical Property	Details	References
1	Molecular weight	908–2150	Gross et al. (1958)
2	Solubility	Insoluble in water	Savy et al. (2015)
3	Melting point	252–259 °C	Han et al. (2019)
4	Soluble in solvent	Soluble in an organic solvent such as THF, acetone, ethylene glycol, and strong acid/base	Sameni et al. (2016)
5	Boiling point	100–124 °C	Aadil et al. (2018a, b)
6	Availability	Natural available	Aadil and Jha (2016), Aadil et al. (2016a, c), Barapatre et al. (2017)
7	Function group	−OH, −COOH	Aadil et al. (2016b)
8	Nature	Biodegradable	Aadil and Jha (2016)
9	Uses	See Fig. 1	Aadil et al. (2014, 2016b), Barapatre et al. (2015, 2016)

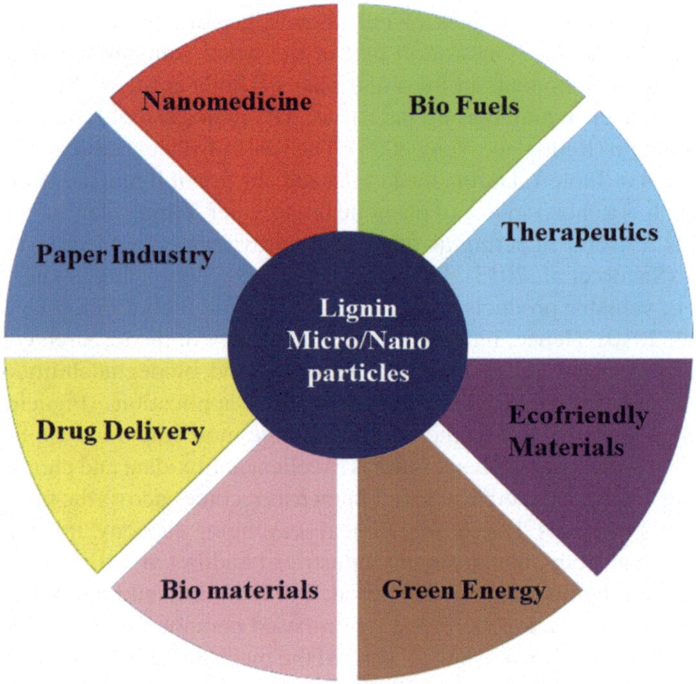

Fig. 1 Simple overview of lignin micro- and nanoparticles for different application

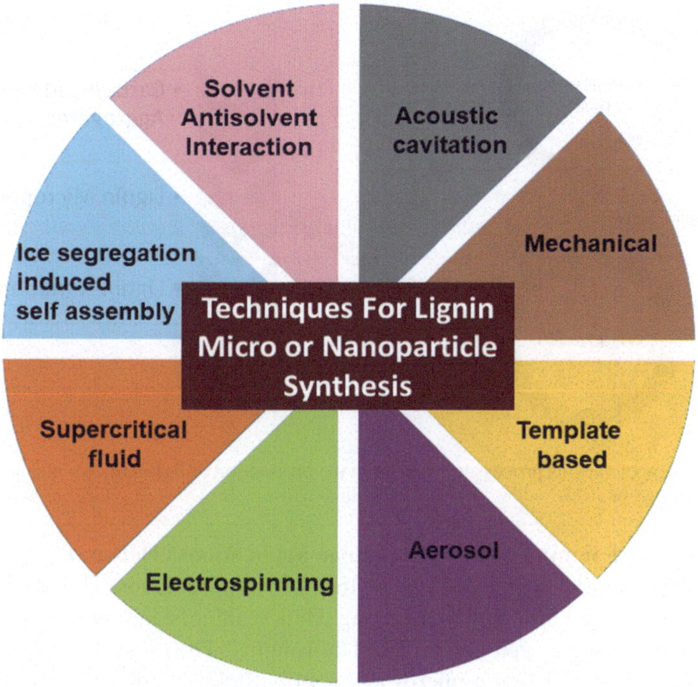

Fig. 2 List of techniques for synthesis of micro- and nanoparticles of lignin

preparation of lignin micro and nanostructures. However, the formulation of lignin micro- or nanoparticles by controlling their physicochemical properties becomes quite challenging. Lignin nanoparticle synthesis has gained importance because of the environment-friendly nature of lignin. Several methods have been developed for the synthesis of lignin nanoparticles (Lievonen et al. 2016). Figure 2 represents the list of some interesting techniques used for the synthesis of lignin micro- and nanoparticles. The details on these techniques are discussed in the following sections.

2 Techniques for Synthesis of Micro- and Nanostructures of Lignin

The properties of lignin or polymer at the nanoscale are more interesting when compared to the bulk materials. There are a large number of techniques developed for the preparation of lignin micro- or nanoparticles (Frangville et al. 2012; Duval and Lawoko 2014; Gilca et al. 2015; Richter et al. 2016; Beisl et al. 2017a, b). Figure 2 presents the schematic view of various processes used for the synthesis of micro- and nanoparticles of lignin. Figure 3 displays the various processes involved in processing of lignin. The different techniques mentioned in Fig. 2 produce a specific type of

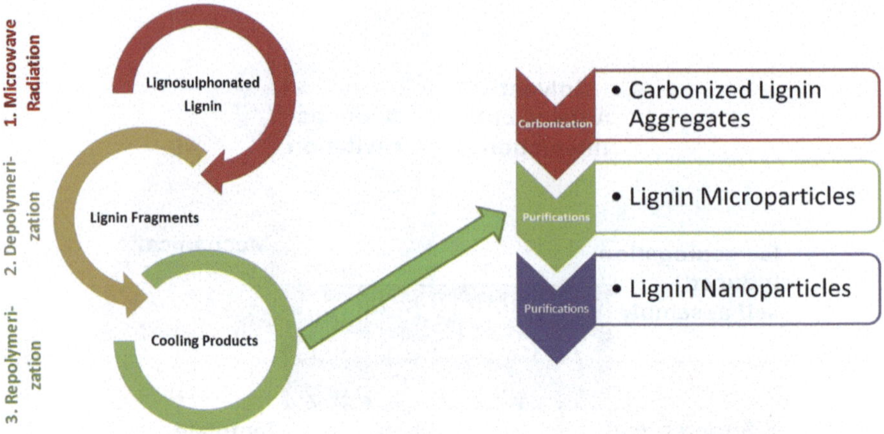

Fig. 3 Overview of various processes involved in the production of lignin micro- and nanoparticles

nanostructures of lignin used for particular applications. Hence, the formulation of lignin is generally based on the application requirements. This makes lignin nanostructures unique and more interesting. Hence, understanding each process and the effect of parameters need to be studied thoroughly. The present chapter discusses the various methods for lignin micro- and nanoparticles synthesis, advantages, and disadvantages of the synthesis processes with suitable examples (Duval and Lawoko 2014; Beisl et al. 2017a, b). All processes represented schematically in Fig. 2 are discussed in detail in the following sections.

2.1 *Mechanical Processes*

Milling and high-pressure homogenization (HPH) techniques are also called a mechanical method or mechanical treatment techniques. Milling method is one of the conventional and traditional methods used for nanoformulations or particle-size reduction. This technique is widely used in pharmaceutical industries because it is simple, cost-effective and easy to scale up. Milling again is of two types, viz. dry milling and wet milling. Another mechanical process is the High-Pressure Homogenization (HPH) method. HPH technique is classified into two types, namely hot HPH and cold HPH depending on the processing temperature. As the time of milling or number of homogenization cycles increase, there is a reduction in particle size. Lignin nanoparticles with particle size 500 nm was prepared by Nair et al. via mechanical treatment process using softwood KL aqueous solution of 5 g/L. The solution was treated for 4 h at 15 K rpm homogenization (Nair et al. 2014). Gilca et al. and Tortora et al. prepared the lignin nanoparticles of size 10–20 nm through mechanical treatment using sonication energy of 20 kHz frequency and 600 W power (Tortora et al. 2014; Gilca et al. 2015). 20 nm size lignin particles were successfully formulated by Zimniewska et al. using ultrasonication-based mechanical process. Shawn et al. and Zhiming et al. reduced lignin particle size to 10 nm through the milling process at room temperature and −18 °C (Zimniewska et al. 2008). The major disadvantage with milling processes is the possibility of contamination and formation of particles with large-size and broad-size distribution.

2.2 *Solvent Shifting Method*

Solvent shifting method is a very simple method used for the preparation of lignin particles with size range from tens of nanometer to hundreds of micrometers. The name of methods itself clears the chemistry involved in this technique. In this method, micro- and nanoparticles of lignin will be created by shifting the suitable lignin substrate from one solvent to another solvent under controlled conditions. The solution of lignin will be subjected to mixing with the second solvent in which lignin is insoluble or immiscible, which results in the precipitation of lignin. Most suitable solvents for lignin synthesis through this technique are Tetrahydrofuran (THF), acetone, acetone/water, and dimethyl sulfoxide (DMSO). At optimum conditions, particles with size range from microns to nanometers are generated. This technique is widely used for the preparation of a wide range of polymeric nanostructures. This technique usually generates spherical-shaped particles. Figueiredo et al. formulated lignin nanoparticles with particle size 221 ± 10 nm using THF as solvent and water as antisolvent (Figueiredo et al. 2017). Lievonen et al. (2016) prepared 200–500 nm sized lignin particles using solvent shifting method. The major drawback of this technique is very low yield and difficult to control the morphology of particles.

2.3 *pH Shifting Technique*

The pH shifting method also has kind of similarities with solvent shifting method. In this method, pH is changed from acidic to basic or vice versa. The pH change mainly depends on the physicochemical properties of the substrate used for the preparation of lignin nano/microparticles. Due to the change in solution pH, lignin changes its solubility or miscibility. Generally, initial pH is the one at which the lignin substrate gets dissolved. For example, low-sulfonated lignin is used as a substrate which is usually insoluble in water at neutral and acidic conditions. However, the solubility increases significantly in ethylene glycol (EG) and water above pH 10. Using chemistry based on pH, two different approaches have been reported for the synthesis of lignin micro/nanoparticles. In one of the processes using ethylene glycol as a solvent, lignin nanoparticles have been prepared by addition of aqueous hydrochloric acid (HCl). In another process, NaOH is used to control the pH of water at 12, followed by the reduction of pH by the addition of aqueous nitric acid (HNO_3) which results in formation of particles of lignin. This method was initially developed by Frangville et al. (2012). Later on, this process has been widely used for preparation of lignin and other polymer nanoparticles (Gupta et al. 2014; Yang et al. 2015a, b, 2016a, b; Richter et al. 2016; Beisl et al. 2017b).

2.4 Template-Based Synthesis Technique

Most of the methods for the preparation of nanomaterials are simple and produce the nanostructures with different sizes and shapes. However, a major challenge still remains is the use of regular and well-known bottom-up methods to produce the nanomaterials with desired size and shape (Felix et al. 2005; Dubey et al. 2013; Gupta et al. 2016b). Since there is a very high demand of specific size and shape of nanomaterials for particular applications, many techniques are able to do so but the major problem is the reproducibility and very difficult to achieve industrial-scale production (Gupta et al. 2015; Sharma et al. 2016; Chawla et al. 2017b, c). Hence, template-based processes are in high demand for the synthesis of specific-shaped nanomaterials with desirable physicochemical properties (Dutt and Siril 2014a; Dutt et al. 2015, 2017; Chawla et al. 2016; Dutt and Sharma 2017). There are usually two different types of templates such as soft templates and hard templates. Soft and hard templates can be further classified into different categories. Figure 4 presents the formulation of hexagonal mesophases also called swollen liquid crystal (SLCs) templates using oil, water, surfactant, and cosurfactant along with the formation of polymeric nanoparticles. Dutt et al. discussed in detail the different template-based techniques for nanomaterials synthesis in their recent review (Dutt et al. 2017). They also thoroughly studied the preparation of different nanomaterials such as polyaniline nanoparticles and nanofibers and nanocomposites of polyaniline with different metals like gold, silver, iron, platinum, and palladium with a focus on controlling the size and shape of the nanomaterials for several applications such as catalysis, dye reduction, and glucose biosensing (Singh et al. 2009; Siril et al. 2012; Dutt and Siril 2014a; Dutt et al. 2015, 2017, 2018; Chaudhary et al. 2015; Chawla et al. 2016; Vats et al. 2016; Vats and Siril 2017; Zhou et al. 2019; Shi et al. 2019; Hosseini et al. 2019). They also employed the swollen liquid crystals as a soft templates for the preparation of various polymeric nanoparticles, nanocomposites with several metals such as palladium, platinum, gold, and iron (Dutt and Siril 2014a, b, 2015; Dwivedi et al. 2014; Dutt et al. 2015, 2017; Dutt and Sharma 2017). Vats et al. further explored the swollen liquid crystals for the synthesis of carbon nanomaterials such as graphene, graphene oxide, reduced graphene oxides, and their nanocomposites with palladium for catalytic applications (Vats et al. 2016; Gupta and Nandi 2017; Vats and Siril 2017; Dutt et al. 2018). Chawla et al. used a similar template-based synthesis approach for synthesizing pure metallic nanostructures of palladium, platinum, and bimetallic nanomaterials for catalytic applications (Gupta et al. 2016a; Chawla et al. 2016, 2017a, b, c, 2018). However, template-based synthesis techniques not well studied for lignin micro- and nanostructures formulations. Interestingly, lignin-based materials are used as template for the synthesis of other nanomaterials. Chen

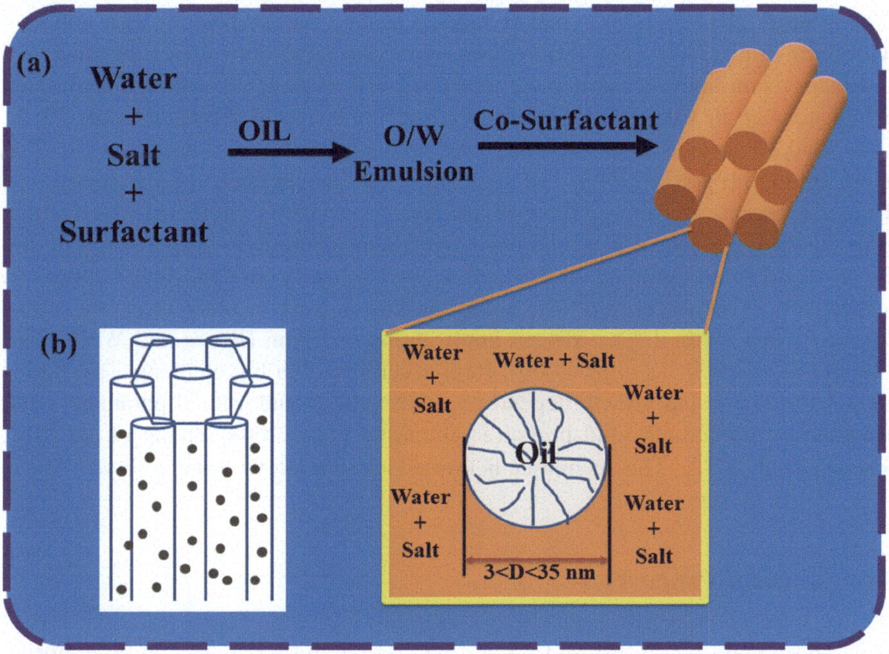

Fig. 4 **a** Simple view of generation of soft templates (swollen hexagonal mesophase) using water, surfactant, oil, and cosurfactant and **b** formation of nanoparticles of polymers and metals in soft templates

et al. utilized un-modified lignin as template for the synthesis of mesoporous TiO_2 nanoparticles. Formulated m-TiO_2 showed excellent photocatalytic performance (Chen et al. 2016). Wang et al. used lignin amine (LA) as template for solid-phase synthesis of mesoporous ZnO nanoparticles (Wang et al. 2014). Zhong et al. formulated in situ lignin-based silver nanoparticle composite using lignin as template and sodium borohydrate and ascorbic acids used as reducing agents (Zhong et al. 2015). Hence, further studies on preparation of lignin nanoparticles using templates are needed to be explored. The major drawbacks of the template-based synthesis process are use of toxic chemicals, removability of template materials, purification, and toxicity due to surfactant (e.g., CTAB or SDS) used in templates. However, soft templates are good choice over hard templates.

2.5 *Ice-Segregation-Induced Self-assembly Method*

In Ice-Segregation-Induced Self-Assembly (ISISA) technique, monomer solution is subjected to freezing. The freezing of solution results in the formation of ice crystals and growing these ice crystals separates the polymer phase in the control process

(Zhang et al. 2011). The polymer is formed around the ice crystals as a template (Gutiérrez et al. 2008). The different experimental conditions affect the final size and shape. Nanofibers of lignin have been produced by using ice-segregation method by Spender et al. using alkali lignin as a starting material (Spender et al. 2012). They used 300 rpm as drum rotation speed and cooled the reaction mixture to 77 K using liquid nitrogen (liq. N_2). The rate of cooling, temperature, cooling conditions, drum size, speed of drum, and concentration of precursors are the parameters that affect the physicochemical properties of produced nanofibers. Caicedo et al. and Ten et al. prepared lignin nanotubes and nanowires using template-based approach, respectively (Caicedo et al. 2012; Ten et al. 2014). Several other researchers also employed this method for different applications of lignin (Bruce and West 1989; Hatfield et al. 1994; Li et al. 2009). Reproducibility is a big challenge, as the ice crystal formation and growth depend on many crucial and critical parameters and sensitive to temperature and pressure (Gutiérrez et al. 2008; Zhang et al. 2011; Spender et al. 2012). This technique is limited application with scalability.

2.6 *Aerosol Process*

The aerosol technique is a simple and single-step method (Eerikäinen et al. 2003). The major advantage of this technique is great reproducibility and excellent control over size and shape of the particles (Buesser and Pratsinis 2012). This process consists of an atomizer which is working continuously, and generates aerosol. The formed aerosol is passed through a hot tube connected to a low-pressure impactor where the collection and fractionation of formed dry particles occur. Ago et al. successfully formulated lignin nanoparticles with size range 30–2000 nm (Ago et al. 2016). Good product yield, low cost, and no by-products in liquid form make this technique more useful and interesting (Buesser and Pratsinis 2012). This method has been used to very limited extent for the synthesis of lignin nanoparticles. However, there is a huge scope for further studies to explore the possibilities that the process offers and make it an interesting approach.

2.7 *Electrospinning Method*

Electrospinning is a well known and widely used process in the field of nanomaterials. This technique is popular for making the nano- and microfibers and nanocomposites of natural and synthetic polymers (Antman-Passig and Shefi 2016; Baranes et al. 2016; Antman-Passig et al. 2017; Marcus et al. 2017; Aadil et al. 2018a, b). The basic schematic experimental setup of electrospinning used for fabrication of micro- or nanofibers of lignin is shown in Fig. 5. In this method, a solution of the polymer in a suitable solvent pumped through a very small nozzle. The widely preferred nozzles are with the inner diameter of 100 μm. It works as an electrode and counter electrode

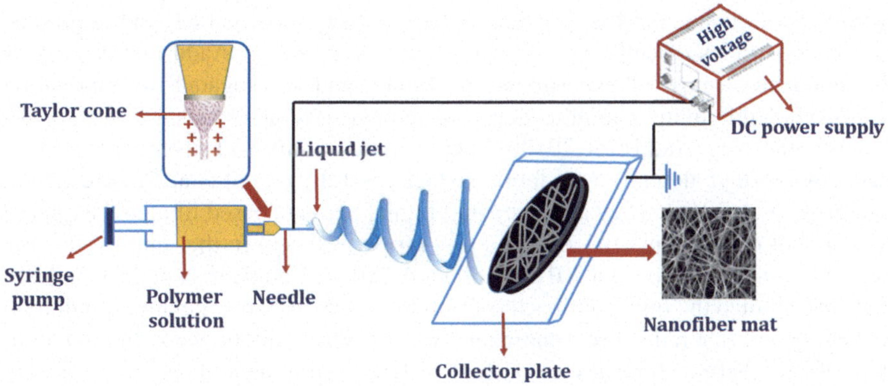

Fig. 5 Schematic overview of preparation of lignin micro- or nanofibers through electrospinning technique

simultaneously. Usually, 100–500 kV/m electrical field is used. The produced fibers are collected between the nozzle and the counter electrode. Electrospinning generally produces polymeric nanofibers range between 10 and 25 cm in a typical laboratory setup. Nanofibers are excellent materials and their usage in a wide range applications is growing monotonically. Nanofibers are promising materials and have been widely investigated for their applications in drug delivery, tissue engineering, and biomedical applications (Ahmed et al. 2006; Seidlits et al. 2008; Jin et al. 2011; Shah et al. 2014). Nanofibers also have great importance in manufacturing and automobile industries. Shefi and coworkers have applied nanofibers for controlling the growth of neuron directions and formulation of various types of scaffolds for tissue engineering and regenerative medicine (Antman-Passig and Shefi 2016). A number of daily need products are composed of polymer nanofibers. However, the use of nanofibers of lignin is limited due to difficulties in their processing. This limitation arises due to the limitation in physicochemical properties and solubility of lignin. Initial investigation suggested that, due to branched structure, heterogeneous nature, and ability of degradation due to low molecular weight, formulation of nanofibers of lignin in a continuous process like electrospinning is difficult. Hence, it was suggested that electrospraying is a more favorable technique for lignin as compared to electrospinning. Dallmeyer et al. thoroughly investigated the feasibilities of electrospinning of lignin using seven different kinds of lignins (Dallmeyer et al. 2010). They also studied the effect of different solvents. However, only electrospraying produced fibers, but electrospinning failed to generate fiber-type structures of lignin. Finally, poly(ethylene oxide) (PEO) with 1 wt% addition successfully produced lignin fibers with diameters 700–1600 nm using different lignin types (Klossner et al. 2008; Varesano et al. 2008; Nie et al. 2009). The diameters can be tuned by controlling the lignin concentration. Later, many researchers prepared the lignin fibers and their nanocomposites through electrospinning and applied these for different applications (Aadil et al. 2018a, b). Ago et al. prepared the lignin nanofibers of diameters 61 ± 3 to 500 ± 3 4 nm using softwood KL and reinforced with nanoparticles of cellulose (Ago et al. 2012a, b, 2013). Aadil and coworkers also conducted range of experiments on lignin, and investigated the fabrication of biocompatible lignin nanofibers and nanocomposites nanofibers as scaffolds for tissue engineering (Aadil et al. 2018b). Electrospun nanofibers of silver nanoparticles loaded poly(vinyl alcohol) were fabricated successfully (Aadil et al. 2016b, 2018a; Barapatre et al. 2016). They also prepared lignin–alginate based film, and explored the effect of different plasticizers. The film properties were further improved using Acacia lignin (Aadil and Jha 2016; Aadil et al. 2016c; Barapatre et al. 2017). Aadil et al. also studied the lignin antibacterial activities, use of lignin–gelatin film in food packaging and free radical scavenging activity of Acacia nilotia wood lignin (Aadil et al. 2014b, 2016a; Barapatre et al. 2015, 2016). Lignin nanofibers are promising materials for biomedical and tissue engineering applications.

2.8 Supercritical Fluid Processes

Supercritical fluid (SCF) technique is more interesting and widely used in the pharmaceutical industry and drug delivery. Generally, carbon dioxide is used as a supercritical fluid due to its favorable physicochemical properties such as abundance nature, nonflammability, nontoxicity, and low cost. The critical temperature 304.3 K and critical pressure 7.4 MPs have made carbon dioxide as excellent material for the supercritical fluid process (Mukhopadhyay and Dalvi 2004; Dalvi and Mukhopadhyay 2009). Figure 6 presents the list of several important supercritical techniques used for the preparation of polymeric micro- and nanoparticles. Based on the technique, experimental setup, and processing conditions, supercritical fluids techniques are generally classified into supercritical antisolvent (SAS) method, supercritical gas antisolvent (GAS) method, rapid expansion of supercritical solution (RESS), and rapid expansion of supercritical solution and co-solvent (RESS-SC) method (Martín and Cocero 2008). Several critical reviews reported and discussed a range of supercritical fluid-based methods in detail (Martín and Cocero 2008a). This technique has gained significant interest in the last 20 years for the production of various types of nanomaterials, microstructures (Ye and Wai 2003), and nanocomposites (Reverchon and Adami 2006). Various researchers have thoroughly investigated and successfully prepared the micro- and nanoparticles of active pharmaceutical ingredients (M. Charoenchaitrakool et al. 2000; Pathak et al. 2004), dyes (Reverchon et al. 2005), pigments (Montero et al. 2000), and polymers (Meziani et al. 2004; Sun et al. 2005) using this technique. Further, many researchers also studied the use of other solvents as supercritical solvents for the fabrication of nanoparticles and nanocomposites of several compounds.

Synthesis of lignin nanostructures using supercritical fluids (SCF) or supercritical carbon dioxide (SC-CO_2) has a major advantage of industrial scale up when compared to other processes discussed so far. This technique has a very precise ability to control the size and shape of the nanomaterials easily, simply by tuning the experimental

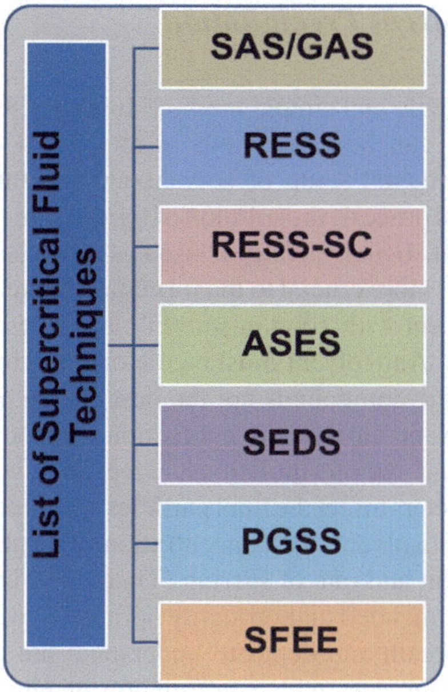

Fig. 6 Various supercritical processes well established and widely used for preparation of polymeric micro- and nanoparticles

parameters. Lu et al. prepared the 144 nm lignin nanoparticles through supercritical method using lignin (OS), supercritical carbon dioxide, and acetone at 30 MPs and 35 °C pressure and temperature, respectively, were used for the synthesis (Lu et al. 2012). Similarly, Myint and coworkers also prepared nano-lignin using KL and dimethyl formamide (DMF) as solvent (Myint et al. 2016). Effect of different parameters has also been reported widely, which concluded that aggregation of nanostructures decreases with increase of pressure. Particle size increases with reducing the solution flow rate and increasing the concentration of lignin in solution. (Lu et al. 2012; Myint et al. 2016). The major advantage of the process is that product yields of ~90% can be achieved. The major limitation of this technique is the solubility of compounds in supercritical carbon dioxide. However, to overcome this, another solvent like methanol or ethanol along with carbon dioxide is generally used. There are number of experimental parameters that affect the size and shape of the synthesized nanomaterials and hence, there is a need to optimize the experimental conditions. The major limitations of this technique are expertise to run, sophisticated laboratory facility is also needed, and maintenance is slightly expensive.

2.9 Solvent Antisolvent Precipitation

Nanoprecipitation is a facile and simple method. It is easy to perform with minimal experimental setup and can be carried out in almost all chemistry laboratories. The very simple experimental setup of solvent–antisolvent interaction method is presented in Fig. 7. In this process, the solution of lignin in an organic solvent is slowly added into an antisolvent. Usually, water is used as an antisolvent. The selection of solvent and antisolvent is more crucial to the formation of nanoparticles (Kumar et al. 2015a, 2019c). The solvent with good miscibility is the best choice for solvent and antisolvent combination. Antisolvent must be the one in which lignin is not soluble or poorly soluble. Organic compounds are the best choice of substances to prepare nanoparticles using solvent–antisolvent interaction method (Kumar et al. 2015b). In the organic compound nanoformulations, water is a widely used antisolvent. However, various parameters affect the final particles size such as solvent, antisolvent, the ratio of solvent to antisolvent, the concentration of a substance in a solvent, the temperature of antisolvent and rate of stirring (Thorat and Dalvi 2012; Kumar et al. 2014). Hence, it is recommended to thoroughly optimize the conditions. Kumar et al. conducted the studies breath and depth to understand the phenomena of solvent–antisolvent precipitation process with a correlation of physicochemical properties of solvent and antisolvent (Kumar et al. 2014). They developed the evaporation-assisted solvent–antisolvent interaction (EASAI) method for the preparation of nanoparticles of organic compounds (Kumar and Siril 2014; Kumar 2019b, 2020). They also found that acetone and water are the best combinations. Acetone has boiling point ~56 °C and, once the acetone solution of substrate injected into water at a temperature above the boiling point of acetone, it results in the quick evaporation of acetone from the surface of the hot antisolvent leading to the fast supersaturation, nucleation, and formation of ultrafine nanoparticles (Kumar and Siril

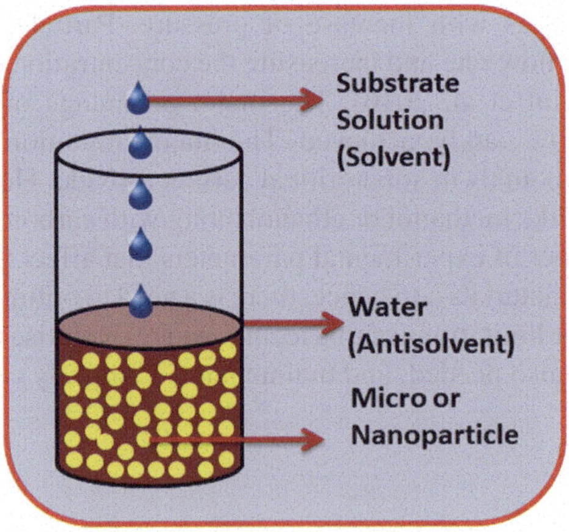

Fig. 7 Simple view of solvent–antisolvent interaction method

2018). Kumar et al. also explored the effect of different solvents and interestingly, found that the solvents with good miscibility and lower boiling points result in the formation of smaller size spherical-shaped nanoparticles (Kumar et al. 2015b), while solvent with poor miscibility with antisolvent and higher boiling points result in the particles with uncontrolled morphology and larger size particles. They also controlled the growth of the particles to enhance solid loading using a high concentration of drug solution in acetone using stabilizers' aqueous solution as antisolvent. Stabilizers form a monolayer on the surface of the particles immediately after formation and control the further growth of particles (Kumar and Siril 2016). Polymeric stabilizers, polyvinylpyrrolidone (PVP), polyvinyl alcohol (PVA) and hydroxypropyl methylcellulose (HPMC) were used (Kumar and Siril 2015). Kumar and coworkers further incorporated the advanced microfluidics technology with evaporation-assisted solvent–antisolvent interaction method to transform the batch process into a continuous process with high solid loading. They developed different approaches using microfluidics for nanoparticles synthesis such as drop-by-drop addition method and drop-to-drop solvent–antisolvent interaction method for nanocrystallization using microfluidic droplet technology. They successfully prepared the nanoparticles of high energetic organic compounds such as RDX, HMX, and TNT; and APIs (Sulfamethoxazole) through the drop-by-drop method and drop-to-drop method (Kumar et al. 2019b; Kumar and Siril 2020). Through this process, they were able to prepare nanoparticles with small size, continuous, and high solid loading. Hence, the process was made feasible for industrial applications. Nanoparticles of lignin were also prepared using solvent–antisolvent interaction method. Hardwood dioxane lignin (DL) and softwood (alkali lignin) nanoparticles are successfully prepared by Yearla and Padmasree (Yearla and Padmasree 2016). They used acetone solution (10 mg/mL) and water 9:1 (v/v)% under stirring of 300 rpm injected using a microsyringe at 20 °C. The lignin particles with size range from 80 to 104 nm were reported. Both nanoparticles showed good antioxidant and UV protectant potential (Yearla and Padmasree 2016). Further investigation of lignin solubility in range of solvent and precipitation of nano- and microparticles at optimum condition need to be investigate. This can be a promising approach to explore the application and performance of lignin at micro- and nanoscale level in short time span. The major limitation of solvent–antisolvent interaction method is controlling the shape of the particles, removability of solvent, and colloidal stability.

2.10 Acoustic Cavitation Assisted Nanoparticles Preparation

Ultrasound is one of the greener energy sources. Due to the unique experimental conditions provided by ultrasound, it is gaining more and more interest (Pilloni et al. 2018; Kumar et al. ; Rogowska-Tylman et al. 2019). In liquid medium, sound waves generate "bubbles" which are growing with time. At a critical size, suddenly the bubbles collapse resulting in the release of very high amount of energy. (Kumar et al. 2018a, 2019a). This released energy can increase local temperature and pressure up to

5000 K and 1000 psi, respectively, which helps in the formulations of nanoparticles (Suslick and Price 1999; Gedanken 2004). Initially, Suslick et al. developed and thoroughly explained the sonication process, mechanism of cavitation formation, acoustic cavitation phenomena, temperature and pressure during sonication (Suslick et al. 1986, 1999; Yuri T. Didenko et al. 1999; McNamara et al. 1999; Flannigan and Suslick 2005). Then, Gedanken and his colleagues extensively investigated the use of ultrasonication in fabrication of nanomaterials (Topaz et al. 2002; Rana et al. 2002; Gedanken 2004; Kumar et al. 2020a). Muthupandian et al. significantly contributed in explanation of cavitation in liquid and well explained the effect of range of parameters on cavitation as well as particle size and shape (Ashokkumar et al. 2010; Babu et al. 2019; Singh et al. 2019; Hafeez et al. 2019; Son et al. 2019). This technique has been widely employed for the preparation of a wide variety of nanoparticles and nanomaterials (Xu et al. 2013). Ultrasound is also gaining more interest in the biomedical applications and controlled and targeted drug delivery (Kumar 2016; Kumar et al. 2019d, e, 2020b). Ultrasound can also be used to breakdown the single strand of DNA molecules (Furusawa et al. 2012). Kumar et al. employed the ultrasound for the synthesis of nanoparticles and drug delivery system for controlled and sustained delivery of non-steroidal anti-inflammatory poor water soluble active pharmaceutical ingredients (Kumar et al. 2018a). Acoustic cavitation assisted solvent–antisolvent method was also developed for the preparation of drug loaded solid lipid nanoparticles (Kumar 2019a). Schematic experimental setup of acoustic cavitation assisted the synthesis of micro- and nanoparticles presented in Fig. 8. Zimniewska et al. reported the successful preparation for lignin nanoparticles using acoustic cavitation (Zimniewska et al. 2008). The prepared nanoparticles of lignin were employed for textile fiber treatments. Schilling and Gilca et al. also achieved the submicron size particle of lignin by hydroxy methylation (Gilca et al. 2015). It was incorporated into water-based ink. Epoxidation is also used for lignin size reduction. Sonochemical synthesis of lignin nanoparticles is one of the best green process.

3 Conclusion and Future Prospective

Lignin is clearly recognized as a major source of natural renewable biopolymer for various applications. Lignin is promising biopolymer with excellent physicochemical and biological properties. Micro- and nanoparticles of lignin possess several advantages due to unique and novel physicochemical properties at nanoscale. Several nanotechnology-based methods are developed for the formulation of lignin micro- or nanoparticles. Among all the techniques, milling method is suitable to prepare smallest nanoparticles (10 nm) of lignin. THF, acetone, and DMSO are the most suitable solvent to prepare nano-lignin through solvent shifting method. Mesoporous TiO_2 and ZnO are formulated using lignin as template. Swollen liquid crystal is good choice of templates for the synthesis of nanoparticles and nanocomposites of lignin. Compare to other method, ice-segregation-induced self-assembly method is most sensitive method and suitable for preparation of nanotubes, nanowires, and nanofibers of lignin. Lignin nanofibers are formulated using

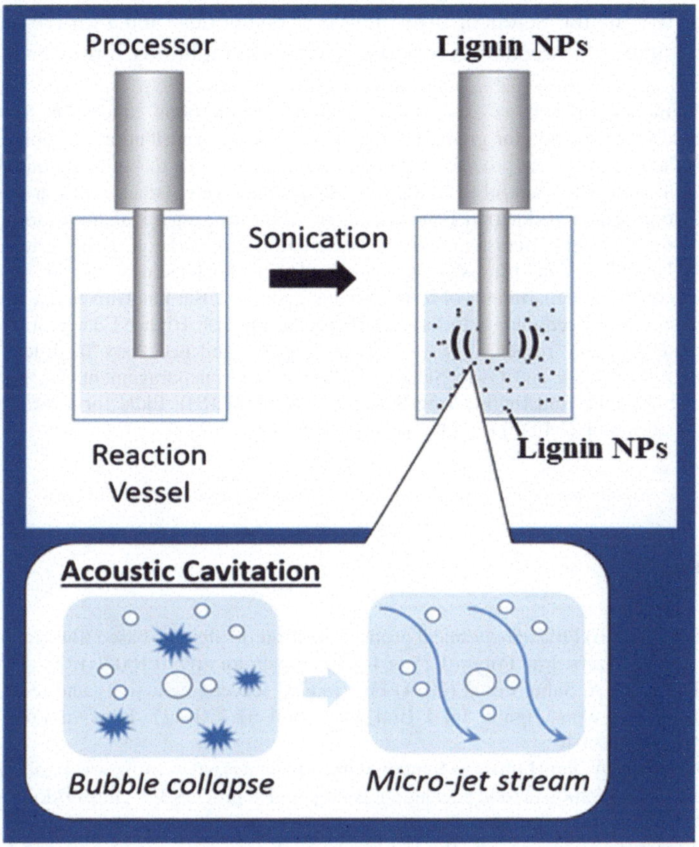

Fig. 8 Schematic presentation of preparation of lignin nanoparticles using acoustic cavitation assisted method

electrospinning with 1% PEO. The physicochemical and antibacterial properties of lignin film improved further by using Acacia lignin. SCF process is most feasible for industrial application and successfully formulated lignin nanoparticles with particles size 144 nm. Antisolvent precipitation method is most simple and easy to prepare lignin nanoparticles. Microfluidic-based antisolvent techniques are more suitable for control and continuous formulation of lignin nanoparticles. Sonochemical approach is one of the green approaches for reduction of particle size of lignin. It is concluded that compared to bottom-up techniques, top-down process offers more flexibility to control the particle size and shape. Simple solvent reprecipitation technique in combination of microfluidics, milling, and high-pressure homogenization, supercritical fluids-based nanoformulations techniques are the promising methods with the feasibility of industrial applications. Lignin nanoparticles or microparticles synthesis have great economical, social, and health impact. The large-scale production may also effect the application of lignin in the field of environmental science, drug delivery, cosmetics, and biofuels. This may completely change the scenarios of biopolymers use including lignin.

Acknowledgements All authors express their gratitude to the book editors Dr. Swati Sharma and Dr. Ashok Kumar Nadda for giving an opportunity for the contribution of book chapter and reviewer for their kind review process, comments, and suggestions to uplift the quality of book chapter. Dr. Raj Kumar is thankful to Council of Higher Education, Israel, for awarding Planning and Budget Committee Postdoctoral Fellowship. Dr. Abhishek Gupta is thankful to postdoctoral fellowships from the Jacob Blaustein Center for Scientific Cooperation, Ben-Gurion University of the Negev, Israel. Dr. Vijay Bhooshan Kumar, Dr. Abhishek Chaudhary, and Dr. Sunil Dutt are thankful to directors and departments of respective institutes viz., Bar Ilan University, Israel; Jaypee University Institute of Technology, India; and Himachal Pradesh Higher Education Department, Himachal Pradesh, India, respectively, for proving support and necessary facilities. Dr. Mohit Chawla thanks IIT Mandi and Gyan Sankul, Kullu for kind encouragement. Dr. Keshaw Ram Aadil thanks to Science and Engineering Research Board (SERB), India for awarding National Postdoctoral Fellowship (NPDF) (NPDF/2016/001156).

References

Aadil KR, Jha H (2016) Physico-chemical properties of lignin–alginate based films in the presence of different plasticizers. Iran Polym J 25:661–670. https://doi.org/10.1007/s13726-016-0449-1

Aadil KR, Barapatre A, Sahu S et al (2014) Free radical scavenging activity and reducing power of Acacia nilotica wood lignin. Int J Biol Macromol 67:220–227. https://doi.org/10.1016/J.IJBIOMAC.2014.03.040

Aadil KR, Barapatre A, Jha H (2016a) Synthesis and characterization of Acacia lignin-gelatin film for its possible application in food packaging. Bioresour Bioprocess 3:27. https://doi.org/10.1186/s40643-016-0103-y

Aadil KR, Barapatre A, Meena AS, Jha H (2016b) Hydrogen peroxide sensing and cytotoxicity activity of Acacia lignin stabilized silver nanoparticles. Int J Biol Macromol 82:39–47. https://doi.org/10.1016/J.IJBIOMAC.2015.09.072

Aadil KR, Prajapati D, Jha H (2016c) Improvement of physcio-chemical and functional properties of alginate film by Acacia lignin. Food Packag Shelf Life 10:25–33. https://doi.org/10.1016/J.FPSL.2016.09.002

Aadil KR, Mussatto SI, Jha H (2018a) Synthesis and characterization of silver nanoparticles loaded poly(vinyl alcohol)-lignin electrospun nanofibers and their antimicrobial activity. Int J Biol Macromol 120:763–767. https://doi.org/10.1016/J.IJBIOMAC.2018.08.109

Aadil KR, Nathani A, Sharma CS et al (2018b) Fabrication of biocompatible alginate-poly(vinyl alcohol) nanofibers scaffolds for tissue engineering applications. Mater Technol 33:507–512. https://doi.org/10.1080/10667857.2018.1473234

Ago M, Jakes JE, Johansson L-S et al (2012a) Interfacial properties of lignin-based electrospun nanofibers and films reinforced with cellulose nanocrystals. ACS Appl Mater Interfaces 4:6849–6856. https://doi.org/10.1021/am302008p

Ago M, Okajima K, Jakes JE et al (2012b) Lignin-based electrospun nanofibers reinforced with cellulose nanocrystals. Biomacromol 13:918–926. https://doi.org/10.1021/bm201828g

Ago M, Jakes JE, Rojas OJ (2013) Thermomechanical properties of lignin-based electrospun nanofibers and films reinforced with cellulose nanocrystals: a dynamic mechanical and nanoindentation study. ACS Appl Mater Interfaces 5:11768–11776. https://doi.org/10.1021/am403451w

Ago M, Huan S, Borghei M et al (2016) High-throughput synthesis of lignin particles (~30 nm to ~2 μm) via Aerosol Flow Reactor: Size Fractionation and Utilization in Pickering Emulsions. ACS Appl Mater Interfaces 8:23302–23310. https://doi.org/10.1021/acsami.6b07900

Ahmed I, Liu H-Y, Mamiya PC et al (2006) Three-dimensional nanofibrillar surfaces covalently modified with tenascin-C-derived peptides enhance neuronal growthin vitro. J Biomed Mater Res Part A 76A:851–860. https://doi.org/10.1002/jbm.a.30587

Anderson JM, Hiltner A, Schodt K, Woods R (1972) Biopolymers as biomaterials: mechanical properties of?-benzyl-L-glutamate-L-Leucine copolymers. J Biomed Mater Res 6:25–35. https://doi.org/10.1002/jbm.820060405

Anderson DG, Levenberg S, Langer R (2004) Nanoliter-scale synthesis of arrayed biomaterials and application to human embryonic stem cells. Nat Biotechnol 22:863–866. https://doi.org/10.1038/nbt981

Antman-Passig M, Shefi O (2016) Remote magnetic orientation of 3D collagen hydrogels for directed neuronal regeneration. Nano Lett 16:2567–2573. https://doi.org/10.1021/acs.nanolett.6b00131

Antman-Passig M, Levy S, Gartenberg C et al (2017) Mechanically oriented 3D collagen hydrogel for directing neurite growth. Tissue Eng Part A 23:403–414. https://doi.org/10.1089/ten.tea.2016.0185

Ashokkumar M, Lee J, Iida Y et al (2010) Spatial distribution of acoustic cavitation bubbles at different ultrasound frequencies. ChemPhysChem 11:1680–1684. https://doi.org/10.1002/cphc.200901037

Atkins M, Damidot D, Glasser FP (1993) Performance of cementitious systems in the repository. MRS Proc 333:315. https://doi.org/10.1557/PROC-333-315

Babu SG, Karthik P, John MC et al (2019) Synergistic effect of sono-photocatalytic process for the degradation of organic pollutants using CuO-TiO2/rGO. Ultrason Sonochem 50:218–223. https://doi.org/10.1016/J.ULTSONCH.2018.09.021

Baranes K, Shevach M, Shefi O, Dvir T (2016) Gold nanoparticle-decorated scaffolds promote neuronal differentiation and maturation. Nano Lett 16:2916–2920. https://doi.org/10.1021/acs.nanolett.5b04033

Barapatre A, Aadil KR, Tiwary BN, Jha H (2015) In vitro antioxidant and antidiabetic activities of biomodified lignin from Acacia nilotica wood. Int J Biol Macromol 75:81–89. https://doi.org/10.1016/J.IJBIOMAC.2015.01.012

Barapatre A, Aadil KR, Jha H (2016) Synergistic antibacterial and antibiofilm activity of silver nanoparticles biosynthesized by lignin-degrading fungus. Bioresour Bioprocess 3:8. https://doi.org/10.1186/s40643-016-0083-y

Barapatre A, Aadil KR, Jha H (2017) Biodegradation of Malachite Green by the Ligninolytic Fungus *Aspergillus flavus*. Clean Soil Air Water 45:1600045. https://doi.org/10.1002/clen.201600045

Beisl S, Friedl A, Miltner A et al (2017a) Lignin from micro- to nanosize: applications. Int J Mol Sci 18:2367. https://doi.org/10.3390/ijms18112367

Beisl S, Miltner A, Friedl A et al (2017b) Lignin from micro- to nanosize: production methods. Int J Mol Sci 18:1244. https://doi.org/10.3390/ijms18061244

Bruce RJ, West CA (1989) Elicitation of lignin biosynthesis and isoperoxidase activity by pectic fragments in suspension cultures of castor bean. Plant Physiol 91:889–897. https://doi.org/10.1104/PP.91.3.889

Buesser B, Pratsinis SE (2012) Design of nanomaterial synthesis by aerosol processes. Annu Rev Chem Biomol Eng 3:103–127. https://doi.org/10.1146/annurev-chembioeng-062011-080930

Caicedo HM, Dempere LA, Vermerris W (2012) Template-mediated synthesis and bio-functionalization of flexible lignin-based nanotubes and nanowires. Nanotechnology 23:105605. https://doi.org/10.1088/0957-4484/23/10/105605

Charoenchaitrakool M, Dehghani F, Foster NR, Chan HK (2000) Micronization by rapid expansion of supercritical solutions to enhance the dissolution rates of poorly water-soluble pharmaceuticals. Ind Eng Chem Res 39:4794–4802. https://doi.org/10.1021/IE000151A

Chaudhary A, Dwivedi C, Gupta A, Nandi CK (2015) One pot synthesis of doxorubicin loaded gold nanoparticles for sustained drug release. RSC Adv 5:97330–97334. https://doi.org/10.1039/C5RA12892G

Chawla M, Kumar R, Siril PF (2016) High catalytic activities of palladium nanowires synthesized using liquid crystal templating approach. J Mol Catal A: Chem 423:126–134. https://doi.org/10.1016/J.MOLCATA.2016.06.014

Chawla M, Dubey R, Singh G et al (2017a) Controlling the morphology of layered double hydroxides of Mn and Co and their exceptional catalytic activities. Thermochim Acta 654:130–139. https://doi.org/10.1016/J.TCA.2017.05.017

Chawla M, Randhawa JK, Siril PF (2017b) Calcination temperature as a probe to tune the non-enzymatic glucose sensing activity of Cu–Ni bimetallic nanocomposites. New J Chem 41:4582–4591. https://doi.org/10.1039/C6NJ03920K

Chawla M, Sharma V, Randhawa JK (2017c) Facile one pot synthesis of CuO nanostructures and their effect on Nonenzymatic Glucose Biosensing. Electrocatalysis 8:27–35. https://doi.org/10.1007/s12678-016-0337-7

Chawla M, Kumari A, Siril PF (2018) Exceptional catalytic activities and sensing performance of palladium decorated anisotropic gold nanoparticles. ChemistrySelect 3:9071–9083. https://doi.org/10.1002/slct.201801426

Chen X, Kuo D-H, Lu D et al (2016) Synthesis and photocatalytic activity of mesoporous TiO2 nanoparticle using biological renewable resource of un-modified lignin as a template. Microporous Mesoporous Mater 223:145–151. https://doi.org/10.1016/J.MICROMESO.2015.11.005

Cui H, Hanus R, Kessler MR (2013) Degradation of ROMP-based bio-renewable polymers by UV radiation. Polym Degrad Stab 98:2357–2365. https://doi.org/10.1016/J.POLYMDEGRADSTAB.2013.08.003

Dallmeyer I, Ko F, Kadla JF (2010) Electrospinning of technical lignins for the production of fibrous networks. J Wood Chem Technol 30:315–329. https://doi.org/10.1080/02773813.2010.527782

Dalton L (2002) Nonlinear optical polymeric materials: from chromophore design to commercial applications. Polymers for photonics applications I. Springer, Berlin, Heidelberg, pp 1–86

Dalvi SV, Mukhopadhyay M (2009) A novel process for precipitation of ultra-fine particles using sub-critical CO2. Powder Technol 195:190–195. https://doi.org/10.1016/J.POWTEC.2009.05.029

Dang JM, Leong KW (2006) Natural polymers for gene delivery and tissue engineering. Adv Drug Deliv Rev 58:487–499. https://doi.org/10.1016/J.ADDR.2006.03.001

Davin LB, Lewis NG (2005) Lignin primary structures and dirigent sites. Curr Opin Biotechnol 16:407–415. https://doi.org/10.1016/J.COPBIO.2005.06.011

De Oliveira W, Glasser WG (1994) Multiphase materials with lignin. XII. Blends of poly(vinyl chloride) with lignin–caprolactone copolymers. J Appl Polym Sci 51:563–571. https://doi.org/10.1002/app.1994.070510320

del Agua I, Marina S, Pitsalidis C et al (2018) Conducting polymer scaffolds based on Poly(3,4-ethylenedioxythiophene) and Xanthan Gum for live-cell monitoring. ACS Omega 3:7424–7431. https://doi.org/10.1021/acsomega.8b00458

Didenko YT, McNamara WB, Suslick KS (1999) Hot spot conditions during cavitation in water. https://doi.org/10.1021/ja9844635

Dubey R, Chawla M, Siril PF, Singh G (2013) Bi-metallic nanocomposites of Mn with very high catalytic activity for burning rate enhancement of composite solid propellants. Thermochim Acta 572:30–38. https://doi.org/10.1016/J.TCA.2013.09.005

Dutt S, Sharma R (2017) Controlling the polypyrrole microstructures using swollen liquid crystals as structure directing agent. Mater Res Express 4:105305. https://doi.org/10.1088/2053-1591/aa9192

Dutt S, Siril PF (2014a) A novel approach for the synthesis of polyaniline nanostructures using swollen liquid crystal templates. Mater Lett 124:50–53. https://doi.org/10.1016/J.MATLET.2014.03.068

Dutt S, Siril PF (2014b) Morphology controlled synthesis of polyaniline nanostructures using swollen liquid crystal templates. J Appl Polym Sci 131. https://doi.org/10.1002/app.40800

Dutt S, Siril PF (2015) Controlling the morphology of polyaniline–platinum nanocomposites using swollen liquid crystal templates. Synth Met 209:82–90. https://doi.org/10.1016/J.SYNTHMET.2015.07.012

Dutt S, Kumar R, Siril PF (2015) Green synthesis of a palladium–polyaniline nanocomposite for green Suzuki-Miyaura coupling reactions. RSC Adv 5:33786–33791. https://doi.org/10.1039/C5RA05007C

Dutt S, Siril PF, Remita S (2017) Swollen liquid crystals (SLCs): a versatile template for the synthesis of nano structured materials. RSC Adv 7:5733–5750. https://doi.org/10.1039/C6RA26390A

Dutt S, Vats T, Siril PF (2018) Synthesis of polyaniline–magnetite nanocomposites using swollen liquid crystal templates for magnetically separable dye adsorbent applications. New J Chem 42:5709–5719. https://doi.org/10.1039/C7NJ04637E

Duval A, Lawoko M (2014) A review on lignin-based polymeric, micro- and nano-structured materials. React Funct Polym 85:78–96. https://doi.org/10.1016/J.REACTFUNCTPOLYM.2014.09.017

Dwivedi C, Gupta A, Chaudhary A, Nandi CK (2014) Gold nanoparticle chitosan composite hydrogel beads show efficient removal of methyl parathion from waste water. RSC Adv 4:39830. https://doi.org/10.1039/C4RA03870C

Edgar KJ, Buchanan CM, Debenham JS et al (2001) Advances in cellulose ester performance and application. Prog Polym Sci 26:1605–1688. https://doi.org/10.1016/S0079-6700(01)00027-2

Eerikäinen H, Watanabe W, Kauppinen EI, Ahonen PP (2003) Aerosol flow reactor method for synthesis of drug nanoparticles. Eur J Pharm Biopharm 55:357–360. https://doi.org/10.1016/S0939-6411(03)00005-5

Felix SP, Singh G, Sikder AK, Aggrawal JP (2005) Studies on energetic compounds: Part 33: thermolysis of keto-RDX and its plastic bonded explosives containing thermally stable polymers. Thermochim Acta 426:53–60. https://doi.org/10.1016/J.TCA.2004.06.020

Figueiredo P, Lintinen K, Kiriazis A et al (2017) In vitro evaluation of biodegradable lignin-based nanoparticles for drug delivery and enhanced antiproliferation effect in cancer cells. Biomaterials 121:97–108. https://doi.org/10.1016/J.BIOMATERIALS.2016.12.034

Figueiredo P, Lintinen K, Hirvonen JT et al (2018) Properties and chemical modifications of lignin: towards lignin-based nanomaterials for biomedical applications. Prog Mater Sci 93:233–269. https://doi.org/10.1016/J.PMATSCI.2017.12.001

Fitz Patrick M, Champagne P, Cunningham MF, Whitney RA (2010) A biorefinery processing perspective: treatment of lignocellulosic materials for the production of value-added products. Bioresour Technol 101:8915–8922. https://doi.org/10.1016/J.BIORTECH.2010.06.125

Flannigan DJ, Suslick KS (2005) Plasma formation and temperature measurement during single-bubble cavitation. Nature 434:52–55. https://doi.org/10.1038/nature03361

Francis Suh J-K, Matthew HW (2000) Application of chitosan-based polysaccharide biomaterials in cartilage tissue engineering: a review. Biomaterials 21:2589–2598. https://doi.org/10.1016/S0142-9612(00)00126-5

Frangville C, Rutkevičius M, Richter AP et al (2012) Fabrication of environmentally biodegradable lignin nanoparticles. ChemPhysChem 13:4235–4243. https://doi.org/10.1002/cphc.201200537

Furusawa Y, Fujiwara Y, Campbell P et al (2012) DNA double-strand breaks induced by cavitational mechanical effects of ultrasound in cancer cell lines. PLoS One 7:e29012. https://doi.org/10.1371/journal.pone.0029012

Gedanken A (2004) Using sonochemistry for the fabrication of nanomaterials. Ultrason Sonochem 11:47–55. https://doi.org/10.1016/J.ULTSONCH.2004.01.037

Ghosh D, Khastgir D (2018) Degradation and stability of polymeric high-voltage insulators and prediction of their service life through environmental and accelerated aging processes. ACS Omega 3:11317–11330. https://doi.org/10.1021/acsomega.8b01560

Gilca IA, Popa VI, Crestini C (2015) Obtaining lignin nanoparticles by sonication. Ultrason Sonochem 23:369–375. https://doi.org/10.1016/J.ULTSONCH.2014.08.021

Gross SK, Sarkanen K, Schuerch C (1958) Determinations of molecular weight of lignin degradation products by three methods. Anal Chem 30:518–521. https://doi.org/10.1021/ac60136a020
Gupta A, Nandi CK (2017) PC12 live cell ultrasensitive neurotransmitter signaling using high quantum yield sulphur doped carbon dots and its extracellular Ca2 + ion dependence. Sensors Actuators B Chem 245:137–145. https://doi.org/10.1016/J.SNB.2017.01.145
Gupta AK, Mohanty S, Nayak SK (2014) Synthesis, characterization and application of Lignin Nanoparticles (LNPs). Mater Focus 3:444–454. https://doi.org/10.1166/mat.2014.1217
Gupta A, Chaudhary A, Mehta P et al (2015) Nitrogen-doped, thiol-functionalized carbon dots for ultrasensitive Hg(ii) detection. Chem Commun 51:10750–10753. https://doi.org/10.1039/C5CC03019F
Gupta A, Verma NC, Khan S et al (2016a) Paper strip based and live cell ultrasensitive lead sensor using carbon dots synthesized from biological media. Sensors Actuators B Chem 232:107–114. https://doi.org/10.1016/J.SNB.2016.03.110
Gupta A, Verma NC, Khan S, Nandi CK (2016b) Carbon dots for naked eye colorimetric ultrasensitive arsenic and glutathione detection. Biosens Bioelectron 81:465–472. https://doi.org/10.1016/J.BIOS.2016.03.018
Gutiérrez MC, Ferrer ML, del Monte F (2008) Ice-templated materials: sophisticated structures exhibiting enhanced functionalities obtained after unidirectional freezing and ice-segregation-induced self-assembly. Chem Mater 20:634–648. https://doi.org/10.1021/cm702028z
Hafeez HY, Lakhera SK, Ashokkumar M, Neppolian B (2019) Ultrasound assisted synthesis of reduced graphene oxide (rGO) supported InVO4-TiO2 nanocomposite for efficient hydrogen production. Ultrason Sonochem 53:1–10. https://doi.org/10.1016/J.ULTSONCH.2018.12.009
Han T, Sophonrat N, Tagami A et al (2019) Characterization of lignin at pre-pyrolysis temperature to investigate its melting problem. Fuel 235:1061–1069. https://doi.org/10.1016/J.FUEL.2018.08.120
Hatfield RD, Jung H-JG, Ralph J et al (1994) A comparison of the insoluble residues produced by the Klason lignin and acid detergent lignin procedures. J Sci Food Agric 65:51–58. https://doi.org/10.1002/jsfa.2740650109
Hosseini M, Fazelian N, Fakhri A et al (2019) Preparation, and structural of new NiS-SiO2 and Cr2S3-TiO2 nano-catalyst: photocatalytic and antimicrobial studies. J Photochem Photobiol B Biol 194:128–134. https://doi.org/10.1016/J.JPHOTOBIOL.2019.03.016
Huang L, McMillan RA, Apkarian RP et al (2000) Generation of synthetic elastin-mimetic small diameter fibers and fiber networks. https://doi.org/10.1021/ma991858f
Huebsch N, Mooney DJ (2009) Inspiration and application in the evolution of biomaterials. Nature 462:426–432. https://doi.org/10.1038/nature08601
Iden R, Schrof W, Hadeler J, Lehmann S (2003) Combinatorial materials research in the polymer industry: speed versus flexibility. Macromol Rapid Commun 24:63–72. https://doi.org/10.1002/marc.200390019
Jin G-Z, Kim M, Shin US, Kim H-W (2011) Neurite outgrowth of dorsal root ganglia neurons is enhanced on aligned nanofibrous biopolymer scaffold with carbon nanotube coating. Neurosci Lett 501:10–14. https://doi.org/10.1016/J.NEULET.2011.06.023
Jung YJ, Kar S, Talapatra S et al (2006) Aligned Carbon nanotube – polymer hybrid architectures for diverse flexible electronic applications. https://doi.org/10.1021/nl052238x
Kabanov AV (2006) Polymer genomics: an insight into pharmacology and toxicology of nanomedicines. Adv Drug Deliv Rev 58:1597–1621. https://doi.org/10.1016/J.ADDR.2006.09.019
Kogan G, Šoltés L, Stern R, Gemeiner P (2006) Hyaluronic acid: a natural biopolymer with a broad range of biomedical and industrial applications. Biotechnol Lett 29:17–25. https://doi.org/10.1007/s10529-006-9219-z
Košíková B, Gregorová A, Osvald A, Krajčovičová J (2007) Role of lignin filler in stabilization of natural rubber–based composites. J Appl Polym Sci 103:1226–1231. https://doi.org/10.1002/app.24530

Klossner RR, Queen HA, Coughlin AJ, Krause WE (2008) Correlation of chitosan's rheological properties and its ability to electrospin. Biomacromolcules, 2009, 9(10):2947–2953. https://doi.org/10.1021/bm800738u
Kumar R (2016) Repositioning of Non-Steroidal Anti Inflammatory Drug (NSAIDs) for cancer treatment: promises and challenges. J Nanomed Nanotechnol 7:e140. https://doi.org/10.4172/2157-7439.1000e140
Kumar R (2019a) Lipid-based nanoparticles for drug-delivery systems. Nanocarriers Drug Deliv 249–284. https://doi.org/10.1016/B978-0-12-814033-8.00008-4
Kumar R (2019b) Nanotechnology based approaches to enhance aqueous solubility and bioavailability of griseofulvin: a literature survey. J Drug Deliv Sci Tec 53:101221. https://doi.org/10.1016/j.jddst.2019.101221
Kumar R (2020) Solubility and Bioavailability of Fenofibrate Nanoformulations. ChemistrySelect, 5(4):1478–1490. https://doi.org/10.1002/slct.201903647
Kumar R, Siril PF (2014) Ultrafine carbamazepine nanoparticles with enhanced water solubility and rate of dissolution. RSC Adv 4:48101–48108. https://doi.org/10.1039/C4RA08495K
Kumar R, Siril PF (2015) Controlling the size and morphology of griseofulvin nanoparticles using polymeric stabilizers by evaporation-assisted solvent–antisolvent interaction method. J Nanoparticle Res 17:256. https://doi.org/10.1007/s11051-015-3066-6
Kumar R, Siril PF (2016) Preparation and characterization of polyvinyl alcohol stabilized griseofulvin nanoparticles. Mater Today Proc 3:2261–2267. https://doi.org/10.1016/J.MATPR.2016.04.135
Kumar R, Siril PF (2018) Enhancing the solubility of fenofibrate by nanocrystal formation and encapsulation. AAPS PharmSciTech 19:284–292. https://doi.org/10.1208/s12249-017-0840-z
Kumar R, Siril PF (2020) Drop-by-drop solvent hot antisolvent interaction method for engineering nanocrystallization of sulfamethoxazole to enhanced water solubility and bioavailablility. J Drug Deliv Sci Tec 55:101359. https://doi.org/10.1016/j.jddst.2019.101359
Kumar R, Siril PF, Soni P (2014) Preparation of Nano-RDX by evaporation assisted solvent antisolvent interaction. Propellants Explos Pyrotech 39:383–389. https://doi.org/10.1002/prep.201300104
Kumar R, Siril PF, Soni P (2015a) Tuning the particle size and morphology of high energetic material nanocrystals. Def Technol 11:382–389. https://doi.org/10.1016/J.DT.2015.07.002
Kumar R, Siril PF, Soni P (2015b) Optimized synthesis of HMX nanoparticles using antisolvent precipitation method. J Energ Mater 33:277–287. https://doi.org/10.1080/07370652.2014.988774
Kumar R, Siril PF, Javid F (2016) Unusual anti-leukemia activity of nanoformulated naproxen and other non-steroidal anti-inflammatory drugs. Mater Sci Eng C 69:1335–1344. https://doi.org/10.1016/j.msec.2016.08.024
Kumar R, Singh A, Garg N, Siril PF (2018a) Solid lipid nanoparticles for the controlled delivery of poorly water soluble non-steroidal anti-inflammatory drugs. Ultrason Sonochem 40:686–696. https://doi.org/10.1016/J.ULTSONCH.2017.08.018
Kumar VB, Kumar R, Gedanken A, Shefi O (2018b) Fluorescent metal-doped carbon dots for neuronal manipulations. Ultrason Sonochem, 2019, 52:205–213. https://doi.org/10.1016/j.ultsonch.2018.11.017
Kumar R, Kumar VB, Marcus M (2019a) Elements (B, N, P) doped carbon dots interaction with neural cells: promising results and future prospective. In: Proceedings SPIE 10892, colloidal nanoparticles for biomedical applications XIV, p 1089214. https://doi.org/10.1117/12.2509610
Kumar R, Soni P, Siril PF (2019b) Engineering the morphology and particle size of high energetic compounds using drop-by-drop and drop-to-drop solvent-antisolvent interaction methods. ACS Omega 4:5424–5433. https://doi.org/10.1021/acsomega.8b03214
Kumar VB, Kumar R, Friedman O et al (2019c) One-pot hydrothermal synthesis of elements (B, N, P)-doped fluorescent Carbon dots for cell labelling, differentiation and outgrowth of neuronal cells. ChemistrySelect 4:4222–4232. https://doi.org/10.1002/slct.201900581

Kumar R, Singh A, Garg N (2019d) Acoustic cavitation-assisted formulation of solid lipid nanoparticles using different Stabilizers. ACS Omega 4 (8):13360–13370. https://doi.org/10.1021/acsomega.9b01532
Kumar R, Singh A, Garg N (2019e) Acoustic cavitation assisted hot melt mixing technique for solid lipid nanoparticles formulation, characterization, and controlled delivery of poorly water soluble drugs. J Drug Deliv Sci Tec 54:101277. https://doi.org/10.1016/j.jddst.2019.101277
Kumar R, Kumar VB, Gedanken A (2020a) The sonochemical synthesis of carbon dots: Synthetic route, effect of parameters, and catalytic, energy, biomedical and tissue engineering applications. Ultrason Sonochem 105009. https://doi.org/10.1016/j.ultsonch.2020.105009
Kumar R, Singh A, Sharma K, Dhasmana D, Garg N, Siril PF (2020b) Preparation, characterization and in vitro cytotoxicity of Fenofibrate and Nabumetone loaded solid lipid nanoparticles. Mater Sci Eng: C 106:110184. https://doi.org/10.1016/j.msec.2019.110184
Li Y, Sarkanen S (2005) Miscible blends of kraft lignin derivatives with low-Tg polymers. Macromolecules, 2005, 38(6):2296–2306. https://doi.org/10.1021/ma047546g
Li J, Gellerstedt G, Toven K (2009) Steam explosion lignins; their extraction, structure and potential as feedstock for biodiesel and chemicals. Bioresour Technol 100:2556–2561. https://doi.org/10.1016/J.BIORTECH.2008.12.004
Lievonen M, Valle-Delgado JJ, Mattinen M-L et al (2016) A simple process for lignin nanoparticle preparation. Green Chem 18:1416–1422. https://doi.org/10.1039/C5GC01436K
Lopes MS, Jardini AL, Filho RM (2012) Poly (Lactic Acid) production for tissue engineering applications. Procedia Eng 42:1402–1413. https://doi.org/10.1016/J.PROENG.2012.07.534
Lu Q, Zhu M, Zu Y et al (2012) Comparative antioxidant activity of nanoscale lignin prepared by a Supercritical Antisolvent (SAS) process with non-nanoscale lignin. Food Chem 135:63–67. https://doi.org/10.1016/J.FOODCHEM.2012.04.070
Marcus M, Baranes K, Park M et al (2017) Interactions of Neurons with physical environments. Adv Healthc Mater 6:1700267
Martín A, Cocero MJ (2008) Micronization processes with supercritical fluids: fundamentals and mechanisms. Adv Drug Deliv Rev 60:339–350. https://doi.org/10.1016/J.ADDR.2007.06.019
McNamara WB, Didenko YT, Suslick KS (1999) Sonoluminescence temperatures during multi-bubble cavitation. Nature 401:772–775. https://doi.org/10.1038/44536
Meziani MJ, Pathak P, Hurezeanu R et al (2004) Supercritical-fluid processing technique for nanoscale polymer particles. Angew Chemie Int Ed 43:704–707. https://doi.org/10.1002/anie.200352834
Mogoşanu GD, Grumezescu AM (2014) Natural and synthetic polymers for wounds and burns dressing. Int J Pharm 463:127–136. https://doi.org/10.1016/J.IJPHARM.2013.12.015
Montero GA, Smith CB, Hendrix WA, Butcher DL (2000) Supercritical fluid technology in textile processing: an overview. Ind Eng Chem Res 39:4806–4812. https://doi.org/10.1021/IE0002475
Mukhopadhyay M, Dalvi SV (2004) Partial molar volume fraction of solvent in binary (CO2–solvent) solution for solid solubility predictions. J Supercrit Fluids 29:221–230. https://doi.org/10.1016/S0896-8446(03)00087-1
Myint AA, Lee HW, Seo B et al (2016) One pot synthesis of environmentally friendly lignin nanoparticles with compressed liquid carbon dioxide as an antisolvent. Green Chem 18:2129–2146. https://doi.org/10.1039/C5GC02398J
Naik SN, Goud VV, Rout PK, Dalai AK (2010) Production of first and second generation biofuels: a comprehensive review. Renew Sustain Energy Rev 14:578–597. https://doi.org/10.1016/J.RSER.2009.10.003
Nair SS, Sharma S, Pu Y et al (2014) High shear homogenization of lignin to nanolignin and thermal stability of Nanolignin-Polyvinyl Alcohol blends. Chemsuschem 7:3513–3520. https://doi.org/10.1002/cssc.201402314
Nie H, He A, Wu W et al (2009) Effect of poly(ethylene oxide) with different molecular weights on the electrospinnability of sodium alginate. Polymer (Guildf) 50:4926–4934. https://doi.org/10.1016/J.POLYMER.2009.07.043

Nitz H, Semke H, Mülhaupt R (2001) Influence of lignin type on the mechanical properties of lignin based compounds. Macromol Mater Eng 286:737. https://doi.org/10.1002/1439-2054(20011201)286:12%3c737:AID-MAME737%3e3.0.CO;2-2

Pathak P, Meziani MJ, Desai T, Sun YP (2004) Nanosizing drug particles in supercritical fluid processing. J Am Chem Soc 126:10842–10843. https://doi.org/10.1021/JA046914T

Paul DR, Robeson LM (2008) Polymer nanotechnology: nanocomposites. Polymer (Guildf) 49:3187–3204. https://doi.org/10.1016/J.POLYMER.2008.04.017

Pilloni M, Kumar VB, Ennas G et al (2018) Formation of metallic silver and copper in non-aqueous media by ultrasonic radiation. Ultrason Sonochem 47:108–113. https://doi.org/10.1016/J.ULTSONCH.2018.04.018

Rana RK, Mastai Y, Gedanken A (2002) Acoustic cavitation leading to the morphosynthesis of mesoporous silica vesicles. Adv Mater 14:1414–1418. https://doi.org/10.1002/1521-4095(20021002)14:19%3c1414:AID-ADMA1414%3e3.0.CO;2-F

Reddy N, Yang Y (2005) Biofibers from agricultural byproducts for industrial applications. Trends Biotechnol 23:22–27. https://doi.org/10.1016/J.TIBTECH.2004.11.002

Reddy N, Reddy R, Jiang Q (2015) Crosslinking biopolymers for biomedical applications. Trends Biotechnol 33:362–369. https://doi.org/10.1016/J.TIBTECH.2015.03.008

Reverchon E, Adami R (2006) Nanomaterials and supercritical fluids. J Supercrit Fluids 37:1–22. https://doi.org/10.1016/J.SUPFLU.2005.08.003

Reverchon E, Adami R, De Marco I et al (2005) Pigment Red 60 micronization using supercritical fluids based techniques. J Supercrit Fluids 35:76–82. https://doi.org/10.1016/J.SUPFLU.2004.10.010

Richter AP, Bharti B, Armstrong HB et al (2016) Synthesis and characterization of biodegradable lignin nanoparticles with tunable surface properties. Langmuir 32:6468–6477. https://doi.org/10.1021/acs.langmuir.6b01088

Rogowska-Tylman J, Locs J, Salma I et al (2019) In vivo and in vitro study of a novel nanohydroxyapatite sonocoated scaffolds for enhanced bone regeneration. Mater Sci Eng C 99:669–684. https://doi.org/10.1016/J.MSEC.2019.01.084

Sahoo S, Misra M, Mohanty AK (2013) Effect of compatibilizer and fillers on the properties of injection molded lignin-based hybrid green composites. J Appl Polym Sci 127:4110–4121. https://doi.org/10.1002/app.37667

Sameni J, Krigstin S, Sain M (2016) Characterization of Lignins isolated from industrial residues and their beneficial uses. BioResources 11(4):8435–8456. https://doi.org/10.15376/biores.11.4.8435-8456

Sarkanen S, Chen Y, Wang Y-Y (2016) Journey to polymeric materials composed exclusively of simple lignin derivatives. ACS Sustain Chem Eng 4:5223–5229. https://doi.org/10.1021/acssuschemeng.6b01700

Savy D, Nebbioso A, Mazzei P et al (2015) Molecular composition of water-soluble lignins separated from different non-food biomasses. Fuel Process Technol 131:175–181. https://doi.org/10.1016/J.FUPROC.2014.11.011

Seidlits SK, Lee JY, Schmidt CE (2008) Nanostructured scaffolds for neural applications. Nanomedicine 3:183–199. https://doi.org/10.2217/17435889.3.2.183

Shah S, Yin PT, Uehara TM et al (2014) Guiding stem cell differentiation into oligodendrocytes using graphene-nanofiber hybrid scaffolds. Adv Mater 26:3673–3680. https://doi.org/10.1002/adma.201400523

Sharma V, Sinha N, Dutt S et al (2016) Tuning the surface enhanced Raman scattering and catalytic activities of gold nanorods by controlled coating of platinum. J Colloid Interface Sci 463:180–187. https://doi.org/10.1016/J.JCIS.2015.10.036

Shi X, Karachi A, Hosseini M et al (2019) Ultrasound wave assisted removal of Ceftriaxone sodium in aqueous media with novel nano composite g-C3N4/MWCNT/Bi2WO6 based on CCD-RSM model. Ultrason Sonochem. https://doi.org/10.1016/J.ULTSONCH.2019.01.018

Singh G, Kapoor IPS, Dubey S, Siril PF (2009) Preparation, characterization and catalytic activity of transition Metal Oxide nanocrystals. J Sci Conf Proc 1:11–17. https://doi.org/10.1166/jcp.2009.002

Singh K, Gupta SP, Kumar A, Kumar A (2019) The effect of High Intensity Ultrasound (HIU) on the kinetics of crystallization of sucrose: elimination of latent period. Ultrason Sonochem 52:19–24. https://doi.org/10.1016/J.ULTSONCH.2018.05.030

Siril PF, Lehoux A, Ramos L et al (2012) Facile synthesis of palladium nanowires by a soft templating method. New J Chem 36:2135. https://doi.org/10.1039/c2nj40342k

Son Y, Lee D, Lee W et al (2019) Cavitational activity in heterogeneous systems containing fine particles. Ultrason Sonochem 58:104599. https://doi.org/10.1016/J.ULTSONCH.2019.05.016

Song M, Seo J, Kim H, Kim Y (2017) Flexible thermal sensors based on organic field-effect transistors with polymeric channel/gate-insulating and light-blocking layers. ACS Omega 2:4065–4070. https://doi.org/10.1021/acsomega.7b00494

Spender J, Demers AL, Xie X et al (2012) Method for production of polymer and carbon nanofibers from water-soluble polymers. Nano Lett 12:3857–3860. https://doi.org/10.1021/nl301983d

Stewart D (2008) Lignin as a base material for materials applications: Chemistry, application and economics. Ind Crops Prod 27:202–207. https://doi.org/10.1016/J.INDCROP.2007.07.008

Sun Y-P, Meziani MJ, Pathak P, Qu L (2005) Polymeric nanoparticles from rapid expansion of supercritical fluid solution. Chem A Eur J 11:1366–1373. https://doi.org/10.1002/chem.200400422

Suslick KS, Price GJ (1999) Applications of ultrasound to materials Chemistry. Annu Rev Mater Sci 29:295–326. https://doi.org/10.1146/annurev.matsci.29.1.295

Suslick KS, Hammerton DA, Cline RE (1986) Sonochemical hot spot. J Am Chem Soc 108:5641–5642. https://doi.org/10.1021/ja00278a055

Suslick KS, McNamara WB, Didenko Y (1999) Hot spot conditions during multi-bubble cavitation. Sonochemistry and sonoluminescence. Springer, Netherlands, Dordrecht, pp 191–204

Suyama K, Shirai M (2009) Photobase generators: recent progress and application trend in polymer systems. Prog Polym Sci 34:194–209. https://doi.org/10.1016/J.PROGPOLYMSCI.2008.08.005

Ten E, Ling C, Wang Y et al (2014) Lignin nanotubes as vehicles for gene delivery into human cells. Biomacromol 15:327–338. https://doi.org/10.1021/bm401555p

Thakur VK, Thakur MK (2014) Processing and characterization of natural cellulose fibers/thermoset polymer composites. Carbohydr Polym 109:102–117. https://doi.org/10.1016/J.CARBPOL.2014.03.039

Thakur VK, Thakur MK, Raghavan P, Kessler MR (2014) Progress in green polymer composites from lignin for multifunctional applications: a review. ACS Sustain Chem Eng 2:1072–1092. https://doi.org/10.1021/sc500087z

Thorat AA, Dalvi SV (2012) Liquid antisolvent precipitation and stabilization of nanoparticles of poorly water soluble drugs in aqueous suspensions: recent developments and future perspective. Chem Eng J 181–182:1–34. https://doi.org/10.1016/J.CEJ.2011.12.044

Thunga M, Chen K, Grewell D, Kessler MR (2014) Bio-renewable precursor fibers from lignin/polylactide blends for conversion to carbon fibers. Carbon 68:159–166. https://doi.org/10.1016/J.CARBON.2013.10.075 (N Y)

Topaz M, Motiei M, Assia E et al (2002) Acoustic cavitation in phacoemulsification: chemical effects, modes of action and cavitation index. Ultrasound Med Biol 28:775–784. https://doi.org/10.1016/S0301-5629(02)00514-8

Tortora M, Cavalieri F, Mosesso P et al (2014) Ultrasound driven assembly of lignin into microcapsules for storage and delivery of hydrophobic molecules. Biomacromol 15:1634–1643. https://doi.org/10.1021/bm500015j

Uggla C, Mellerowicz EJ, Sundberg B et al (1998) Indole-3-Acetic acid controls cambial growth in scots pine by positional signaling. Plant Physiol 117:113–121. https://doi.org/10.1104/pp.117.1.113

Varesano A, Aluigi A, Vineis C, Tonin C (2008) Study on the shear viscosity behavior of keratin/PEO blends for nanofibre electrospinning. J Polym Sci Part B: Polym Phys 46:1193–1201. https://doi.org/10.1002/polb.21452

Vats T, Siril PF (2017) A dataset for preparing pristine graphene-palladium nanocomposites using swollen liquid crystal templates. Sci Data 4:170196. https://doi.org/10.1038/sdata.2017.196

Vats T, Dutt S, Kumar R, Siril PF (2016) Facile synthesis of pristine graphene-palladium nanocomposites with extraordinary catalytic activities using swollen liquid crystals. Sci Rep 6:33053. https://doi.org/10.1038/srep33053

Wang X, Zhang Y, Hao C et al (2014) Solid-phase synthesis of Mesoporous ZnO using Lignin-Amine template and its photocatalytic properties. Ind Eng Chem Res 53:6585–6592. https://doi.org/10.1021/ie404179f

Xu H, Zeiger BW, Suslick KS (2013) Sonochemical synthesis of nanomaterials. Chem Soc Rev 42:2555–2567. https://doi.org/10.1039/C2CS35282F

Yang W, Fortunati E, Dominici F et al (2015a) Effect of processing conditions and lignin content on thermal, mechanical and degradative behavior of lignin nanoparticles/polylactic (acid) bionanocomposites prepared by melt extrusion and solvent casting. Eur Polym J 71:126–139. https://doi.org/10.1016/J.EURPOLYMJ.2015.07.051

Yang W, Kenny JM, Puglia D (2015b) Structure and properties of biodegradable wheat gluten bionanocomposites containing lignin nanoparticles. Ind Crops Prod 74:348–356. https://doi.org/10.1016/J.INDCROP.2015.05.032

Yang W, Fortunati E, Dominici F et al (2016a) Synergic effect of cellulose and lignin nanostructures in PLA based systems for food antibacterial packaging. Eur Polym J 79:1–12. https://doi.org/10.1016/J.EURPOLYMJ.2016.04.003

Yang W, Owczarek JS, Fortunati E et al (2016b) Antioxidant and antibacterial lignin nanoparticles in polyvinyl alcohol/chitosan films for active packaging. Ind Crops Prod 94:800–811. https://doi.org/10.1016/J.INDCROP.2016.09.061

Ye X, Wai CM (2003) Making nanomaterials in supercritical fluids: a review. J Chem Educ 80:198. https://doi.org/10.1021/ed080p198

Yearla SR, Padmasree K (2016) Preparation and characterisation of lignin nanoparticles: evaluation of their potential as antioxidants and UV protectants. J Exp Nanosci 11:289–302. https://doi.org/10.1080/17458080.2015.1055842

Zhang X, Li C, Luo Y (2011) Aligned/unaligned conducting polymer Cryogels with three-dimensional macroporous architectures from ice-segregation-induced self-assembly of PEDOT-PSS. Langmuir 27:1915–1923. https://doi.org/10.1021/la1044333

Zhang C, Gao Q, Zhou B, Bhargava G (2017) Preparation, characterization, and surface conductivity of nanocomposites with hollow graphitic carbon nanospheres as fillers in polymethylmethacrylate matrix. J Nanoparticle Res 19:269. https://doi.org/10.1007/s11051-017-3976-6

Zhao W, Liu L, Zhang F et al (2019) Shape memory polymers and their composites in biomedical applications. Mater Sci Eng C 97:864–883. https://doi.org/10.1016/J.MSEC.2018.12.054

Zhong J-F, Xu L, Qin X-L (2015) Efficient antibacterial silver nanoparticles composite using lignin as a template. J Compos Mater 49:2329–2335. https://doi.org/10.1177/0021998314545192

Zhou L, Kamyab H, Surendar A et al (2019) Novel Z-scheme composite Ag2CrO4/NG/polyimide as high performance nano catalyst for photoreduction of CO2: design, fabrication, characterization and mechanism. J Photochem Photobiol A Chem 368:30–40. https://doi.org/10.1016/J.JPHOTOCHEM.2018.09.006

Zimniewska M, Kozłowski R, Batog J (2008) Nanolignin modified linen fabric as a multifunctional product. Mol Cryst Liq Cryst 484:43[409]–50[416]. https://doi.org/10.1080/15421400801903395

Synthesis and Applications of Lignin-Derived Hydrogels

Sanjam Chandna, Sanjeev Kumar Bhardwaj, Shatabdi Paul and Jayeeta Bhaumik

Abstract Lignin, owing to its randomly cross-linked networks of reactive functional groups such as methoxy groups and reductive phenolic hydroxyl groups, is being utilized for the preparation of versatile materials. These functional moieties also offer various properties, such as adhesiveness, mechanical strength, antioxidant, antimicrobial and UV-protection properties to its derivatives. Hydrogels are three-dimensional highly hydrated networks of natural or synthetic polymeric chains, which have enormous potential for various applications. Adaption of natural polymers in modified form for the preparation of hydrogels provides several advantages over synthetic polymers, such as low toxicity, biodegradability, biocompatibility, eco-friendly properties and susceptibility to enzymatic degradation. Lignin-derived hydrogels are prepared via different methods including hydrothermal, cross-linking, ultrasonication methods. These hydrogels are currently being exploited for multiple applications, including therapeutic agents, adhesives, adsorbents and other miscellaneous applications, such as conductive carbon microfibers, supercapacitors and biosensors. Although the applications of lignin-derived hydrogels are still emerging, the employment of lignin in hydrogel preparation has come up as a green and sustainable alternative against harsh chemical-based hydrogels.

Keywords Lignin · Hydrogels · Therapeutic · Adhesives · Adsorbents

---

S. Chandna · S. K. Bhardwaj · S. Paul · J. Bhaumik (✉)
Department of Nanomaterials and Application Technology, Center of Innovative and Applied Bioprocessing (CIAB), Sector 81 (Knowledge City), S.A.S. Nagar, Punjab 140306, India
e-mail: jayeeta@ciab.res.in

S. Chandna
Department of Microbial Biotechnology, Panjab University, South Campus, Sector 25, Chandigarh 160036, India

S. Paul
Regional Centre for Biotechnology, Faridabad, India

S. Sharma and A. Kumar (eds.), *Lignin*, Springer Series on Polymer and Composite Materials, https://doi.org/10.1007/978-3-030-40663-9_8

1 Introduction

Lignin is one of the most abundant biopolymers derived from plants, which is considered as major waste material from the paper and pulp industry. It is present in the cell walls of plants, providing them chemical and mechanical protection (Chandna et al. 2019; Larrañeta et al. 2018). It has the potential to replace the conventional synthetic materials due to its industrially important properties such as adhesiveness, stability and antibacterial activity (Kaur et al. 2020). Lignin is an aromatic heteropolymer possessing a network of multiple reactive and randomly cross-linked functional groups including methoxy and reductive phenolic hydroxyls groups. These functional groups are also possibly responsible for their antioxidant, antimicrobial and UV-protection properties of lignin (Gan et al. 2019). Recently, use of lignin for water purification, as binder in lithium batteries and as a biomedical material has also been reported (Domínguez-Robles et al. 2017; Dong et al. 2011; Kai et al. 2015, 2016). Despite all these important properties, less than 2% of the total lignin produced by paper industry is reutilized for specialty products, such as vanillin, catechols, fuels, etc. while the remainder being used as a low grade burning fuel or considered as waste (Kai et al. 2016; Raschip et al. 2013).

Hydrogels are three-dimensional highly hydrated networks of natural or synthetic polymeric chains, which have enormous potential for various applications. They have become a research hotspot in the recent years due to their application-oriented interesting properties, such as hydrophilicity, resilience, close resemblance to the natural extracellular matrix (ECM) and soft consistency (Fonseca Silva et al. 2012; Gao et al. 2015; Jin et al. 2018; Passauer et al. 2015). Hydrogels are normally formed of hydrophilic polymers, which become water-insoluble due to cross-linking. However, despite being insoluble polymer networks, hydrogels are competent of absorbing and retaining a large amount of solvent, which allows them to acquire the shape of their surroundings upon confining (Zerpa et al. 2018). This phenomenon is caused due to surface absorption, physical filling of the open pores of the cross-linked polymer by capillary pressure and osmotic pressure (Zerpa et al. 2018). Due to these interesting properties possessed by hydrogels, they can be used in different applications such as therapeutic agents and scavengers, which can lead to the development of industrially relevant products.

2 Lignin as a Potential Material for Hydrogel Development

Approximately, 50 million tons of lignin is generated as a by-product from paper pulp industries every year (Graça et al. 2012). In recent times, this heteropolymer has been valorized to different products even in nanoscale (Chandna et al. 2019; Shen et al. 2014). Lignin possesses biocompatibility, biodegradability, non-toxicity and promising thermal properties, which makes it a promising candidate for the synthesis of hydrogels for their applications in versatile sectors (Ciolacu and Cazacu 2018; Raschip et al. 2013).

Hydrogels can be derived from both synthetic and natural biopolymers (Hasan and Fatehi 2018; Raschip et al. 2011; Zerpa et al. 2018). The commercially available hydrogels are mainly made up of polyacrylic acids (PAA), polyvinyl alcohol (PVA) and polyvinyl pyrrolidones (PVP) for biomedical or technical applications. In current years, hydrogels with superabsorbent properties are being developed by grafting of polymerizable ethylene type monomers onto polysaccharides. These superabsorbent hydrogels can be used in various applications ranging from biomedicine to supercapacitors (Passauer 2012; Peng et al. 2018). Mostly stable and swellable hydrogels are obtained by cross-linking between various groups by thermal or electron beam-induced cross-linking (Schupp et al. 2019). In some cases, thermo-irreversible hydrogels have been formed arising from poly(ethylene glycol)-based blocks having injectable and moldable properties (Hu et al. 2018). Recently, biopolymer-derived hydrogels have gained substantial research consideration (Jin et al. 2010; Qu et al. 2010). These biopolymers such as lignin, cellulose, xylan and chitosan can be potentially used to prepare multifunctional hydrogels with good market value (Agnihotri et al. 2012; Fonseca Silva et al. 2012; Peng et al. 2018).

Hydrogel development has been shifted from the utilization of synthetic polymers to the natural polymers in modified form, due to low toxicity, biodegradability, biocompatibility, eco-friendly properties and susceptibility to enzymatic degradation of the later ones (Feng et al. 2017; Figueiredo et al. 2018; Kalinoski and Shi 2019; Yamamoto et al. 1999; Yao et al. 2014). Therefore, production of biorenewable polymer-based hydrogels has come up as a green and sustainable alternative against harsh chemical-based hydrogels (Meng et al. 2019). As such, lignin-based hydrogels can be promising biomaterials for hydrogel development due to the aforementioned characteristics of this biomolecule (Jiang et al. 2018; Thakur et al. 2017). In the last decade, lignin has been used in different types of hydrogels to facilitate their preparation and to advance their properties (Ciolacu et al. 2013; Feng et al. 2014; Grishechko et al. 2013; Ji et al. 2017; Jin et al. 2018; Nakasone and Kobayashi 2016; Sathawong et al. 2018; Teng et al. 2017). Hydrogels have been used to produce wound dressings, contact lenses, adhesives, adsorbents and for tissue engineering as well as drug delivery (Caló and Khutoryanskiy 2015; Hunt et al. 2014; Osti 2006; Vashist et al. 2014; Wichterle and Lím 1960).

3 Synthetic Procedures for Lignin-Derived Hydrogels

Lignin-derived hydrogels have been synthesized through a variety of methods, such as hydrothermal methods, ultrasonic polymerization, wet spinning, ultrasound and various cross-linking methods, such as esterification reaction, copolymerization with other polymers such as acrylic acid (Fig. 1) (Feng et al. 2012; Park et al. 2015a, b; Passauer et al. 2011; Shen et al. 2016; Zmejkoski et al. 2018). Synthetic techniques to prepare lignin-derived hydrogels are described below.

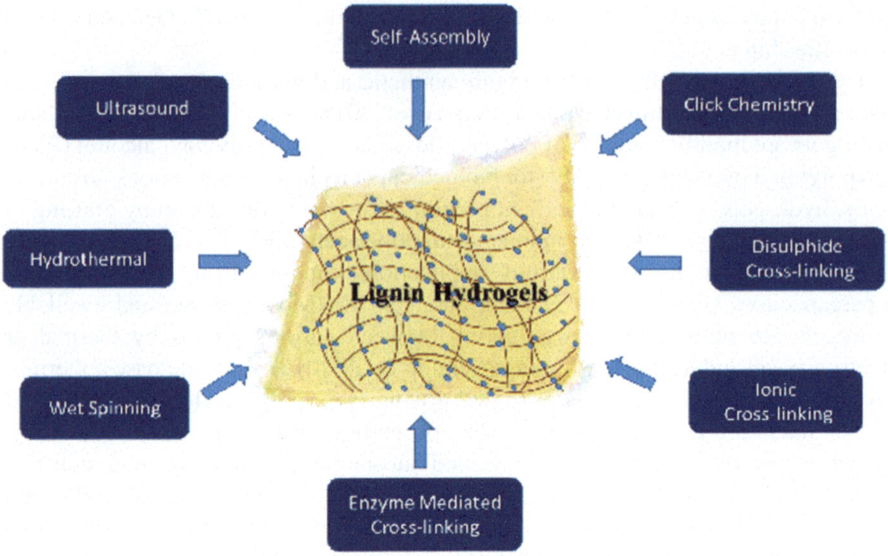

Fig. 1 Different approaches to the synthesis of lignin hydrogels

3.1 Cross-Linking Method

Cross-linking is one of the most prevalent methods to synthesize lignin-derived hydrogels by using various cross-linking agents. In a study, chemical cross-linking method was followed, in which epichlorohydrin (ECH) and poly(ethylene glycol) PEG were used in different concentrations for cross-linking with lignin (Shen et al. 2016). They have also designed a xylan-containing lignin/epoxide-terminated PEG (ETPEG)-based hydrogels by cross-linking xylan and lignin with ETPEG. The cross-linking properties of hydroxyl (OH) functional groups present in the lignin were also investigated. (Li et al. 2018). Glycerol diglycidyl ether (GDE), being non-toxic in nature, was found to be the most effective cross-linker for lignin polymerization to develop lignin–epoxy resins. These resins can find applications as wood adhesives. In another study, lignin hydrogels were synthesized by using gantrez (GAN) as a cross-linker (Larrañeta et al. 2018). The presence of excessive alcohol groups in lignin makes it suitable for cross-linking with GAN. PEG was also used as a plasticizer and to maximize cross-linking, leading to the development of hydrogels. Notably, a hydrophilic polyether-based polyurethane (HPU) cross-linker was utilized for the synthesis of lignin hydrogels, which enhanced their mechanical properties as well as toughness due to hydrogen bonding (Oveissi et al. 2018).

Lignin-based hydrogel can also be synthesized by combining lignin polymer on the matrix of the polyacrylic acid network (Sun et al. 2016a, b). For this purpose, lignin and acrylic acid were cross-linked in the presence of *N*,*N*′-methylenebisacrylamide (MBAm) while ammonium persulfate (APS) acted as an initiator for the polymerization.

3.2 ATRP and RAFT Methods

Lignin-derived hydrogels with controlled structures have been synthesized by using atom transfer radical polymerization (ATRP) and reversible addition-fragmentation chain transfer (RAFT) methods (Chung et al. 2013; Liu and Chung 2017). ATRP is a 'graft from' and RAFT polymerization is a 'graft onto' method to fabricate hydrogels in an aqueous or ionic liquid medium. The mechanisms of ATRP and RAFT methods were utilized to develop active sites onto the backbone of lignin polymers. In the case of ATRP method, lignin-based organic halide acted as an initiator to react with acrylonitrile monomers (Chung et al. 2013). In RAFT method, lignin served as a core and azobisisobutyronitrile (AIBN) as an initiator for polymerization of lignin core through radical reaction (Liu and Chung 2017).

3.3 Copolymerization Method

Lignin-derived hydrogels have also been prepared by copolymerizing lignin with various other polymers by using enzyme-based initiators, cross-linkers or techniques such as graft copolymerization. It has been found that polymers, such as acrylic acid, can be copolymerized with lignosulfonates to synthesize a series of lignosulfonate-*g*-acrylic acid (LS-*g*-AA) hydrogels in the presence of laccase/t-BHP as an initiator and N,N'-methylene-bis-acrylamide (MBA) as a cross-linker, as demonstrated in a study by Yu et al. (2016).

In another study, green synthesis of lignin-based hydrogels was accomplished by using graft copolymerization technique in the presence of acrylamide monomer and poly(vinyl alcohol) (PVA) along with alkaline or kraft lignin, followed by mixing with acrylamide monomer. In this method, the prepared hydrogel showed high swelling properties (El-Zawawy 2005).

3.4 Ultrasonication Method

Ultrasonication method follows the principle of cavitation induced polymerization reaction for the synthesis of hydrogels. Ultrasonication leads to the generation of a compression cycle of ultrasonic wave followed by the rarefaction cycle. Therefore, when ultrasonic waves pass through the reaction media, a large number of cavitation bubbles are generated which mediates the synthesis of hydrogels. A method of ultrasound synthesis has been employed in preparing pH-sensitive sodium lignosulphonate-grafted poly(acrylic acid-co-poly(vinyl pyrrolidone)) hydrogel (Wang et al. 2016).

3.5 Hydrothermal Method

Hydrothermal method owing to its ease of operation is a major choice for the synthesis of hydrogels (Xu et al. 2010). A simple hydrothermal treatment was used to produce pressure-sensitive lignosulfonate/single-walled carbon nanotube-HNO_3 (Lig/SWCNTHNO_3) hydrogel and cellulose-based hydrogel via phase-inversion method (Peng et al. 2018). In this procedure, carboxyl and hydroxyl groups of lignin were treated with nitric acid followed by dispersion in SWCNTs. During the hydrothermal process of hydrogel synthesis, Lig/SWCNTHNO_3 3D network structure was attained within a trace of nitric acid, which resulted due to π-π stacking and hydrogen bonding between lignin and SWCNTs.

3.6 Wet Spinning Method

Wet spinning, which is an eco-friendly process, has been successfully applied for microfiber hydrogel production (Wang et al. 2019). In this technique, hydrogel synthesis takes place in aqueous calcium chloride solution, which results in fast aggregation of the hydrogel precursors comprised of wood-derived lignin and TEMPO-oxidized cellulose nanofibrils (TNF). Wang and coworkers developed carbon microfibers by carrying out the wet spinning of bicomponent precursors containing lignin and cellulose nanofibrils, which was followed by one-step carbonization at 900 °C (Wang et al. 2019).

A summary of synthesis, characterization and applications of lignin-derived hydrogels is described in Table 1.

4 Applications of Lignin-Derived Hydrogels

Hydrogels have similar physical characteristics as soft biological tissues, which allow them to accommodate a large volume of water via capillary effect or surface tension. This key feature makes them more important in a range of fields varying from academic research to industrial applications (Appel et al. 2012; Babu et al. 2014; Choi et al. 2013; Li et al. 2017; Rudnitskaya et al. 2013; Rybtchinski 2011; Zhang and Ma 2013). Key applications of lignin-derived hydrogels including therapeutics, adhesives, adsorbents, coating materials, conductive carbon microfibers, supercapacitors and biosensors are described in Fig. 2.

Table 1 Summary of synthesis routes, methods of characterization and applications of lignin-based hydrogels

S. No.	Components of hydrogel	Synthetic procedure	Characterization	Applications	References
1	Lignin/SWCNTHNO_3	Hydrothermal	SEM, TEM, BET, XRD, XPS, FTIR	Flexible supercapacitors (FSCs)	Peng et al. (2018)
2	Lignin with poly(ethylene glycol), poly(methyl vinyl ether-co-maleic acid)	Microwave irradiation	FTIR, SEM, ^{13}C NMR	Drug-eluting hydrogels	Larrañeta et al. (2018)
3	Alkaline/kraft lignin, acrylamide (AM) and poly(vinyl alcohol) (PVA)	Copolymerization	UV-vis, SEM, FTIR	Absorbent	El Zawawy et al. (2005)
4	Acrylic acid (AA), lignosulfonate (LS) and *N,N′*-methylene -*bis*-acrylamide	Free radical polymerization	FTIR	Superabsorbent	Yu et al. (2016)
5	Lignin, hemicellulose and poly(acrylic acid)	Free radical polymerization reaction	FTIR, SEM, ^{1}H NMR, UV-Vis, Thermogravimetric method	Adsorption of dyes	Song et al. (2016)
6	Lignin and 2,2,6,6-tetra-methylpiperidine-1-oxyl (TEMPO)-oxidized cellulose nanofibrils (TOCNF)	Wet spinning	SEM, Rheology, Wide-angle X-ray scattering (WAXS), TGA, Raman	Conductive carbon microfibers	Wang et al. (2019)
7	Sodium lignosulfonate and poly (acrylic acid-*co*-poly(vinyl pyrrolidone))	Ultrasonication	FTIR, SEM, TGA	Drug delivery	Wang et al. (2016)
8	Ag-lignin nanoparticles, pectin and polyacrylic acid	Redox catechol chemistry	Electron spin-resonance spectroscopy, SEM, TEM	Adhesive	Gan et al. (2019)

(continued)

Table 1 (continued)

S. No.	Components of hydrogel	Synthetic procedure	Characterization	Applications	References
9	Lignin model dehydrogenate polymer (DHP), alginate (Alg)	Suspension	Antimicrobial	Wound healing	Spasojević et al. (2016)
10	Bacterial cellulose, lignin (dehydrogenative polymer of coniferyl alcohol)	Suspension	HPLC-MS, SEM, FTIR, AFM	Wound healing	Zmejkoski et al. (2018)
11	Kraft lignin and glycerol	Cross-linking	TGA, FTIR	Adhesive	Li et al. (2018)
12	IL-lignin/epoxide-terminated poly(ethylene glycol) [ETPEG]	Cross-linking	SEM, water vapor transmission (WVT)	Food packaging or wound dressings	Shen et al. (2016)
13	Lignin and hydrophilic polyether-based polyurethane (HPU)	Cross-linking	Rheology	Fabricating applications, such as fiber spinning, casting and 3D printing	Oveissi et al. (2018)
14	Lignin, starch and acrylamide	Cross-linking	SEM, FTIR	Adsorbent	Peñaranda-Avila and Sabino 2010
15	Acetic acid lignin and NCO-terminated polyurethane ionomers	Cross-linking	TGA, FTIR	Coating material	Peng and Chen (2011)
16	Lignosulfonate and graphene	Oxidation	AFM, EDX, SEM, BET, XPS, TGA, FTIR, XRD	Adsorbent	Li et al. (2017)
17	Lignin and silicate	Self-assembly	Cyclic voltammetry and DC potential amperometry (DCPA), FTIR	Biosensing	Burrs et al. (2013)

(continued)

Table 1 (continued)

S. No.	Components of hydrogel	Synthetic procedure	Characterization	Applications	References
18	Lignin, cellulose and xylan	Stirring	SEM, pH stability, thermal stability, UV-vis	Enzyme immobilization	Park et al. (2015a, b)
19	Lignin and cellulose	Stirring	FTIR, UV-vis, pH stability, thermal stability	Enzyme immobilization	Borisenkov et al. (2016)
20	Fractionated kraft lignin and PVA	Freeze-thawing	TGA, SEM	Tissue engineering	Rejmontová et al. (2019)

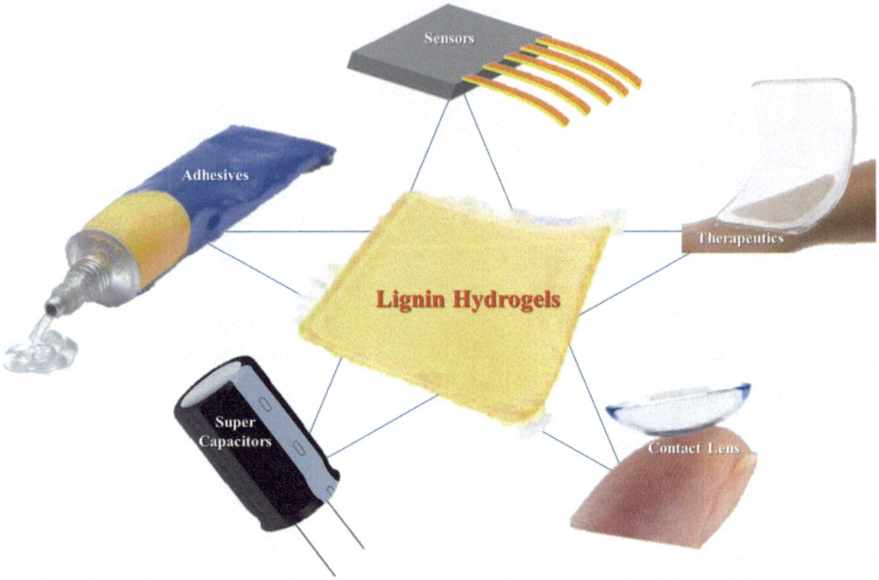

Fig. 2 Applications of lignin-derived hydrogels in various fields

4.1 Therapeutic Agents

Microbial infections severely restrict the use of healthcare materials due to the emergence of drug-resistant microbes. In this respect, hydrogels demonstrate tunable and reversible behavior of their supramolecular interactions, in addition to the capability to incorporate therapeutic agents (Farhat et al. 2017). These properties make hydrogels a potential candidate to achieve efficient antimicrobial therapy (Hu et al. 2018).

In a study, curcumin (a model hydrophobic drug) was loaded in lignin-derived hydrogels and was sustainably delivered for up to four days (Larrañeta et al. 2018). Notably, those hydrogels displayed water uptake capacities up to 500% with lignin contents ranging between 40 and 24%. On in vitro microbiological assessment, it was found that the antimicrobial activity of the lignin-based hydrogels against *Staphylococcus aureus* and *Proteus mirabilis* was commendable since a logarithmic reduction in the adherence of microbes by orders of 5.0 was observed compared to poly(vinyl chloride) (PVC), which is a commonly employed medical material.

In another study, the lignin model polymer dehydrogenate polymer (DHP) in alginate (ALG) hydrogel (Alg) was synthesized and its antimicrobial and antibiofilm activity was evaluated against various bacterial strains (Spasojević et al. 2016). It was found that DHP–Alg has strong antibacterial potential against several bacterial strains, i.e., Gram-negative and Gram-positive bacteria and there was no toxic effect of biofilms on human epithelial cells. The obtained results demonstrated that the

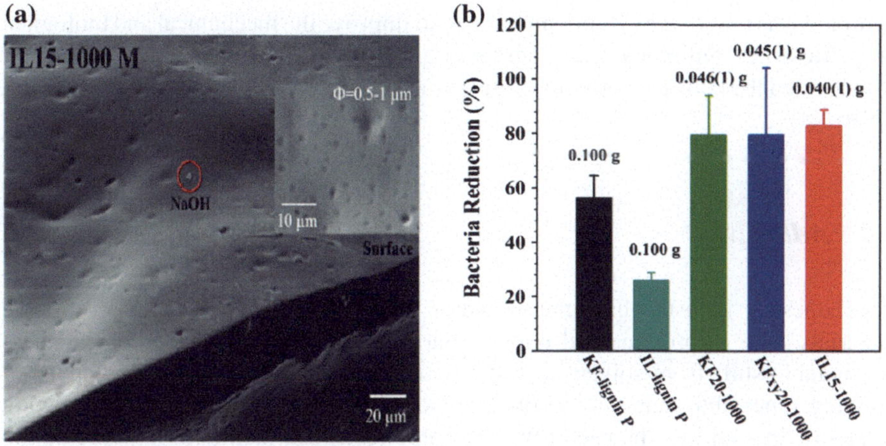

Fig. 3 **a** SEM images of lignin/ETPEG1000 hydrogel membranes after air-drying with lignin; **b** Antimicrobial activity of lignin powder (P) and lignin/ETPEG hydrogel membranes (9 mm diameter) determined using the shake flask method. The numbers in the figure represent the dry masses of the samples used, KFxy represents KF-lignin-xylan; reproduced with permission from Ref. Shen et al. (2016), Royal Society of Chemistry

prepared lignin model polymer DHP-alginate hydrogels could be used as wound healing agents.

Hydrogels with excellent water absorbency were derived from epoxide-terminated PEG (ETPEG). Figure 3a demonstrates SEM images of membranes developed from lignin/ETPEG1000 hydrogel after air-drying (Shen et al. 2016). Figure 3b represents the antimicrobial activity of the lignin powder (P) and the prepared lignin/ETPEG hydrogel membranes. The developed membranes were found to possess good antioxidant, antimicrobial and water vapor transmission, indicating their potential to be used as a wound dressing and food packaging materials.

Zmejkoski et al. have shown that bacterial cellulose–lignin composite hydrogel can be utilized as a potential agent in chronic wound healing (Zmejkoski et al. 2018). The developed composite designed as a dressing which displayed bactericidal effect against laboratory strains and biofilm-forming bacteria. It was observed that during the first 24 h, the highest amount of the active substance, i.e., dehydrogenative polymer of coniferyl alcohol (DHP) was released, keeping the constant elevated concentration of the antimicrobial compounds for the next 72 h. Furthermore, the synthesized hydrogel exhibited high absorbing capacity confirming it's applicability in chronic wound healing.

In a similar study, a pH-sensitive lignosulfonate-based hydrogel was synthesized to incorporate drug and check its release performance (Wang et al. 2016). The researchers used amoxicillin as a model drug and studied its release properties in enzyme-free simulated gastrointestinal fluids. Interestingly, the prepared hydrogel exhibited good pH responsivity, high water swelling property and sustained release performance in the intestine. Similarly, a fractionated kraft lignin-derived PVA hydrogel was reported by Rejmontová et al. to improve the mechanical and biological properties (i.e., antioxidant and antimicrobial activities) of PVA-based scaffolds, which aided the cells to grow and survive within its internal pores (Rejmontová et al. 2019).

4.2 Adhesives

The adhesive industry has manifested a growing interest in natural adhesive materials such as proteins and polysaccharides (Heinrich 2019). Moreover, there is a notable shift from solvent to water-based adhesives. One of the incentives of utilizing renewable materials to produce adhesives is moving towards a growing economy. In addition, the use of biorenewable feedstock facilitates in reducing the carbon footprint. Moreover, renewable materials such as lignin, cellulose, starch or polyhydroxyalkanoates have higher intrinsic biodegradability than synthetic materials such as polyethylene and polypropylene (Heinrich 2019). Formaldehyde-containing condensation resins are still used as wood adhesives both for solid wood and wood composites. Therefore, another reason that shifts the search for substitute adhesive solutions is the recent categorization of formaldehyde as a harmful substance. All these factors have led to acceleration in research to develop renewable and greener alternatives to the existing chemically derived adhesives. To develop efficient adhesives, physical cross-linking of polymers is required, for which natural biopolymer lignin can be utilized as it is a low-cost material with numerous polar sites on its backbone. In a study by Li and coworkers, they have reported a formaldehyde-free adhesive formulation, which comprised of kraft lignin (paper and pulp) and glycerol (biodiesel) (Li et al. 2018). The adhesion potential of the formulation was elucidated by wood failure rates and ultimate shear stress. The lignin-derived resin revealed similar characteristics as formaldehyde-based resins, such as fast curing and water-tolerance. Moreover, unlike most of the formaldehyde-free adhesives, the lignin–epoxy formulation did not require the utilization of a caustic base or any known carcinogens. Since lignin can be efficiently used for the preparation of adhesives, lignin-based hydrogels are also being engineered for their potential application as adhesives.

Recently, a catechol chemistry-based plant-inspired hydrogel has been prepared, which was reported to possess long-term adhesiveness and high toughness (Gan et al. 2019). The hydrogel was prepared from an aqueous precursor solution containing pectin, acrylic acid and Ag-lignin nanoparticles under an ambient environment. The hydrogel was found to possess excellent mechanical properties due to the interpenetrating network formed by pectin and polyacrylic acid (PAA) with multi-cross-linking of non-covalent and covalent bonds (Fig. 4). The continuous generation of catechol groups in the inner hydrogel network by Ag-lignin nanoparticles provided repeatable and long-term adhesiveness to the hydrogels.

In another study, lignin was used as an additional cross-linking agent in hydrophilic polyether-based polyurethane hydrogels, which significantly improved

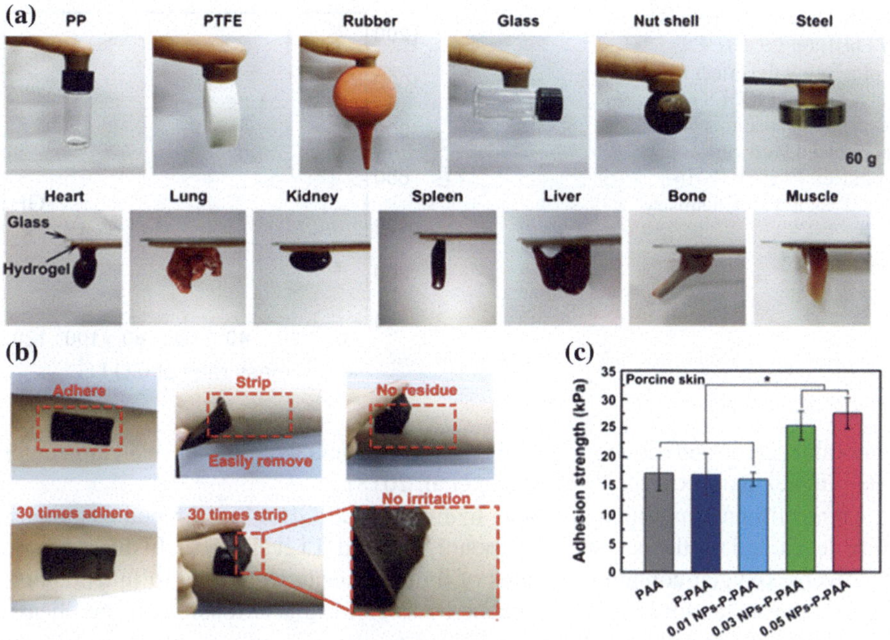

Fig. 4 Adhesive properties of the NPs-P-PAA hydrogel. **a** The hydrogel adhered to various material surfaces and tissues. **b** The hydrogel repeatedly adhered on the skin of the author. After peeling off, no residue or irritation on the skin was found. **c** The adhesive strength of various hydrogels to porcine skin. The numbers in the figure represent the content of the nanoparticles (gm L^{-1}); reproduced with permission from Ref. Gan et al. (2019), Springer Nature

their mechanical properties as well as processability (Oveissi et al. 2018). Addition of 2.5 wt% lignin significantly increased Young's modulus, fracture energy and lap shear adhesiveness of the prepared hydrogels. It was also found that the hydrophilic polyether-based polyurethane hydrogels were processable by a range of techniques, such as fiber spinning and 3D printing.

4.3 *Adsorbents*

Three-dimensional (3D) porous materials with high surface area and interconnected pores are promising candidates for capture and adsorption of pollutants and small molecules (Huang et al. 2012; Li et al. 2016; Parajuli et al. 2006; Parajuli et al. 2005; Song et al. 2016; Sun et al. 2016a, b). Lignin-derived materials offer various oxygen-containing groups such as carbonyl, carboxyl, hydroxyl and methoxyl groups, acting as adsorbing sites, which are ideal for developing low-cost adsorbents (Ma et al. 2017; Rajan et al. 2018; Wang et al. 2017). Therefore, efficient utilization of numerous significant binding sites offered by lignin derivatives is essential to develop

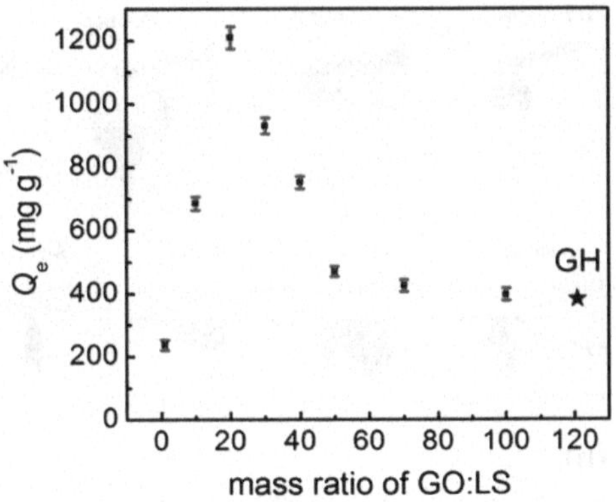

Fig. 5 Adsorption behaviors of Pb(II) on LS-GH representing the effect of the mass ratio of GO: LS on adsorption capacities; reproduced with permission from Ref. Li et al. (2016), Royal Society of Chemistry

renewable, green and economic adsorbents for toxic metals and other small molecules (Domínguez-Robles et al. 2018; Zhang et al. 2018).

Lignosulfonate-modified graphene hydrogels (LS-GHs) based metal adsorbents were developed via the one-step eco-friendly method (Li et al. 2016). In this network, the diverse oxygen-containing groups of lignosulfonate were used to functionalize the 3-D porous architecture of graphene hydrogel. Lignosulfonate adsorbed heavy metal ions on adequate active sites provided by the network and graphene offered multileveled porous skeleton. Different mass ratios of graphene oxide (GO) and lignosulfonate (LS) were utilized to check the adsorption behaviors of Pb(II) on lignosulfonate-modified graphene hydrogels (Fig. 5). The synthesized (LS-GHs) can be used as a column-packing device, due to their flexible and free-standing nature. The findings provided a competitive pathway for fast and efficient removal of Pb(II) with very high adsorption capacity.

Lignosulfonate-based hydrogels were synthesized with superabsorbent properties. In this material, there was enhanced interaction between the cationic dyes and lignin-based absorbent hydrogel due to the presence of carboxylic acid moieties in the lignosulfonate backbone (Yu et al. 2016). The superabsorbent properties of the prepared hydrogels were evaluated by targeting their removal efficiency towards methylene blue. Overall, this study demonstrated a new approach to manufacture lignin-derived hydrogel-based superabsorbents for the absorption and removal of wastewater contaminants.

Peñaranda and Sabino used interpenetrated lignin and peat inside acrylamide–starch matrix to develop IPN hydrogels (Peñaranda-Avila and Sabino 2010). The hydrogels prepared by the researchers were characterized and their absorption capacity was measured against metal ions, i.e., nickel (Ni^{2+}) and copper (Cu^{2+}) present in contaminated water. Wang et al. used ultrasonic polymerization to synthesize novel superabsorbent hydrogels based on sodium lignosulfonate (SLS) (Wang et al. 2017). Due to the properties such as favorable adsorption capacity and low cost, the SLS-hydrogel exhibited promising adsorbent properties to remove Ni(II) from aqueous solution. Different experimental conditions which impact the adsorption capacity of the prepared hydrogels were investigated. The experimental conditions included variation in optimization of temperature, pH, contact time and initial concentration of the Ni^{2+} ions in the solution. The adsorption behavior was found to be endothermic and spontaneous.

Similarly, a lignin-containing hemicellulose-based hydrogel was prepared for adsorption of dyes contaminants (Song et al. 2016). The adsorption behavior of the hydrogel was checked toward the removal of methylene blue, a model dye. Both the swelling ratio and the adsorption capacity were dependent on sodium lignosulfonate dosage. It was found that the developed hydrogel exhibited approximately 80% recycle efficiency after four cycles.

4.4 *Conductive Carbon Microfibers*

Carbon fibers are materials of high interest the demand is increasing annually with the rate of 10% which is expected to touch 89,000 tons by 2020 (Wang et al. 2019). Such a high demand has been fulfilled by petroleum-based precursors, namely polyacrylonitrile (PAN) and mesoporous pitch, accounting for 90% of the market requirements. The major bottleneck in carbon fiber production is excessive dependence on fossil precursors as well as processing costs. As such, alternative sources are highly required owing to the high and fluctuating cost of petroleum-based precursors along with the risk of shortage in long-term supplies. This leads to critical urgency for converting renewable materials to carbon fibers in the near future. In this respect, relatively inexpensive, renewable and abundant precursors such as lignin and cellulose can be considered as potential alternative precursors to polyacrylonitrile and mesoporous pitch. In this context, Wang and group have synthesized carbon microfibers comprising lignin and cellulose nanofibrils (Wang et al. 2019). Lignin was loaded up to 70 weight% to produce filaments from the respective hydrogel dopes, without the use of any harsh conditions, such as high temperature and strong or volatile antisolvents. The morphology and properties of the bicomponent microfibers such as mechanical strength, fibril orientation and thermal stability were significantly controlled by lignin. In addition, no stabilization treatment was given to carbonize the microfibers, and circular cross sections without any signs of inter-fiber fusion were preserved in the obtained carbon microfibers. The study demonstrated that abundant lignin can be efficiently processed and used to develop efficient conductive carbon microfibers and other conducting materials (Wang et al. 2018, 2019).

4.5 *Supercapacitors*

Flexible supercapacitors (FSCs) are potential candidates for energy storage devices in wearable/flexible electronics due to their attractive characteristics, such as high power density, extraordinary mechanical flexibility which endures

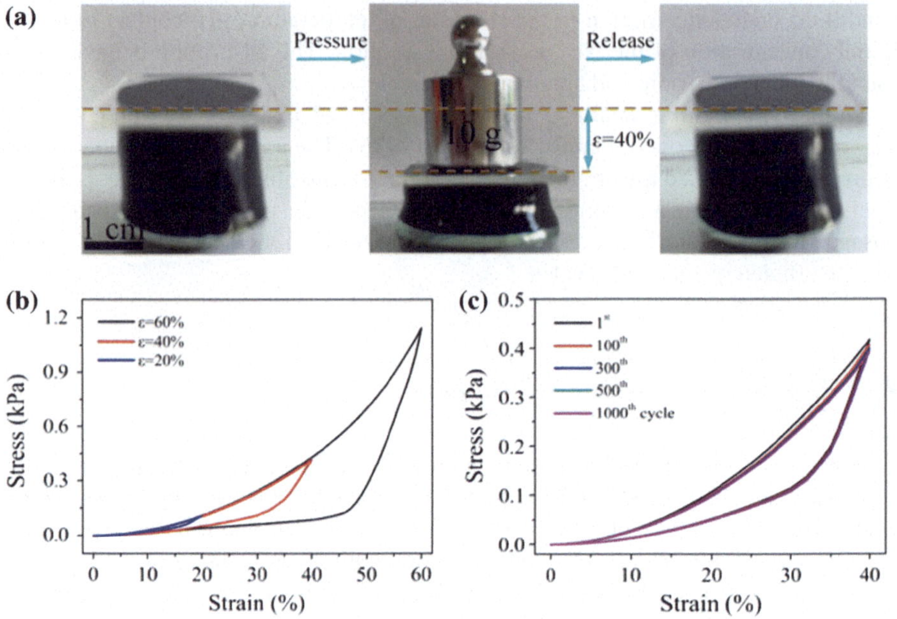

Fig. 6 **a** The image of compression-recovery processes for Lig/SWCNTHNO$_3$ hydrogel. The compressive cyclic stress-strain curves of the Lig/SWCNTHNO$_3$ hydrogel with **b** different strains and **c** different cycles; reproduced with permission from Ref. Peng et al. (2018), American Chemical Society

bending, folding or twisting and ultra-long life cycle (Peng et al. 2018). To develop high-performance flexible electrode materials, remarkable efforts have been dedicated. Recently, sustainable, low-cost and environmentally friendly lignin-derived composite hydrogel electrodes derived FSCs have been achieved for the first time. High specific capacitance, outstanding energy density and excellent rate capability were possessed by the assembled symmetric FSCs. Furthermore, in 1000 bending cycles for 90° angle, 98% of the capacitance was retained in the developed FSCs showing their exceptional electrochemical stability (Fig. 6) (Peng et al. 2018). This study opened a pathway for the utilization of biomass-based materials to design biocompatible and environment friendly FSCs and provides important opportunities for the progress of sustainable energy storage systems.

4.6 Sustained Release Materials

Lignin-derived hydrogels can also be used to develop tough coating materials by cross-linking lignin with other strong and biocompatible polymers, such as polyurethane. Such materials can be used for controlled release of various agents, such as pesticides, fertilizers and aroma compounds (Appel et al. 2012). In a study conducted by Peng et al., acetic acid lignin (AAL) was chemically cross-linked with isothiocyanate(NCO)-terminated polyurethane ionomers (IPUI) to develop hydrogels (Peng and Chen 2011). It was observed that the developed hydrogels were pH-sensitive as the swelling ratio increased linearly with the pH. The introduction of lignin in the hydrogels improved the thermal stability, as revealed by the results of the thermogravimetric analysis. The prepared hydrogels were used as coating materials for ammonium sulfate for the development of the slow-release fertilizers. Release experiments for ammonium sulfate demonstrated that the prepared hydrogels can be used as potential coating materials for developing slow-release fertilizers.

In another study, lignin and polyacrylic acid were used to prepare a novel composite porous hydrogel (LBPAA) (Sun et al. 2016a, b). The developed hydrogel was found to be porous, to facilitate slow release of pesticides. The LBPAA was examined for its properties, such as water swelling and slow release of entrapped pesticides. The porous LBPAA hydrogel was shown to demonstrate improvised properties when compared to polyacrylic acid. Hence, this type of material can be considered as a potential candidate for utilization in controlled release systems in the agricultural sector.

4.7 Biosensors

Simple, cost-effective and user-friendly nanomaterial-based sensing membranes which are easy to develop in laboratories have always been in demand (Burrs et al. 2013). Lignin-derived hydrogels doped with nanomaterials can be potentially used in sensing applications. The processes used for lignin extraction, such as pulping or de-lignification, determine the sensing properties of lignin-based materials by defining the resulting lignin molecular weight, structure and the available functional groups. In a study by Burrs et al., they have demonstrated the use of lignin, silicate and cellulose (derived from waste pulp sludge) to prepare hydrogels (Burrs et al. 2013). The combination of these prepared hydrogels with graphene-based nanocomposites was utilized to form a biocompatible film on platinum electrodes for developing potentiometric electrochemical biosensors.

4.8 Enzyme Immobilization

Lignocellulosic biomass-based hydrogels can be used to immobilize various enzymes such as lipase, cellulase, amylase, etc. for their potential applications in the field of biocatalysis to generate industrially relevant products. Lipase has many industrial applications, such as in the fields of food, pharmaceuticals, bioremediation and cosmetics. Park et al. used lipase from *Candida rugosa*, for entrapment in beads made of wood mimetic composite hydrogels, which were composed of cellulose and lignin (Park et al. 2015a, b). The stability and activity of the entrapped lipase were also investigated as a function of composition and source of each component in the cellulose/lignin composite hydrogel. Higher activity and stability were found in the lipase immobilized on cellulose/lignin beads than that of the lipase which was immobilized on pure cellulose beads. Interestingly, lignin content in the cellulose/lignin beads was found to affect the stability and activity of the immobilized lipase in an incremental manner. A similar study was done by Borisenkov et al., to immobilize beta-glucuronidase and estrogen inside the matrix of pectin/lignin hydrogel (Borisenkov et al. 2016).

5 Conclusion

To summarize, lignin-derived hydrogels are an upcoming material of interest owing to the abundance of lignin in nature and its various inherent properties, such as biodegradability, adhesive nature, presence of numerous functional groups and so on. Syntheses of lignin-derived hydrogels were achieved by different techniques, namely ultrasonication, polymerization, hydrothermal method, etc. Hydrogels prepared using lignin have been demonstrated to find utility in numerous applications, including, therapeutic agents, adhesives, adsorbents, biosensors, etc. It has been observed that various materials can be accompanied by lignin, such as graphene or cellulose to develop lignin-derived composite hydrogels with enhanced activities, such as mechanical strength, swelling and adsorbent properties. The lignin-derived hydrogels have shown to possess high water absorbency, antimicrobial activity and good mechanical properties. Therefore, it can be concluded that lignin can serve as a potential material to develop green hydrogels, which can find applications in food, pharmaceutical and other industries.

Acknowledgements J.B. is grateful to the Department of Science and Technology for funding support via ECRA scheme. S.C. is thankful to Indian Council of Medical Research for the senior research fellowship and Department of Microbial Biotechnology, Punjab University, Chandigarh for Ph.D. registration. S.P. acknowledges junior research fellowship support from CSIR and she is registered in the Ph.D. program of Regional Center for Biotechnology (DBT) at CIAB. For postdoctoral fellowship, S.K.B. is grateful to CIAB, DBT.

References

Agnihotri S, Mukherji S, Mukherji S (2012) Antimicrobial chitosan–PVA hydrogel as a nanoreactor and immobilizing matrix for silver nanoparticles. Appl Nanosci 2(3):179–188

Appel EA, del Barrio J, Loh XJ, Scherman OA (2012) Supramolecular polymeric hydrogels. Chem Soc Rev 41(18):6195–6214

Babu SS, Praveen VK, Ajayaghosh A (2014) Functional π-Gelators and their applications. Chem Rev 114(4):1973–2129

Borisenkov MF, Karmanov AP, Kocheva LS, Markov PA, Istomina EI, Bakutova LA, Litvinets SG, Martinson EA, Durnev EA, Vityazev FV, Popov SV (2016) Adsorption of β-glucuronidase and estrogens on pectin/lignin hydrogel particles. Int J Polym Mater Polym Biomater 65(9):433–441
Burrs SL, Jairam S, Vanegas D, Tong Z, McLamore E (2013) Lignin and silicate based hydrogels for biosensor applications. Progr Biomed Opt Imag Proc SPIE 8719:87190H. https://doi.org/10.1117/12.2016285
Caló E, Khutoryanskiy VV (2015) Biomedical applications of hydrogels: a review of patents and commercial products. Eur Polymer J 65:252–267
Chandna S, Thakur NS, Reddy YN, Kaur R, Bhaumik J (2019) Engineering lignin stabilized bimetallic nanocomplexes: structure, mechanistic elucidation, antioxidant and antimicrobial potential. ACS Biomateri Sci Eng 5(7):3212–3227
Choi M, Choi JW, Kim S, Nizamoglu S, Hahn SK, Yun SH (2013) Light-guiding hydrogels for cell-based sensing and optogenetic synthesis in vivo. Nat Photonics 7:987
Chung H, Al-Khouja A, Washburn NR (2013) Lignin-based graft copolymers *via* ATRP and Click Chemistry, Green Polymer Chemistry: biocatalysis and materials II. American Chemical Society, pp 373–391
Ciolacu D, Cazacu G (2018) New green hydrogels based on Lignin. J Nanosci Nanotechnol 18(4):2811–2822
Ciolacu D, Doroftei F, Cazacu G, Cazacu M (2013) Morphological and surface aspects of cellulose-lignin hydrogels. Cellulose Chem Technol 47(5–6):377–386
Domínguez-Robles J, Peresin MS, Tamminen T, Rodríguez A, Larrañeta E, Jääskeläinen A-S (2018) Lignin-based hydrogels with "super-swelling" capacities for dye removal. Int J Biol Macromol 115:1249–1259
Domínguez-Robles J, Sánchez R, Díaz-Carrasco P, Espinosa E, García-Domínguez MT, Rodríguez A (2017) Isolation and characterization of lignins from wheat straw: application as binder in lithium batteries. Int J Biol Macromol 104:909–918
Dong X, Dong MD, Lu YJ, Turley A, Jin T, Wu CQ (2011) Antimicrobial and antioxidant activities of lignin from residue of corn stover to ethanol production. Ind Crops Prod 34(3):1629–1634
El-Zawawy WK (2005) Preparation of hydrogel from green polymer. Polym Adv Technol 16(1):48–54
Farhat W, Venditti R, Mignard N, Taha M, Becquart F, Ayoub A (2017) Polysaccharides and lignin based hydrogels with potential pharmaceutical use as a drug delivery system produced by a reactive extrusion process. Int J Biol Macromol 104:564–575
Feng Q, Chen F, Zhou X (2012) Preparation of thermo-sensitive hydrogels from Acrylated Lignin and N-Isopropylacrylamide through photocross linking
Feng Q, Cheng H, Xie Y, Chen F (2017) Development of Sulfomethylated Lignin-based adsorbent for Methylene Blue removal. J Biobased Mater Bioenergy 11(6):655–661
Feng Q, Li J, Cheng H, Chen F, Xie Y (2014) Synthesis and characterization of porous hydrogel based on lignin and polyacrylamide. BioResources 9(3):4369–4381
Figueiredo P, Lintinen K, Hirvonen JT, Kostiainen MA, Santos HA (2018) Properties and chemical modifications of lignin: towards lignin-based nanomaterials for biomedical applications. Prog Mater Sci 93:233–269
Fonseca Silva TC, Filpponen I, Habibi Y, Colodette JL, Lucia LA (2012) A facile approach for the synthesis of Xylan-Derived Hydrogels, functional materials from renewable sources. American Chemical Society, pp 257–270
Gan D, Xing W, Jiang L, Fang J, Zhao C, Ren F, Fang L, Wang K, Lu X (2019) Plant-inspired adhesive and tough hydrogel based on Ag-Lignin nanoparticles-triggered dynamic redox catechol chemistry. Nat Commun 10(1):1487
Gao G, Xu WZ, Kadla JF (2015) Reversible pH-responsive hydrogels of softwood kraft lignin and poly [(2-dimethylamino) ethyl methacrylate]-based polymers. J Wood Chem Technol 35(1):73–90

Graça MPF, Rudnitskaya A, Faria C, Evtuguin FAC, Gomes DV, Oliveira JA, Costa LC (2012) Electrochemical impedance study of the lignin-derived conducting polymer. Electrochimica Acta 76:69–76

Grishechko L, Amaral-Labat G, Szczurek A, Fierro V, Kuznetsov B, Pizzi A, Celzard A (2013) New tannin–lignin aerogels. Ind Crops Prod 41:347–355

Hasan A, Fatehi P (2018) Synthesis and characterization of lignin–poly (acrylamide)–poly (2-methacryloyloxyethyl) trimethyl ammonium chloride copolymer. J Appl Polym Sci 135(23):46338

Heinrich LA (2019) Future opportunities for bio-based adhesives—advantages beyond renewability. Green Chem 21(8):1866–1888

Hu B, Owh C, Chee PL, Leow WR, Liu X, Wu Y-L, Guo P, Loh XJ, Chen X (2018) Supramolecular hydrogels for antimicrobial therapy. Chem Soc Rev 47(18):6917–6929

Huang Y, Zeng M, Ren J, Wang J, Fan L, Xu Q (2012) Preparation and swelling properties of graphene oxide/poly(acrylic acid-co-acrylamide) super-absorbent hydrogel nanocomposites

Hunt JA, Chen R, van Veen T, Bryan N (2014) Hydrogels for tissue engineering and regenerative medicine. J Mater Chem B 2(33):5319–5338

Ji X, Zhang Z, Chen J, Yang G, Chen H, Lucia LA (2017) Synthesis and characterization of alkali lignin-based hydrogels from ionic liquids. BioResources 12(3):5395–5406

Jiang P, Sheng X, Yu S, Li H, Lu J, Zhou J, Wang H (2018) Preparation and characterization of thermo-sensitive gel with phenolated alkali lignin. Scientif Rep 8(1):14450

Jin C, Song W, Liu T, Xin J, Hiscox WC, Zhang J, Liu G, Kong Z (2018) Temperature and pH responsive hydrogels using methacrylated lignosulfonate cross-linker: synthesis, characterization, and properties. ACS Sustain Chem Eng 6(2):1763–1771

Jin Z, Yu Y, Shao S, Ye J, Lin L, Iiyama K (2010) Lignin as a cross-linker of acrylic acid-grafted carboxymethyl lignocellulose. J Wood Sci 56(6):470–476

Kai D, Low ZW, Liow SS, Abdul Karim A, Ye H, Jin G, Li K, Loh XJ (2015) Development of Lignin Supramolecular Hydrogels with mechanically responsive and self-healing properties. ACS Sustain Chem Eng 3(9):2160–2169

Kai D, Tan MJ, Chee PL, Chua YK, Yap YL, Loh XJ (2016) Towards lignin-based functional materials in a sustainable world. Green Chem 18(5):1175–1200

Kalinoski RM, Shi J (2019) Hydrogels derived from lignocellulosic compounds: evaluation of the compositional, structural, mechanical and antimicrobial properties. Ind Crops Prod 128:323–330

Kaur R, Thakur NS, Chandna S, Bhaumik J (2020) Development of agri-biomass based lignin derived zinc oxide nanocomposites as promising UV protectant-cum-antimicrobial agents. J Mater Chem B, 8(2):260–269

Larrañeta E, Imízcoz M, Toh JX, Irwin NJ, Ripolin A, Perminova A, Domínguez-Robles J, Rodríguez A, Donnelly RF (2018) Synthesis and characterization of Lignin Hydrogels for potential applications as drug eluting antimicrobial coatings for medical materials. ACS Sustain Chem Eng 6(7):9037–9046

Li F, Wang X, Sun R (2017) A metal-free and flexible supercapacitor based on redox-active lignosulfonate functionalized graphene hydrogels. J Mater Chem A 5(39):20643–20650

Li F, Wang X, Yuan T, Sun R (2016) A lignosulfonate-modified graphene hydrogel with ultrahigh adsorption capacity for Pb(ii) removal. J Mater Chem A 4(30):11888–11896

Li RJ, Gutierrez J, Chung Y-L, Frank CW, Billington SL, Sattely ES (2018) A lignin-epoxy resin derived from biomass as an alternative to formaldehyde-based wood adhesives. Green Chem 20(7):1459–1466

Liu H, Chung H (2017) Lignin-based polymers via graft copolymerization. J Polym Sci, Part A: Polym Chem 55(21):3515–3528

Ma Y, Lv L, Guo Y, Fu Y, Shao Q, Wu T, Guo S, Sun K, Guo X, Wujcik EK (2017) Porous lignin based poly (acrylic acid)/organo-montmorillonite nanocomposites: swelling behaviors and rapid removal of Pb (II) ions. Polymer 128:12–23

Meng Y, Lu J, Cheng Y, Li Q, Wang H (2019) Lignin-based hydrogels: a review of preparation, properties, and application. Int J Biol Macromol 135:1006–1019

Nakasone K, Kobayashi T (2016) Cytocompatible cellulose hydrogels containing trace lignin. Mater Sci Eng, C 64:269–277
Osti E (2006) Cutaneous burns treated with hydrogel (burnshield) and a semipermeable adhesive film. JAMA Surgery 141(1):39–42
Oveissi F, Naficy S, Le TYL, Fletcher DF, Dehghani F (2018) Tough and processable hydrogels based on Lignin and hydrophilic Polyurethane. ACS Appl Bio Mater 1(6):2073–2081
Parajuli D, Adhikari CR, Kuriyama M, Kawakita H, Ohto K, Inoue K, Funaoka M (2006) Selective recovery of gold by novel lignin-based adsorption gels. Ind Eng Chem Res 45(1):8–14
Parajuli D, Inoue K, Ohto K, Oshima T, Murota A, Funaoka M, Makino K (2005) Adsorption of heavy metals on crosslinked lignocatechol: a modified lignin gel. React Funct Polym 62(2):129–139
Park S, Kim SH, Kim JH, Yu H, Kim HJ, Yang Y-H, Kim H, Kim YH, Ha SH, Lee SH (2015a) Application of cellulose/lignin hydrogel beads as novel supports for immobilizing lipase. J Mol Catal B Enzym 119:33–39
Park S, Kim SH, Won K, Choi JW, Kim YH, Kim HJ, Yang Y-H, Lee SH (2015b) Wood mimetic hydrogel beads for enzyme immobilization. Carbohyd Polym 115:223–229
Passauer L (2012) Highly swellable Lignin hydrogels: novel materials with interesting properties, functional materials from renewable sources. American Chemical Society, pp 211–228
Passauer L, Fischer K, Liebner F (2011) Preparation and physical characterisation of strongly swellable oligo(oxyethylene) lignin hydrogels
Passauer L, Hallas T, Bäucker E, Ciesielski G, Lebioda S, Hamer U (2015) Biodegradation of hydrogels from oxyethylated lignins in model soils. ACS Sustain Chem Eng 3(9):1955–1964
Peñaranda-Avila J, Sabino M (2010) Effect of the presence of lignin or peat in IPN hydrogels on the sorption of heavy metals
Peng Z, Chen F (2011) Synthesis and Properties of Lignin-Based Polyurethane Hydrogels. Int J Polym Mater Polym Biomater 60(9):674–683
Peng Z, Zou Y, Xu S, Zhong W, Yang W (2018) High-performance biomass-based flexible solid-state supercapacitor constructed of pressure-sensitive lignin-based and cellulose hydrogels. ACS Appl Mater Interfaces 10(26):22190–22200
Qu Y, Tian Y, Zou B, Zhang J, Zheng Y, Wang L, Li Y, Rong C, Wang Z (2010) A novel mesoporous lignin/silica hybrid from rice husk produced by a sol–gel method. Biores Technol 101(21):8402–8405
Rajan K, Mann JK, English E, Harper DP, Carrier DJ, Rials TG, Labbé N, Chmely SC (2018) Sustainable hydrogels based on lignin-methacrylate copolymers with enhanced water retention and tunable material properties. Biomacromol 19(7):2665–2672
Raschip IE, Hitruc EG, Vasile C (2011) Semi-interpenetrating polymer networks containing polysaccharides. II. Xanthan/lignin networks: a spectral and thermal characterization. High Perform Polym 23(3):219–229
Raschip IE, Hitruc GE, Vasile C, Popescu M-C (2013) Effect of the lignin type on the morphology and thermal properties of the xanthan/lignin hydrogels. Int J Biol Macromol 54:230–237
Rejmontová P, Kovalcik A, Humpolíček P, Capáková Z, Wrzecionko E, Sáha P (2019) The use of fractionated Kraft lignin to improve the mechanical and biological properties of PVA-based scaffolds. RSC Adv 9(22):12346–12353
Rudnitskaya A, Evtuguin DV, Costa LC, Graça MPF, Fernandes AJS, Correia MRP, Gomes MTSR, Oliveira JABP (2013) Potentiometric chemical sensors from lignin–poly(propylene oxide) copolymers doped by carbon nanotubes. Analyst 138(2):501–508
Rybtchinski B (2011) Adaptive supramolecular nanomaterials based on strong noncovalent interactions. ACS Nano 5(9):6791–6818
Sathawong S, Sridach W, Techato K-A (2018) Recovery of Kraft Lignin from OPEFB and using for Lignin-Agarose Hydrogel. J Polym Environ 26(8):3307–3315
Schupp DJ, Zhang X, Sun S, Cölfen H (2019) Mineral plastic hydrogels from the cross-linking of polyacrylic acid and alkaline earth or transition metal ions. Chem Commun 55(34):4913–4916

Shen X, Berton P, Shamshina JL, Rogers RD (2016) Preparation and comparison of bulk and membrane hydrogels based on Kraft- and ionic-liquid-isolated lignins. Green Chem 18(20):5607–5620
Shen Z, Luo Y, Wang Q, Wang X, Sun R (2014) High-value utilization of lignin to synthesize Ag nanoparticles with detection capacity for Hg^{2+}. ACS Appl Mater Interf 6(18):16147–16155
Song X, Chen F, Liu S (2016) A Lignin-containing Hemicellulose-based Hydrogel and its adsorption behavior 11(3):15
Spasojević D, Zmejkoski D, Glamočlija J, Nikolić M, Soković M, Milošević V, Jarić I, Stojanović M, Marinković E, Barisani-Asenbauer T, Prodanović R, Jovanović M, Radotić K (2016) Lignin model compound in alginate hydrogel: a strong antimicrobial agent with high potential in wound treatment. Int J Antimicrob Agents 48(6):732–735
Sun Y, Ma Y, Fang G, Li S, Fu Y (2016a) Synthesis of acid hydrolysis lignin-g-poly-(acrylic acid) hydrogel superabsorbent composites and adsorption of lead ions. BioResources 11(3):5731–5742
Sun Y, Ma Y, Fang G, Ren S, Fu Y (2016b) Controlled pesticide release from porous composite Hydrogels based on Lignin and Polyacrylic Acid 11(1)
Teng X, Xu H, Song W, Shi J, Xin J, Hiscox WC, Zhang J (2017) Preparation and properties of hydrogels based on PEGylated lignosulfonate amine. ACS Omega 2(1):251–259
Thakur S, Govender PP, Mamo MA, Tamulevicius S, Mishra YK, Thakur VK (2017) Progress in lignin hydrogels and nanocomposites for water purification: future perspectives. Vacuum 146:342–355
Vashist A, Vashist A, Gupta YK, Ahmad S (2014) Recent advances in hydrogel based drug delivery systems for the human body. J Mater Chem B 2(2):147–166
Wang L, Ago M, Borghei M, Ishaq A, Papageorgiou AC, Lundahl M, Rojas OJ (2019) Conductive carbon microfibers derived from wet-spun Lignin/Nanocellulose Hydrogels. ACS Sustain Chem Eng 7(6):6013–6022
Wang S, Zhang L, Wang A, Liu X, Chen J, Wang Z, Zeng Q, Zhou H-H, Jiang X, Zhang L (2018) Polymer-laden composite Lignin-based electrolyte membrane for high-performance lithium batteries. ACS Sustain Chem Eng 6(11):14460–14469
Wang X, Zhou Z, Guo X, He Q, Chen H, Ge C (2016) Ultrasonic-assisted synthesis of sodium lignosulfonate-grafted poly(acrylic acid-co-poly(vinyl pyrrolidone)) hydrogel for drug delivery
Wang Y, Xiong Y, Wang J, Zhang X (2017) Ultrasonic-assisted fabrication of montmorillonite-lignin hybrid hydrogel: Highly efficient swelling behaviors and super-sorbent for dye removal from wastewater. Colloids Surf, A 520:903–913
Wichterle O, Lím D (1960) Hydrophilic gels for biological use. Nature 185(4706):117–118
Xu Y, Sheng K, Li C, Shi G (2010) Self-assembled graphene hydrogel via a one-step hydrothermal process. ACS Nano 4(7):4324–4330
Yamamoto H, Amaike M, Saitoh H, Sano Y (1999) Gel formation of lignin and biodegradation of the lignin gels by microorganisms. Mater Sci Eng, C 7(2):143–147
Yao Q, Xie J, Liu J, Kang H, Liu Y (2014) Adsorption of lead ions using a modified lignin hydrogel. J Polym Res 21(6):465
Yu C, Wang F, Zhang C, Fu S, Lucia LA (2016) The synthesis and absorption dynamics of a lignin-based hydrogel for remediation of cationic dye-contaminated effluent. React Funct Polym 106:137–142
Zerpa A, Pakzad L, Fatehi P (2018) Hardwood Kraft Lignin-based Hydrogels: production and performance. ACS Omega 3(7):8233–8242
Zhang J, Ma PX (2013) Cyclodextrin-based supramolecular systems for drug delivery: recent progress and future perspective. Adv Drug Deliv Rev 65(9):1215–1233
Zhang L, Lu H, Yu J, Fan Y, Yang Y, Ma J, Wang Z (2018) Synthesis of lignocellulose-based composite hydrogel as a novel biosorbent for Cu^{2+} removal. Cellulose 25(12):7315–7328
Zmejkoski D, Spasojević D, Orlovska I, Kozyrovska N, Soković M, Glamočlija J, Dmitrović S, Matović B, Tasić N, Maksimović V, Sosnin M, Radotić K (2018) Bacterial cellulose-lignin composite hydrogel as a promising agent in chronic wound healing. Int J Biol Macromol 118:494–503

Lignin Composites for Biomedical Applications: Status, Challenges and Perspectives

Pınar Terzioğlu, Fatma Nur Parın and Yusuf Sıcak

Abstract The interest is rapidly growing to use green and renewable polymers for engineering applications. Herein, biopolymers have shown great potential as environmentally friendly alternatives. Lignin is one of the most abundant biopolymers obtained from lignocellulosic biomass. Lignin constitutes approximately 15–35% of the lignocellulosic biomass. It has unique properties, such as high carbon content, high thermal stability, adhesive properties, biodegradability, biocompatibility and biological activities. Lignin has also the benefit of cost-effective processing with low energy requirements as a by-product of agricultural, paper and pulp industries. Taking account of these properties, a significant amount of research has been carried out toward the fabrication of advanced lignin-based materials with functional properties to be evaluated in several industrial applications for the last few decades. In recent years, lignin composites have gained popularity in the biomedical applications including tissue engineering scaffolds, wound dressing materials, drug delivery carriers and other medical applications. The studies have shown the suitability of lignin composites for multiple biomedical proposes. The present chapter presents a widespread overview of current achievements, limitations and future perspectives of lignin composites in biomedical field. In this chapter, recent advances in the fabrication, characterization and potential biomedical applications of lignin-based composites will be analyzed and discussed in detail.

Keywords Biomaterial · Composite · Lignin · Scaffold · Tissue engineering

P. Terzioğlu (✉) · F. N. Parın
Department of Fiber and Polymer Engineering, Faculty of Engineering and Natural Sciences, Bursa Technical University, Bursa, Turkey
e-mail: pinar.terzioglu@btu.edu.tr

Y. Sıcak
Department of Herbal and Animal Production, Köyceğiz Vocational School, Muğla Sıtkı Koçman University, Muğla, Turkey

S. Sharma and A. Kumar (eds.), *Lignin*, Springer Series on Polymer and Composite Materials, https://doi.org/10.1007/978-3-030-40663-9_9

1 Introduction

The consumption of fossil fuels is increasing continuously with a rapid growth of population and industrialization (Kai et al. 2016a; Duval and Lawoko 2014). Therefore, environmental concerns and run out of fossil fuels are requiring new efforts as evaluation of green and sustainable resources in the twenty-first century (Schutyser et al. 2018). Various biomass resources are considered to have the potential to replace fossil fuels. Besides, biomass energy is one of the candidates and solutions that will be used to ensure sustainable energy supply without causing any environmental pollution (Duval and Lawoko 2014). In the forest ecosystem, woody materials appear the most important biological resources used for energy purposes.

Wood material is a hierarchical structured natural product—lignocellulosic biomass—which contains three main components as cellulose (38–50%), hemicellulose (23–32%) and lignin (12–25%), respectively (Ponnusamy et al. 2018; Carrott and Carrott 2007; Xu et al. 2018). Cellulose and hemicellulose are linked to the lignin with ether and ester bonds (Chio et al. 2019). It indicates why the lignocellulosic structure is so complicated (Duval and Lawoko 2014; Xu et al. 2018; Ten and Vermerris 2015). Lignin is the second greatest inexhaustible biopolymer after cellulose with heterogeneous and amorphous structure (Kai et al. 2016a). It is the main source of aromatic alcohols and also serves as a matrix around the plant cell wall (Kim et al. 2015; Norgren and Edlund 2014). Actually, the main function of lignin is to support the cellulose fibers in plants. Phenylpropanoid-based biopolymer lignin is giving mechanical strength to the plant structure, especially in larger roots and stems as well as providing resistance to microbial attack (Ponnusamy et al. 2018; Spiridon et al. 2018). In addition, it supports the regulation of water and nutrition transport in the cells due to its hydrophobic nature (Kai et al. 2016a; Spiridon et al. 2018).

Lignin is a highly branched, three-dimensional (3D) structure that consists of phenylpropanoid units (Fig. 1), namely p-coumaryl alcohol, coniferyl alcohol and sinapyl alcohol (Fig. 2) (Pandey and Kim 2011). These three monolignols are randomly cross-linked to each other by different chemical bonds such as β-O-4-aryl ether linkages, α-O-4-aryl ether, 4-O-5-diaryl ether, β-5-phenylcoumaran, 5-5-biphenyl, β-1-(1,2-diarylpropane) and β-β-(resinol) (Spiridon et al. 2018). This complex structure of lignin is derived from enzymatic dehydrogenation reactions between monolignols (Carrott and Carrott 2007). Although all of the functional groups and units are identified in lignin, the chemical structure of lignin is not known certainly because of randomly cross-linked linkages.

In the search for a new generation of biomedical materials, lignin has received significant interest due to its promising characteristics. Lignin has been proved to be suitable for biomedical applications such as tissue engineering scaffolds, biocompatible antibacterial agents and release materials in drug delivery systems. The most significant characteristics of lignin and its derivatives can be summarized as antioxidant activity, inherent antibacterial activity, adhesiveness, biodegradability, biocompatibility, favorable stiffness, high thermal stability and low-cost (Naseem et al. 2016). Owing to these unique properties, also there has been a continuously

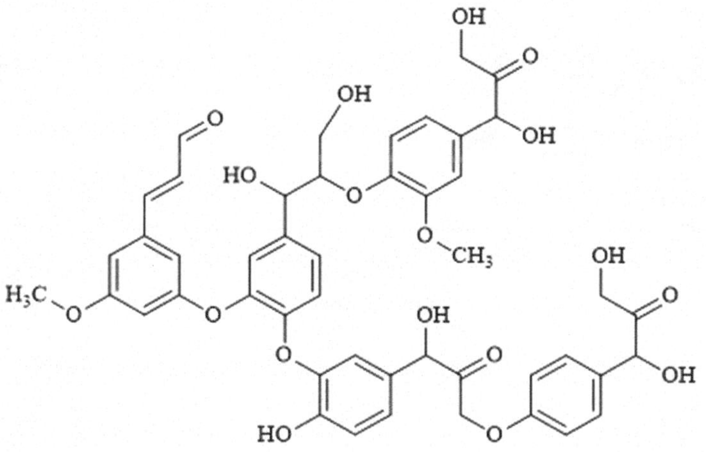

Fig. 1 Common chemical structure of lignin

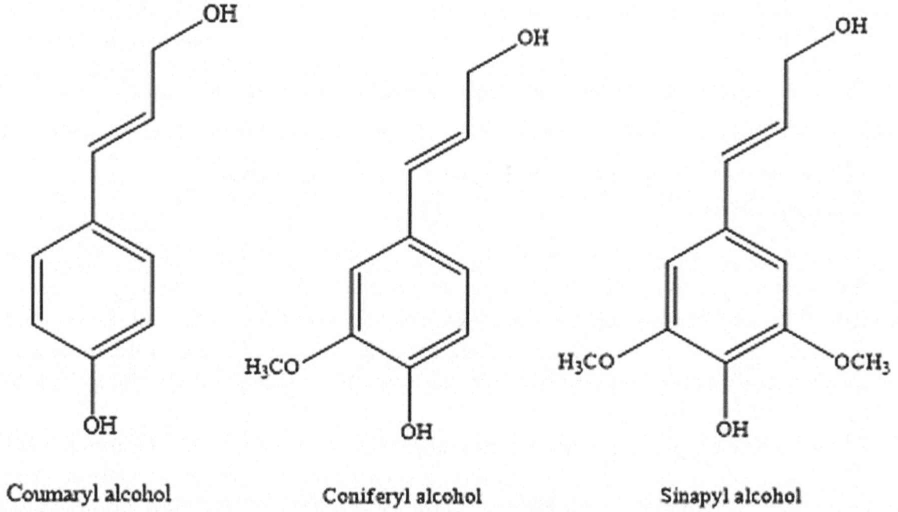

Fig. 2 Chemical structure of phenylpropanoid units of lignin

increasing attention in evaluating lignin-based composites for a wide range of biomedical applications as given in (Fig. 3) Lignin may be used in combination with other polymers to obtain composites with improved functionality.

In this chapter, it is aimed to provide an in-depth overview of lignin-based composites in biomedical applications. Along with the discussion of lignin composites with biological activity, their applicability for scaffold, wound dressing and drug delivery applications is also summarized. Finally, the scope for further research opportunities in the field is also highlighted.

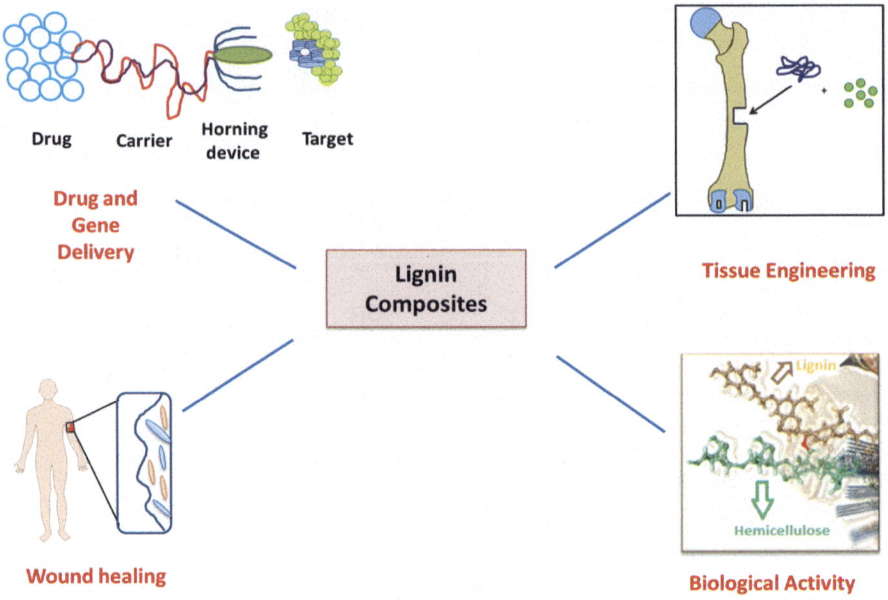

Fig. 3 Schematic illustration of biomedical applications of lignin composites

2 Prominent Properties of Lignin for Biomedical Applications

In the design of polymer-based composite biomaterials, the properties of matrix and reinforcement components have key roles. Especially, the nontoxicity and biocompatibility are the significant features of a biomaterial. Furthermore, materials are expected to meet the requirements of the specification for the desired application (Kaur et al. 2018).

Lignin presents potential for various applications in the field of biomaterials due to its specific structural characteristics (Monteil-Rivera et al. 2013). The biodegradability and biocompatibility of lignin enable its use in the fabrication of bio-based renewable hydrogels (Meng et al. 2019). Lignin has antioxidant, antimicrobial and antiviral properties due to its polyphenolic structure which can be used as a promising natural ingredient in bioactive materials such as biocomposites, drug release agents and sunscreen lotions (Witzler et al. 2018). Among antioxidant activity, lignin with many UV chromophoric groups including double bonds, phenolics, hydroxyl groups and carbonyl groups can be evaluated as an anti-ultraviolet agent in multifunctional hydrogels (Zhu et al. 2018). Moreover, lignin has gaining attention to prepare smart materials such as temperature-sensitive copolymers and hydrogels (Kai et al. 2016b; Diao et al. 2014).

Besides other applications, lignin can be evaluated in tissue engineering scaffolds (e.g., hydrogels and nanofibers) because of providing improved hydrophilicity, mechanical and thermal stability (Witzler et al. 2018; Kai et al. 2016b; Anwer et al. 2015). Lignin composites were suggested as a potential candidate for stem cell-based tissue engineering approaches to guide stem cell differentiation and eventually enhance tissue regeneration process (Witzler et al. 2018). Moreover, lignin-based hydrogel beads were proposed as fine substrates for enzyme immobilization which could be used in multiple biomedical applications (Park et al. 2015). Endowed with unique properties, lignin and its composites seem to be ideal candidates for biomedical applications.

3 Biomedical Applications of Lignin Composites

Polymer composites are broadly used in the biomedical field including drug delivery, dental, tissue engineering (bones, blood vessels, oral tissues, etc.), wound dressing and regenerative medicine applications (Zagho et al. 2018). Through the improvement of the next generation of composite materials, the biopolymer-based ones are of great interest especially due to their eco-efficiency and sustainability. The biopolymers used in these systems stand out by their similarity to biological macromolecules, biodegradability, reduced side effects and nontoxicity (Park et al. 2017). In recent years, lignin and its derivatives have become available which have been starting to be utilized in several biomedical applications in combination with polymers, forming composites. The success of the researches actualized so far and the eventual applications of these new lignin composites have prompted the preparation of biomaterials. In this section, the wound dressing, scaffold, drug delivery system applications of lignin composites, as well as their biological activities are reviewed.

3.1 Lignin Composites with Biological Activity

3.1.1 Antioxidant Activity

Reactive oxygen species (ROSs) which primarily include hydrogen peroxide (H_2O_2), hypochlorite (OCl^-), hydroxyl radical ($\cdot OH$), peroxy nitrite ($ONOO^-$) and superoxide ($O_2^{\cdot -}$) are continuously generated, transformed and consumed by cells during the normal action of the body (Xu et al. 2016). The high concentrations of ROSs lead to various vital molecules (enzymes, lipids, DNA and proteins) damage which can result in disorders in the physiology of tissues and organs (Karaman et al. 2016). Finally, the harmful action of ROSs causes aging, many metabolic diseases including cancer, diabetes mellitus, cardiovascular diseases and degenerative disorders (Albayrak and Aksoy 2019; Abdel-Daim et al. 2018). Therefore, the oxidative balance is a critical and precision issue (Lorenzo et al. 2018). In addition, the reduction of oxidative stress or ROSs is important from the point of view of the biomaterial applications because it can lead to toxicity and inflammatory response for the implantations (Lith et al. 2014). Natural or synthetic antioxidants are widely utilized to eliminate the negative effects of ROSs. Antioxidants are also widely evaluated as additives to retard oxidation in foods and other commercial products (Gu et al. 2017). The materials with antioxidant activity might be favorable for food, pharmaceutical and cosmetic industries.

Lignin, complex aromatic macromolecule containing a variety of hydroxyl and methoxyl functional groups, is a well-known free radical scavenger (Ugartondo et al. 2008; Ugartondo et al. 2009). Studies on lignin have indicated that it could be utilized as antioxidant agents for health care applications (Ge et al. 2014; Wang et al. 2018a). However, the reports on these phenolic biopolymer-based composites with the good antioxidant feature for biomaterial applications are very limited.

Antioxidant lignin–polylactic acid (PLA) composite filaments were prepared using hot-melt extrusion followed by 3D-printing technique (Domínguez-Robles et al. 2019). The antioxidant capacity of filaments was determined by using the 2,2-diphenyl-1-picrylhydrazyl (DPPH) method. The composites were found to have efficient antioxidant activity (Domínguez-Robles et al. 2019). The results of the study indicated that the composites have the potential to be used for health care fused filament fabrication (e.g., wound healing) as well as active food packaging applications.

The importance of lignin evaluation in an antioxidative scaffold for nerve regeneration was the focus of Wang et al. (2018a). The purpose was to prepare polycaprolactone (PCL)/lignin–PCL copolymer composite fibers via electrospinning to support the neuron and Schwann cell growth. The authors stated that PCL/lignin–PCL nanofibers the incorporation of lignin–PCL improved the mechanical properties of PCL nanofibers and also provide the fibrous scaffolds excellent antioxidant activity. Lignin had positive effects on the biocompatibility and antioxidant properties of the nanofibers, which promoted dorsal root ganglion neurons adhesion, proliferation and growth. Moreover, the promoted cell proliferation of the most commonly used seed cells (bone marrow mesenchymal stem and Schwann cells) in nerve tissue engineering and enhanced myelin basic protein expressions of Schwann cells was evident of the great potential of lignin–PCL copolymers incorporated nanofibers for nerve regeneration and cellular oxidative stress concerned other tissues.

Kai et al. (2016a) also synthesized PLA–lignin copolymers and further prepared poly-L-lactide (PLLA)/PLA–lignin nanofibrous composites using electrospinning technique. It was determined that nanofibers had a diameter of 350–500 nm with uniform and bead-free structures. The results of antioxidant assay demonstrated that the lignin copolymers and PLLA/PLA–lignin nanofibers showed excellent DPPH$^{\cdot}$ radical scavenging capacity. In addition, the biocompatibility of lignin and lignin-based materials is a critical point for cosmetic products or biomedical materials. The authors identified the biocompatibility of composites for three different cell types (PC12, human dermal fibroblasts and human mesenchymal stem cells). The composites containing copolymers with 10–30 wt% lignin exhibited higher cell proliferation compared to the neat PLLA showing that the antioxidant activity of the biopolymer may improve the viability of cells. However, composites containing copolymers with 50 wt% lignin exhibited relative low cell proliferation. These findings come to support the potential of lignin-based nanofibers as biomaterials that may locally decrease the oxidative stress-related functional disorders or tissue damages. Nevertheless, it is important to determine the optimal lignin amount to be incorporated into the material matrix to succeed a balance between the antioxidant activity and the cytotoxicity.

Polyvinyl alcohol (PVA)-based hydrogels are important biomaterials (Tamahkar and Özkahraman 2015). However, the low mechanical properties of these hydrogels are the main drawback for biomedical applications. There are efforts to overcome this disadvantage by blending PVA with chitosan. The studies showed that PVA/chitosan hydrogels are suitable candidates for noninvasive transdermal drug delivery system in diabetes chemotherapy (Zu et al. 2012) and wound healing (Yang et al. 2008). A recent study has reported the utilization of lignin nanoparticles in PVA/chitosan hydrogels for drug delivery, food packaging and wound dressing applications (Fig. 4). The lignin incorporation into the network has been proposed as a way to enhance the mechanical, thermal, swelling, antioxidant and antibacterial properties. Due to the synergic effect of chitosan and lignin nanoparticles, efficient antioxidant response was determined by DPPH assay (Yang et al. 2018a). The DPPH scavenging activity of hydrogels raised from 74.3 to 78.6%, while the lignin nanoparticle incorporation ratio increased from 1 wt% to 3 wt%.

Latest studies already focused on the evaluation of lignin as an added-value material for thermoplastic or packaging films to improve sustainable composites. Crouvisier-Urion et al. (2016) pointed out that lignin with antioxidant activity was a good compound to prepare chitosan-based composite films. The researchers prepared chitosan-lignin composite films by the solvent casting method. The antioxidant capacities of films were determined via the method based on the scavenging of the DPPH$^{\cdot}$ radical molecule. Even though neat chitosan films already exhibited a slight natural radical scavenging activity, the incorporation of lignin up to 20

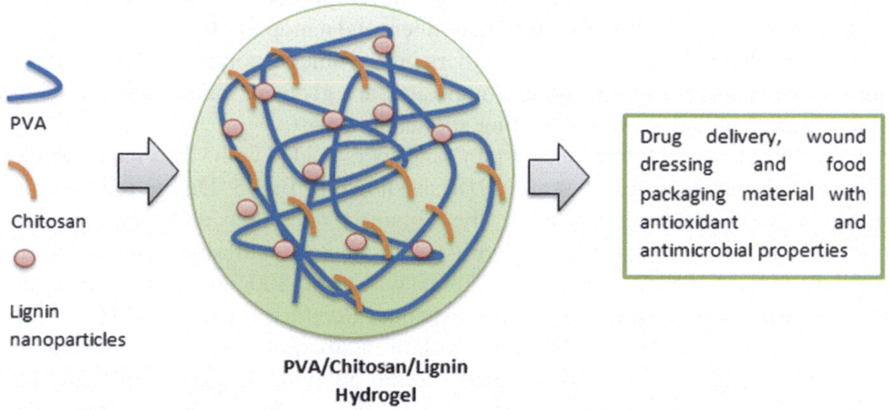

Fig. 4 Schematic diagram of PVA/chitosan/lignin composite hydrogels

wt% significantly enhanced the antioxidant activity of films. However, at high concentration (30 wt%), lignin played as a pro-oxidant. It was concluded that there is not a linear correlation between the antioxidant activity and lignin amount of the films. Furthermore, He et al. (2019) prepared nanocomposite films by blending polyvinyl alcohol (PVA) and citric acid modified lignin nanoparticles. Modification of lignin was applied to overcome the compatibility problem between lignin and PVA. The resulting composite films showed enhanced antioxidant property and transparency. It is stated that lignin nanoparticles are good raw materials for next-generation functional nanocomposites, especially requiring high antioxidant activity and enhanced light absorbance performance.

3.1.2 Antimicrobial Activity

Microbial contamination leads to pathogenic infection which is still one of the most critical complications in various areas, especially in biomedical devices, drugs, health care products, food packaging, textiles and water purification systems (Kenawy et al. 2007; Greenhalgh et al. 2019; Demirci et al. 2018). This situation creates a global health threat. The studies have concentrated on the development of new products with efficient antimicrobial activity.

Investigation of the lignin's antimicrobial capacity has been a long journey of scientific exploration. As a complex polyphenolic polymer, lignin has been proved to exhibit antimicrobial activity against different yeast, gram-negative and gram-positive bacteria including *Bacillus subtilis* (Wang et al. 2018b; Nada et al. 1989), *Bacillus mycoids* (Nada et al. 1989), *Escherichia coli* (Wang et al. 2018b), *Listeria monocytogenes* (Dong et al. 2011), *Salmonella enterica* (Wang et al. 2018b), *Staphylococcus aureus* (Dong et al. 2011) and *Candida lipolytica* (Dong et al. 2011). This raised the attempts to develop lignin-based biomaterials with the purpose to achieve new desirable antimicrobial properties.

To date, researchers have investigated the development of various antimicrobial lignin composites as hydrogels, films, nanofibers and nanoparticles. For this purpose, it has been blended with both biopolymers and synthetic polymers. The literature studies on the antimicrobial lignin composites are given in Table 1. Some of the studies were handled in wound healing applications part.

PVA–Acacia wood lignin electrospun nanofibers loaded with silver nanoparticles were manufactured by Aadil et al. (2018) as shown in Fig. 5. Due to the presence of lignin and silver nanoparticles, the obtained nanofiber mats were yellow. Polydispersed silver nanoparticles with spherical and oval shape are randomly distributed on composite nanofiber. Fibers with 100–300 nm average diameter showed significant antimicrobial activity against *Bacillus circulans* (MTCC 7906) and *Escherichia coli* (MTCC 739) after 24 h of incubation at 37 °C. It is stated that the antimicrobial activity of fibers related to the polyphenolic group of lignin and the presence of silver nanoparticles. The composites can be suitable for preparation of antimicrobial fabrics, membrane filters and wound healing materials.

Table 1 Antimicrobial activities of lignin composites

Material	Test method	In vitro tested efficacy against		Results	Reference
		Gram positive	Gram negative		
PVA/lignin/silver nanofibers	Agar well diffusion	*B. circulans*	*E. coli*	Inhibition zone for *E. coli*: 1.1 ± 0.05 cm *B. circulans* 1.3 ± 0.08 cm	Aadil et al. (2018)
Chitosan/lignosulfonates nanoparticles	Turbidimetric	*S. aureus* *B. subtilis*	*E. coli*	Bacterial growth decrease	Kim et al. (2013)
Lignin/PVA nanocomposite fiber webs	ASTM E 2149-10	*S. aureus*	*E. coli*	99.9% reduction rate against *S. aureus*	Lee et al. (2019)
Lignin/decorated thin multi-walled carbon nanotubes in poly(vinyl alcohol) nanocomposites	Dilution method	*S. aureus*	–	68.7% bacterial growth decrease after 18 h	Lee et al. (2018)
Silver/hydroxyapatite/lignin coatings on titanium	Dilution method	*S. aureus*	–	100% cell reduction after 24 h	Eraković et al. (2013)
Silver/hydroxyapatite/lignin coatings on titanium	Agar diffusion method	*S. aureus*	*E. coli*	100% cell reduction after 24 h	Eraković et al. (2013)
Silver/hydroxyapatite/lignin thin films	Bacterial culture density measurement anti-biofilm	*S. aureus*	*P. aeruginosa* *C. famata*	Protective against the microbial colonization initial phase and mature biofilm development	Janković et al. (2015)
Gelatin/lignin films	Dynamic shake flask method (ASTM E 2149–2010)	*B. subtilis*	–	Excellent bacterial growth inhibition	Mehta and Kumar (2019)
Polyvinyl alcohol/chitosan/lignin hydrogels	–	*S. aureus*	*E. coli*	*E. coli* > 95% *S. aureus* > 85% Cell reduction after 24 h	Yang et al. (2018a)
Polylactide/lignin/silver nanoparticles		*L. monocytogenes*	*E. coli*	100% cell reduction after 9 h	Shankar et al. (2018)

(continued)

Table 1 (continued)

Material	Test method	In vitro tested efficacy against		Results	Reference
		Gram positive	Gram negative		
Artocarpus heterophyllus peel lignin/chitosan biocomposites			*E. coli* *Klebsiella*	Inhibiton zone for *E. coli*: ~0.9 mm *Klebsiella* ~1.1 mm	Jaganathan et al. (2018)
Chitosan/Banana peel composites		*S. aureus*	*E. coli*	*S.aureus* is the most sensitive strain	Rihayat et al. (2019)

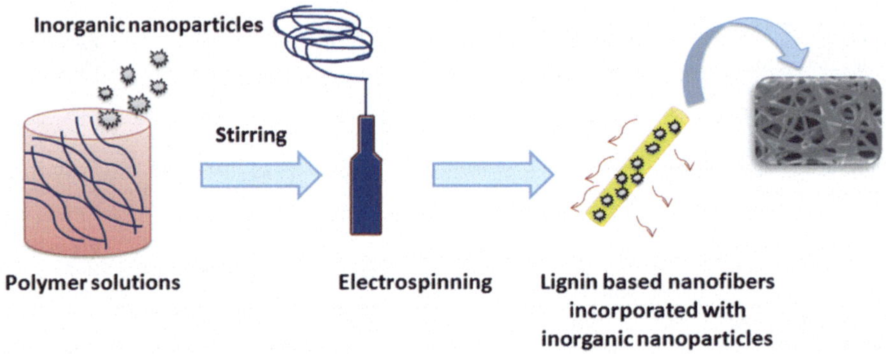

Fig. 5 Schematic representation of the preparation of silver nanoparticles incorporated PVA–lignin nanofibers

Lee et al. (2019) also developed lignin/(PVA) nanocomposite fibers via electrospinning to bring out further possible applications of lignin. According to the antimicrobial evaluation, nanofibers possessed strong antimicrobial activity against *Staphylococcus aureus* but no effect against *Escherichia coli*.

Recently, a research group decided to verify the effect of lignin nanoparticles incorporated PVA/chitosan hydrogels on the bacterial growth of strains representing both gram-positive and gram-negative classes. A drastic reduction in bacterial viability for *Escherichia coli* and *Staphylococcus aureus* bacterial strains was concluded which was associated to the synergetic effect of lignin and chitosan (Yang et al. 2018a).

The studies showed that ternary composites composed of silver/hydroxyapatite/lignin had the ability to inhibit the growth of gram-positive and gram-negative bacteria (Janković et al. 2015; Eraković et al. 2013, 2014). As coating, these composites should provide effective bacterial protection, prevent bacterial infection and the corrosion of an implanted metallic biomaterial (Eraković et al. 2013, 2014). Lignin seems to be a good renewable source for eco-friendly advanced functional materials.

The recent studies reported that besides antimicrobial activity, lignin composites possess UV shielding property (Lee et al. 2019; Shankar et al. 2018; Mehta and Kumar 2019). It is feasible to use lignin in composites as a natural broad-spectrum sun blocker. Based on the outcomes, several antimicrobial composites containing lignin can be a useful candidate for biomaterials and health care products.

3.2 Lignin Composites for Wound Healing Applications

Wound healing is a complex physiological process intending to restore the skin barrier function (Sezer and Cevher 2011, Portela et al. 2019). Healing process is based on the regeneration of dermal and epidermal tissues (Okur et al. 2019). Suitable wound dress materials are the major health care requirements for proper wound healing. The traditional wound dressings such as bandages, gauze, cotton wool and gauze-woven cotton composites are insufficient to provide appropriate conditions for healing. It is expected from an ideal wound dressing to promote a moist environment at the wound surface, protect the wound site from bacterial contamination, allow gaseous exchange, absorb excess exudate and decrease wound pain (Mutlu et al. 2018). Moreover, it should be comfortable, nonallergic, nonadherent, nontoxic and sterile (Mutlu et al. 2018). Therefore, there is extensive research on the development of superior products that heals wounds efficiently. These innovative products are called modern wound dressings which are improved to facilitate the function of the wound rather than only to cover it. Modern wound dressings can be prepared in various types. Hydrogels, hydrocolloid, medicated, semi-permeable foam, semi-permeable film, composite and bioactive wound dressings are among the versatile and convenient types of dressings available in the market (Dhivya et al. 2015).

Biopolymers (alginate, chitosan, cellulose, gelatin, etc.) are considered as promising candidates for modern wound dressing applications. Among the several biopolymers, there has been recent interest in the use of lignin to construct wound dressings. Because lignin has antioxidant, antidiabetic and antimicrobial activities as well as ability to absorb water (Reesi et al. 2018). Many studies have been carried out recently using lignin composites as potential wound dressings.

Ravishankar et al. (2019) suggested that combining the advantages of chitosan and alkali lignin should bring novel materials for promising tissue engineering and regenerative medicine applications. It has been reported that hydrogels and cross-linked films of chitosan and alkali lignin can be successfully obtained by ionotropic cross-linking. The cytotoxicities of the chitosan–alkali lignin hydrogels were investigated using in vitro 3-(4,5-dimethylthiazol-2-yl)-2,5-diphenyl tetrazolium bromide (MTT) assay method, and in vivo toxicity and teratogenicity of gels were determined using a zebra fish (*Danio rerio*) embryo model, to present its suitability as a scaffold for tissue engineering. The gels showed very low cytotoxicities which were considered as nontoxic. Further, the wound healing potential of chitosan–alkali

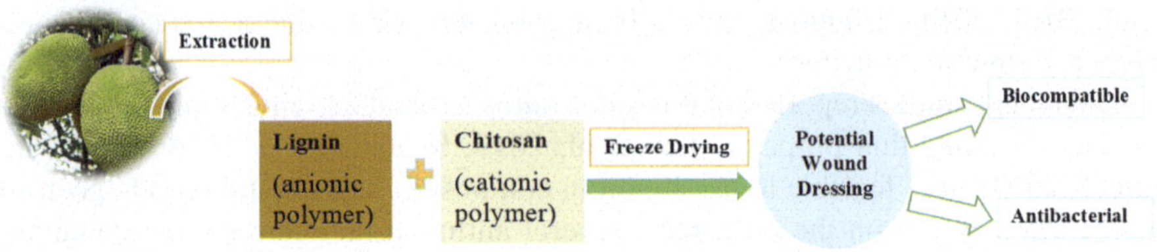

Fig. 6 Schematic diagram of *Artocarpus heterophyllus* peel lignin–chitosan biocomposites for wound healing applications

lignin hydrogels was performed using a scratch assay with mouse fibroblast cell lines (*NIH 3T3*). It was determined that hydrogels significantly enhanced the cells ability to migrate across the scratch. The results showed that the chitosan–alkali lignin hydrogels and films were biocompatible and support the cell attachment and growth with a conductive surface (Ravishankar et al. 2019).

Recently, lignin derived from a tropical fruit (*Artocarpus heterophyllus)* peel was evaluated to prepare lignin–chitosan biocomposites using freezing technique (Jaganathan et al. 2018). The porous structure of a material is one of the key parameters for treatment of wounds. Because porosity has significant effect on the wound tissue infiltration and also cell growth (Lee et al. 2016). According to scanning electron microscope (SEM) results, the morphology of biocomposites was porous-like permeable with a pore width of ~ 100 to 300 μm which should provide quick movement of liquid vapors and gases resulted in better healing effect. The composites were proposed as potential wound healing materials due to being biocompatible and showing superior antibacterial activity against *Escherichia coli* and *Klebsiella* bacteria (Fig. 6).

In another study, it was proposed that banana peel incorporated chitosan films can be used as dressing membrane to promote the process of wound healing. The dressing membranes posed as a synergistic interaction with a wide antimicrobial spectrum against gram-negative (*Escherichia coli*) and gram-positive bacteria (*Staphylococcus aureus*) (Rihayat et al. 2019).

3.3 Lignin Composites for Tissue Engineering Applications

Tissue engineering is an interdisciplinary field that involves the principles of chemistry, engineering, medicine material science, cell and molecular biology. In the light of these disciplines, tissue engineering includes the use of biomaterials, bioactive molecules and cells combinations. Hence, biomaterials are considered as basic components of tissue engineering, which seeks to provide support and scaffolding for cell growth (Keane and Badylak 2014). In other words, providing suitable conditions for the regeneration or replacement of distorted or missing tissue with newly developed functional tissue is expected from a biomaterial.

Extracellular matrix (ECM) is the nature's template for biomaterial which promotes cells into a three-dimensional (3D) structure to guide the growth and generation of the desired tissue (Drury and Mooney 2003). The field extensively based on the use of scaffolds as biomaterials to mimic the ECM. Proper characteristics of scaffolds are required for designing or determining the convenience of a scaffold for tissue engineering applications (O'brien 2011). Behind physical and biological features biocompatibility, proper mechanical properties, porosity, surface structure and topography are the key parameters affecting the regeneration process (Nazeer et al. 2019).

Significant interest and investigations have been focused on the application of biopolymer scaffolds in tissue engineering applications because of their supreme characteristics. Among the biopolymers, lignin-based composites can be taken into account to generate new multifunctional tissue engineering scaffolds. Published data support the use of lignin composites for tissue engineering applications.

Aerogel-based tissue engineering scaffolds are gaining a great deal of attention in the medical field. To date, various polymers (chitosan, polylactic acid, polyurea cross-linked silica, etc.) have been utilized as precursors to develop aerogel-based tissue engineering scaffolds. Alginate is a well-known biopolymer that has commercial biomaterial applications. However, the potential tissue engineering applications of alginate and alginate-based aerogels are limited because of hydrophilic characteristic of the alginate chains which cause decreased protein adsorption and cell adhesion (Rowley et al. 1999; Genes et al. 2004). In alginate–lignin systems, the expectation from lignin is to lessen hydrophilicity of alginate and by the way supply more appropriate environment for cell attachment, growth and differentiation. On the other hand, lignin has good stability which can provide reduction in the scaffold degradation rate and aid to pair it with the new bone tissue regeneration rate (Quraishi et al. 2015). A novel approach toward the fabrication of alginate–lignin hybrid aerogels using ionotropic gelation via pressurized carbon dioxide method was reported by Quraishi et al. (2015). The aerogels were prepared with varying alginate-to-lignin ratios of 2:1, 3:1, 4:1 or 5:1 (w/w). The bulk densities of alginate–lignin aerogels were found to be in the range of 0.03–0.07 g/cm^3. The aerogels present suitable textural and morphological properties, low stiffness, no cytotoxicity and good cell adhesion making them charming candidates for tissue engineering and regenerative medicine applications. Hence, the study opened a way for the preparation of aerogels with controlled porosity which may fill a niche in the biomedical field.

Electrospun PCL/lignin nanocomposites were prepared by Salami et al. (2017). Fiber diameter had an influence on porosity and pore size which have a significant mission on cell adhesion, proliferation and differentiation. The obtained fibers exhibited a suitable pore structure with multilayered micro-sized struts. Lignin incorporation enhanced the porosity and hydrophilic feature of the composite. Overall results showed that lignin-based PCL nanocomposites had enough features to mimic extracellular matrix structure which can be used in the design of soft tissue engineering scaffolds.

Abudula et al. (2018) created hybrid nonwoven scaffolds based on chitin/lignin/poly (glycerol sebacate) matrix using the electrospinning technique. The SEM observations showed that fibers were smooth and uniform. It was determined that the incorporation of poly (glycerol sebacate) greatly enhanced the mechanical performance and antibacterial activity against *S. aureus* and *E. coli* strains of chitin/lignin fibers. This consequence should be ascribed to glycol acid, the acidic degradation product of poly (glycerol sebacate), which promoted the antibacterial performance. Moreover, the scaffolds with an ECM-like fibrous structure have considerable potential to be used in a wide range of applications in the biomedical field.

Another potential scaffold was developed based on lignin/hydroxyapatite and lignin/β-tricalcium phosphate systems (Mansur et al. 2005). Lignin was extracted from Brazilian wood biomass. After sintering at 900 °C, the obtained scaffolds had porosity ranging from 50 to 100 μm. However, further characterization through in vitro and in vivo tests is essential to prove the suitability of this material for bone tissue engineering.

Nano-sized lignin production greatly extends the range of its evaluation in the biomedical field including drug delivery (Zhou et al. 2019), bioink systems (Capecchi et al. 2019), organic antioxidant/antimicrobial materials (Yang et al. 2018a) and UV-blocking composites (Xing et al. 2019; Yang et al. 2018b). A nanocomposite hydrogel was prepared based on polyacrylamide/lignin nanoparticles by simple and easily controlled in situ free radical polymerization method (Chen et al. 2019). The preparation of hydrogels is illustrated in (Fig. 7). The alkali lignin nanoparticles with an average particle size of 200 nm were obtained via an ultrasonic treatment using an ultrasonic homogenizer and further used to prepare hydrogels. Polyacrylamide/lignin hydrogel showed good mechanical properties in terms of high compressive and tensile strength which was the result of the reinforcement behavior of lignin nanoparticles as the co-crosslinker. The performances of polyacrylamide/lignin hydrogels in elasticity, toughness (up to 342 kJ/m^3) and stretchability were determined to be excellent. The elongation at break and tensile stress at fracture of the polyacrylamide/lignin hydrogels were found

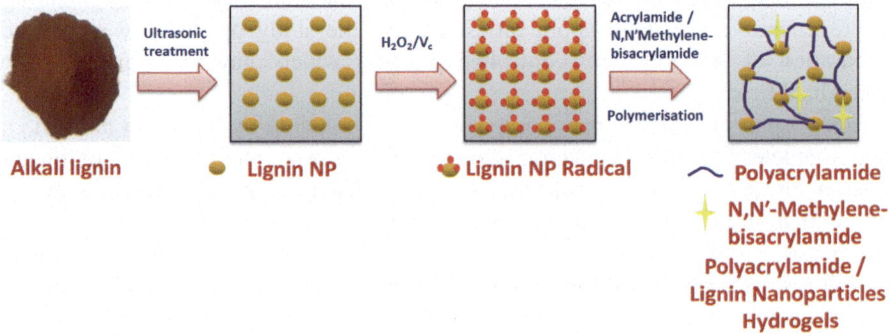

Fig. 7 Schematic diagram of the preparation of polyacrylamide/lignin nanocomposite hydrogel

to be approximately three to four times higher than that of the neat polyacrylamide hydrogel. As candidate biomaterials are to be applied in contact with living tissues, the cytotoxicities of polyacrylamide/lignin hydrogels were determined by MTT assay. The hydrogels were found to be noncytotoxic. It was also reported that the polyacrylamide/lignin hydrogel could enable the formation of 3D aggregated cells as well as assisted survival and proliferation of the cells due to its porous structure. This nanocomposite should be an ideal material for the applications in tissue engineering or regeneration, robust underwater antifouling materials, artificial muscles.

Based on recent research developments, lignin composites emerged as a suitable candidate for tissue engineering materials.

3.4 Lignin Composites for Drug Delivery Systems

Drug delivery systems are engineered technologies used to succeed higher therapeutic effects of a natural compound or pharmaceutical drug in an exact diseased location with minimal toxicological effect (Jacob et al. 2018). Nowadays, natural polymers get attention over synthetic ones. As an abundant biopolymer, lignin gains interest owing to its unique futures as antioxidant activity, antimicrobial behavior, biocompatibility, low toxicity, phenolic nature and UV protection (Zhou et al. 2019). Thereby, researchers endeavor to promote its use as potential drug delivery systems.

Lignin has used in the preparation of composite films for drug delivery systems. Alkaline lignin was extracted from waste hazelnut shells and further used to enhance the mechanical properties of thermoplastic corn starch films (Çalgeris et al. 2012). Lignin was used in concentrations of 1.2, 1.6, 2.0 and 2.4 wt% of total films. Antibiotic drug ciprofloxacin was loaded in composite films at 2 wt%. The obtained films were brown, and the colors darken with the rising amount of lignin. The films containing 2 wt% lignin were used for drug release studies due to presenting better mechanical properties. From the controlled release tests, it was found that films were sensitive to pH. Ciprofloxacin release was accelerated with the decrease in pH. The swelling ratios of composite films were increased with an increment in pH. The drug release mechanism of starch/lignin composite films was determined using Korsmeyer–Peppas model. Films followed a Fickian diffusion controlled drug release mechanism at pH 5.0 and pH 7.4. The starch/lignin biofilms exhibited a quick ciprofloxacin release within 60 min, and then, the drug release rate decelerated and attained to a steady state at the end of tenth day. Therefore, the ciprofloxacin/starch/lignin biofilm might be considered a promising delivery carrier.

Novel drug-loaded PLLA/lignin–poly-D-lactide (PDLA) stereocomplex composite films were developed using solvent volatilization method to improve stability and sustain trans-resveratrol (3,4′,5-*trans*-trihydroxy-stilbene, $C_{14}H_{12}O_3$) release (Liu et al. 2018). The authors reported that addition of lignin–PDLA provides better thermal properties and mechanical characteristic in terms of strength, strain and toughness. The practical application of trans-resveratrol was restricted due to its poor light stability. The PLLA/lignin–PDLA composite films exhibited good

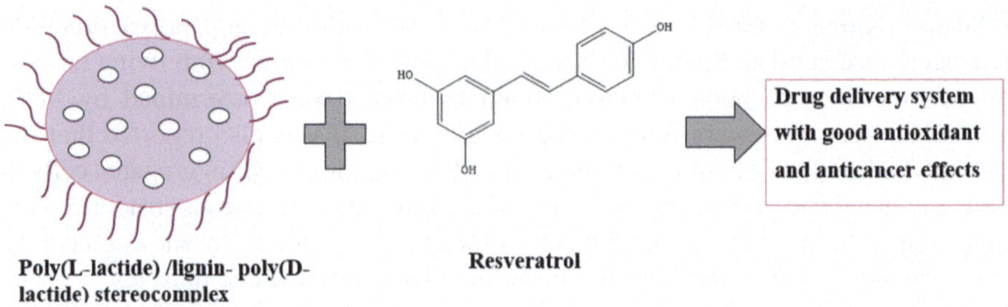

Fig. 8 Schematic diagram of the resveratrol-loaded PLLA/lignin–PDLA stereocomplex

UV-blocking ability, which is a key factor in protecting trans-resveratrol from sunlight. It was also pointed out that the stereocomplex composite films have excellent DPPH radical scavenging capability and anti-tumor activity according to CCK-8 assay (Fig. 8).

In hydrogel form, innovative super-absorbent cellulose–lignin composites were prepared by the chemical cross-linking in the presence of epichlorohydrin to be used in controlled release of polyphenols (Ciolacu et al. 2012). The obtained cellulose–lignin hydrogels showed high swelling capacities. The polyphenol adsorption rate of hydrogels increased as a function of swelling capacity. It was demonstrated that the compositions of the hydrogels have great influence on both the swelling and drug release process. Amount of released polyphenol increased with the raised lignin content.

4 Future Challenges and Perspectives

High-performance composites are required for biomedical applications. The number of articles about lignin-based composites and their evaluation in the biomedical field are increasing day by day. Current research and applications of functional lignin-based biomaterials are mainly focused on the preparation of hydrogel and electrospun scaffolds for tissue engineering as well as drug delivery systems. The positive results confirm its potential to be used in a wide range of applications in the biomedical field. Much attention is needed toward the in vivo (biocompatibility, biodegradability, cell attachment, etc.) evaluation of lignin composites, in order to prove the safety. Meanwhile, the design of novel lignin-based composites with better or distinct properties is also a challenging issue. Therefore, the available knowledge on lignin composites will provide an initial step through the commercial-scale production which can facilitate the usage of these composites in clinical applications in the next future.

It is also important to point out that lignin-based materials can be economically more feasible as compared to synthetic polymer materials because lignin is obtained from abundant natural resources. In this context, along with the evaluation of lignin composites as biomaterials, it is also worth mentioning the use of its composites in different engineering applications, which represents an additional and charming market opportunity. However, implementation of novel techniques including 3D printing, reactive extrusion, thermoforming, melt-electrospinning, 3D braiding and lithography for fabricating lignin composites in different forms beyond hydrogels and fibers is yet needed.

It is predicted that global lignin market will reach from US$ 904.04 to US$ 1021.57 with growth rate of around 2% between 2017 and 2023 years (Alzagameem et al. 2018). In the future perspective, lignin-based composites for biomedical applications are likely to take a great part in this rapidly growing market.

5 Conclusion

Natural polymer-based composites in the biomedical field are an interesting aspect as well as the demand of time. Therefore, the use of biopolymers in biomaterials is increasing steadily. The requirements for biopolymers include low toxicity, biodegradability, biocompatibility and the potential for further chemical modifications. One promising candidate that possesses these characteristics is lignin which plays an important role in the biomedical applications, either independently or with diverse materials. Recent researches highlighted the potential use of lignin in various types of composite preparations. Emerging lignin composites along with the power of several technical advantages have demonstrated a remarkable potential for future drug carriers, wound healing materials and scaffolds for tissue repair and regeneration. However, comprehensive knowledge of their biological, chemical and physical properties is still required to broaden applicability of different formulations of lignin composites in the biomedical field. Because most of the researches are in the initial stage presenting in vitro results. It is important to carry out in vivo studies to encounter with different demands. Additionally, the use of lignin in the development of new composite systems should be the focus of present studies. Hence, there is a need for further research efforts to prepare proper lignin-based composites with the desired properties and functions for specific biomedical applications.

References

Aadil KR, Mussatto SI, Jha H (2018) Synthesis and characterization of silver nanoparticles loaded poly (vinyl alcohol)-lignin electrospun nanofibers and their antimicrobial activity. Int J Biol Macromol 120:763–767

Abdel-Daim MM, Zakhary NI, Aleya L et al (2018) Aging, metabolic, and degenerative disorders: biomedical value of antioxidants. Oxid Med Cell Long. https://doi.org/10.1155/2018/2098123

Abudula T, Gzara L, Simonetti G et al (2018) The effect of poly (glycerol sebacate) incorporation within hybrid chitin-lignin sol–gel nanofibrous scaffolds. Materials 11(3):451

Albayrak S, Aksoy A (2019) Phenolic contents and biological activity of endemic origanum minutiflorum grown in Turkey. Indian J Pharm Edu Res 53(1):160–170

Alzagameem A, Bergs M, Witzler M et al (2018) Lignin-derived biomaterials for drug release and tissue engineering. Molecules 23(8):1885

Anwer MAS, Naguib HE, Celzard A et al (2015) Comparison of the thermal, dynamic mechanical and morphological properties of PLA-Lignin & PLA-Tannin particulate green composites. Compos B Eng 82:92–99

Capecchi ED, Piccinino BM, Bizzarri D et al (2019) Enzymes-lignin nanocapsules are sustainable catalysts and vehicles for the preparation of unique polyvalent bio-inks. Biomacromol 20(5):1975–1988

Carrott P, Carrott MR (2007) Lignin–from natural adsorbent to activated carbon: a review. Biores Technol 98(12):2301–2312

Chen Y, Zheng K, Niu L et al (2019) Highly mechanical properties nanocomposite hydrogels with biorenewable lignin nanoparticles. Int J Biol Macromol 128:414–420

Chio C, Sain M, Qin W (2019) Lignin utilization: a review of lignin depolymerization from various aspects. Renew Sustain Ener Rev 107:232–249

Ciolacu D, Oprea AM, Anghel N et al (2012) New cellulose–lignin hydrogels and their application in controlled release of polyphenols. Mater Sci Eng, C 32(3):452–463

Crouvisier-Urion K, Bodart PR, Winckler P et al (2016) Biobased composite films from chitosan and lignin: antioxidant activity related to structure and moisture. ACS Sustain Chem Eng 4(12):6371–6381

Çalgeris İ, Çakmakçı E, Ogan A et al (2012) Preparation and drug release properties of lignin–starch biodegradable films. Starch-Stärke 64(5):399–407

Demirci F, Yildirim K, Kocer HB (2018) Antimicrobial open-cell polyurethane foams with quaternary ammonium salts. J Appl Polym Sci 135(9):45914

Dhivya S, Padma VV, Santhini E (2015) Wound dressings–a review. Biomedicine 5(4):24–28

Diao BS, Zhang ZX, Zhu JL et al (2014) Biomass-based thermogelling copolymers consisting of lignin and grafted poly(N-isopropylacrylamide), poly(ethylene glycol), and poly(propylene glycol). RSC Adv 4:42996–43003

Domínguez-Robles J, Martin NK, Fong ML et al (2019) Antioxidant PLA composites containing lignin for 3D printing applications: a potential material for healthcare applications. Pharmaceutics 11(4):165

Dong X, Dong M, Lu Y et al (2011) Antimicrobial and antioxidant activities of lignin from residue of corn stover to ethanol production. Ind Crop Prod 34(3):1629–1634

Drury JL, Mooney DJ (2003) Hydrogels for tissue engineering: scaffold design variables and applications. Biomater 24(24):4337–4351

Duval A, Lawoko M (2014) A review on lignin-based polymeric, micro-and nano-structured materials. React Funct Polym 85:78–96

Eraković S, Janković A, Matić IZ et al (2013) Investigation of silver impact on hydroxyapatite/lignin coatings electrodeposited on titanium. Mater Chem Phys 142(2–3):521–530

Erakovic S, Jankovic A, Tsui G et al (2014) Novel bioactive antimicrobial lignin containing coatings on titanium obtained by electrophoretic deposition. Int J Mol Sci 15(7):12294–12322

Ge Y, Wei Q, Li Z (2014) Preparation and evaluation of the free radical scavenging activities of nanoscale lignin biomaterials. BioResources 9(4):6699–6706

Genes NG, Rowley JA, Mooney DJ et al (2004) Effect of substrate mechanics on chondrocyte adhesion to modified alginate surfaces. Arch Biochem Biophys 422(2):161–167

Greenhalgh R, Dempsey-Hibbert NC, Whitehead KA (2019) Antimicrobial strategies to reduce polymer biomaterial infections and their economic implications and considerations. Int Biodeter Biodegr 136:1–14

Gu L, Peng N, Chang C et al (2017) Fabrication of surface-active antioxidant food biopolymers: conjugation of catechin polymers to egg white proteins. Food Biophys 12(2):198–210

He X, Luzi F, Hao X et al (2019) Thermal, antioxidant and swelling behaviour of transparent polyvinyl (alcohol) films in presence of hydrophobic citric acid-modified lignin nanoparticles. Int J Biol Macromol 127:665–676
Jacob J, Haponiuk JT, Thomas S et al (2018) Biopolymer based nanomaterials in drug delivery systems: a review. Mater Today Chem 9:43–55
Jaganathan G, Manivannan K, Lakshmanan S et al (2018) Fabrication and characterization of *Artocarpus heterophyllus* waste derived lignin added chitosan biocomposites for wound dressing application. Sustain Chem Pharm 10:27–32
Janković A, Eraković S, Ristoscu C (2015) Structural and biological evaluation of lignin addition to simple and silver-doped hydroxyapatite thin films synthesized by matrix-assisted pulsed laser evaporation. J Mater Sci Mater Med 26(1):17
Kai D, Ren W, Tian L et al (2016a) Engineering poly (lactide)–lignin nanofibers with antioxidant activity for biomedical application. ACS Sustain Chem Eng 4(10):5268–5276
Kai D, Tan MJ, Chee PL et al (2016b) Towards lignin-based functional materials in a sustainable world. Green Chem 18:1175–1200
Karaman N, Sıcak Y, Taşkın-Tok T et al (2016) New piperidine-hydrazone derivatives: synthesis, biological evaluations and molecular docking studies as AChE and BChE inhibitors. Europ J Medic Chem 124:270–283
Kaur M, Mehta A, Gupta R (2018) Biomedical applications of synthetic and natural biodegradable polymers. In: Ahmed S et al (eds) Green and sustainable advanced materials. Scrivener Publishing LLC, issue no 2, pp 281–310
Keane TJ, Badylak SF (2014) Biomaterials for tissue engineering applications. Semin Pediatr Surg 23(3):112–118
Kenawy ER, Worley S, Broughton R (2007) The chemistry and applications of antimicrobial polymers: a state-of-the-art review. Biomacromol 8(5):1359–1384
Kim S, Fernandes MM, Matamá T et al (2013) Chitosan–lignosulfonates sono-chemically prepared nanoparticles: characterisation and potential applications. Colloid Surf B 103:1–8
Kim YS, Youe WJ, Kim SJ et al (2015) Preparation of a thermoplastic lignin-based biomaterial through atom transfer radical polymerization. J Wood Chem Technol 35(4):251–259
Lee SM, Park IK, Kim YS et al (2016) Physical, morphological, and wound healing properties of a polyurethane foam-film dressing. Biomater Res 20(1):15
Lee ES, Kim YO, Ha YM et al (2018) Antimicrobial properties of lignin-decorated thin multi-walled carbon nanotubes in poly (vinyl alcohol) nanocomposites. Europ Polym J 105:79–84
Lee E, Song Y, Lee S (2019) Crosslinking of lignin/poly (vinyl alcohol) nanocomposite fiber webs and their antimicrobial and ultraviolet-protective properties. Tex Res J 89(1):3–12
Lith R, Gregory EK, Yang J et al (2014) Engineering biodegradable polyester elastomers with antioxidant properties to attenuate oxidative stress in tissues. Biomaterials 35(28):8113–8122
Liu R, Dai L, Zou Z et al (2018) Drug-loaded poly (L-lactide)/lignin stereocomplex film for enhancing stability and sustained release of trans-resveratrol. Int J Biol Macromol 119:1129–1136
Lorenzo JM, Munekata PE, Gomez B et al (2018) Bioactive peptides as natural antioxidants in food products—a review. Trends Food Sci Tech 79:136–147
Mansur HS, Mansur AA, Bicallho SM (2005) Lignin-hydroxyapatite/tricalcium phosphate biocomposites: SEM/EDX and FTIR characterization. Key Eng Mater 284–286:745–748
Mehta MJ, Kumar A (2019) Ionic liquid stabilized gelatin-lignin films: a potential UV-shielding material with excellent mechanical and antimicrobial properties. Chem A Eur J 25(5):1269–1274
Meng Y, Lu J, Cheng Y et al (2019) Lignin-based hydrogels: a review of preparation, properties, and application. Int J Biol Macromol 135:1006–1019
Monteil-Rivera F, Phuong M, Ye M et al (2013) Isolation and characterization of herbaceous lignins for applications in biomaterials. Ind Crop Prod 41:356–364
Mutlu G, Calamak S, Ulubayram K et al (2018) Curcumin-loaded electrospun PHBV nanofibers as potential wound-dressing material. J Drug Deliv Sci Technol 43:185–193
Nada A, El-Diwany A, Elshafei A (1989) Infrared and antimicrobial studies on different lignins. Acta Biotechnol 9(3):295–298

Naseem AS, Tabasum KM, Zia M et al (2016) Lignin-derivatives based polymers, blends and composites: a review. Int J Biol Macromol 93:296–313
Nazeer MA, Yilgor E, Yilgor I (2019) Electrospun polycaprolactone/silk fibroin nanofibrous bioactive scaffolds for tissue engineering applications. Polymer 168:86–94
Norgren M, Edlund H (2014) Lignin: recent advances and emerging applications. Curr Opin Colloid In Sci 19(5):409–416
O'brien FJ (2011) Biomaterials & scaffolds for tissue engineering. Mater Today 14(3):88–95
Okur NÜ, Hökenek N, Okur ME et al (2019) An alternative approach to wound healing field; new composite films from natural polymers for mupirocin dermal delivery. Saudi Pharm J 27(5):738–752
Pandey MP, Kim CS (2011) Lignin depolymerization and conversion: a review of thermochemical methods. Chem Eng Technol 34(1):29–41
Park S, Kim SH, Won K et al (2015) Wood mimetic hydrogel beads for enzyme immobilization. Carbohydr Polym 115:223–229
Park SB, Lih E, Park KS et al (2017) Biopolymer-based functional composites for medical applications. Prog Polym Sci 68:77–105
Ponnusamy VK, Nguyen DD, Dharmaraja J et al (2018) A review on lignin structure, pretreatments, fermentation reactions and biorefinery potential. Biores Technol 271:462–472
Portela R, Leal CR, Almeida PL et al (2019) Bacterial cellulose: a versatile biopolymer for wound dressing applications. Microb Biotechnol. https://doi.org/10.1111/1751-7915.13392
Quraishi SM, Martins A, Barros A et al (2015) Novel non-cytotoxic alginate–lignin hybrid aerogels as scaffolds for tissue engineering. J Supercrit Fluid 105:1–8
Ravishankar K, Venkatesan M, Desingh RP et al (2019) Biocompatible hydrogels of chitosan-alkali lignin for potential wound healing applications. Mater Sci Eng, C 102:447–457
Reesi F, Minaiyan M, Taheri A (2018) A novel lignin-based nanofibrous dressing containing arginine for wound-healing applications. Drug Deliv Trans Res 8(1):111–122
Rihayat T, Siregar JP, Putra AZ et al (2019) Wound dressing based on banana peels waste and chitosan by strengthening lignin as wound healing medicine. IOP Conf Ser Mater Sci Eng 506:1–9. https://doi.org/10.1088/1757-899X/506/1/012056 (IOP Publishing)
Rowley JA, Madlambayan G, Mooney DJ (1999) Alginate hydrogels as synthetic extracellular matrix materials. Biomater 20(1):45–53
Salami MA, Kaveian F, Rafienia M et al (2017) Electrospun polycaprolactone/lignin-based nanocomposite as a novel tissue scaffold for biomedical applications. J Med Signal Sens 7(4):228–238
Schutyser W, Renders T, Bosch SV et al (2018) Chemicals from lignin: an interplay of lignocellulose fractionation, depolymerisation, and upgrading. Chem Soc Rev 47(3):852–908
Sezer AD, Cevher E (2011) Biopolymers as wound healing materials: challenges and new strategies. Biomater Appl Nanomed 19:383–413 (Intech Open)
Shankar S, Rhim JW, Won K (2018) Preparation of poly (lactide)/lignin/silver nanoparticles composite films with UV light barrier and antibacterial properties. Int J Biol Macromol 107:1724–1731
Spiridon I, Poni P, Ghica G et al (2018) Biological and pharmaceutical applications of lignin and its derivatives: a mini-review. Cell Chem Technol 52(7–8):543–550
Tamahkar E, Özkahraman B (2015) Potential evaluation of PVA-based hydrogels for biomedical applications. Hittite J Sci Eng 2(2)
Ten E, Vermerris W (2015) Recent developments in polymers derived from industrial lignin. J Appl Polym Sci 132(24)
Ugartondo V, Mitjans M, Vinardell MP (2008) Comparative antioxidant and cytotoxic effects of lignins from different sources. Biores Technol 99(14):6683–6687
Ugartondo V, Mitjans M, Vinardell MP (2009) Applicability of lignins from different sources as antioxidants based on the protective effects on lipid peroxidation induced by oxygen radicals. Ind Crop Prod 30(2):184–187

Wang J, Tian L, Luo B, Ramakrishna S, Kai D, Loh XJ, Yang IH, Deen GR, Mo X (2018a) Engineering PCL/lignin nanofibers as an antioxidant scaffold for the growth of neuron and Schwann cell. Colloid Surface B 169:356–365

Wang G, Xia Y, Liang B et al (2018b) Successive ethanol–water fractionation of enzymatic hydrolysis lignin to concentrate its antimicrobial activity. J Chem Technol Biotechnol 93(10):2977–2987

Witzler M, Alzagameem A, Bergs M et al (2018) Lignin-derived biomaterials for drug release and tissue engineering. Molecules 23:1885. https://doi.org/10.3390/molecules23081885

Xing Q, Buono P, Ruch D et al (2019) Biodegradable UV-blocking films through core-shell lignin–melanin nanoparticles in poly (butylene adipate-co-terephthalate). ACS Sustain Chem Eng 7(4):4147–4157

Xu Q, He C, Xiao C et al (2016) Reactive oxygen species (ROS) responsive polymers for biomedical applications. Macromol Biosci 16(5):635–646

Xu W, Wang X, Sandler N et al (2018) Three-dimensional printing of wood-derived biopolymers: a review focused on biomedical applications. ACS Sustain Chem Eng 6(5):5663–5680

Yang X, Liu Q, Chen X et al (2008) Investigation of PVA/ws-chitosan hydrogels prepared by combined γ-irradiation and freeze-thawing. Carbohyd Polym 73(3):401–408

Yang W, Fortunati E, Bertoglio F et al (2018a) Polyvinyl alcohol/chitosan hydrogels with enhanced antioxidant and antibacterial properties induced by lignin nanoparticles. Carbohyd Polym 181:275–284

Yang W, Rallini M, Wang DY et al (2018b) Role of lignin nanoparticles in UV resistance, thermal and mechanical performance of PMMA nanocomposites prepared by a combined free-radical graft polymerization/masterbatch procedure. Compos Part A Appl S 107:61–69

Zagho M, Hussein E, Elzatahry A (2018) Recent overviews in functional polymer composites for biomedical applications. Polymers 10(7):739

Zhu W, Lu J, Dai L (2018) Multifunctional pH-responsive sprayable hydrogel based on chitosan and lignin-based nanoparticles. Part Part Syst Charact 35:1800145

Zhou Y, Han Y, Li G et al (2019) Preparation of targeted lignin-based hollow nanoparticles for the delivery of doxorubicin. Nanomater 9(2):188

Zu Y, Zhang Y, Zhao X et al (2012) Preparation and characterization of chitosan–polyvinyl alcohol blend hydrogels for the controlled release of nano-insulin. Int J Biol Macromol 50(1):82–87

Applications of Lignin in the Agri-Food Industry

Rajeev Bhat, Aziz Ahmad and Ivi Jõudu

Abstract Of late, valorization of agri-food industrial wastes and by-products and their sustainable utilization is gaining much contemplation world over. Globally, 'Zero Waste Concept' is promoted with the main emphasis laid toward the generation of minimal wastes and maximal utilization of plant-based agri-food raw materials. One of the wastes/by-products in the agri-food industry is the lignin, which occurs as lignocellulosic biomass. This biomass is deliberated to be an environmental pollutant as they offer resistance to natural biodegradation. Safe disposal of this biomass is often considered a major challenge, especially in low-income countries. Hence, the application of modern technologies to effectively reduce these types of wastes and maximize their potential use/applications is vital in the present-day scenario. Nevertheless, in some of the high-income countries, attempts have been made to efficiently utilize lignin as a source of fuel, as a raw material in the paper industry, as a filler material in biopolymer-based packaging, and for producing bioethanol. However, as of today, agri-food industrial applications remain significantly underexplored. Chemically, lignin is heterogeneous, biopolymeric, and a polyphenolic compound, which is present naturally in plants, providing mechanical strength and rigidity. Reports are available wherein purified lignin is established to possess therapeutic values and are rich in antioxidants, antimicrobial, anti-carcinogenic, anti-diabetic properties, etc. This chapter is divided into four sub-categories focusing on various technological aspects related to isolation and characterization of lignin, established uses of lignin, proved bioactivities and therapeutic potentials of lignin, and finally on identifying the existing research gaps followed by future recommendations for potential use from agri-food industrial wastes.

R. Bhat (✉) · I. Jõudu
ERA Chair for Food (By-) Products Valorisation Technologies, Estonian University of Life Sciences, Fr.R.Kreutzwaldi 1, 51006 Tartu, Estonia
e-mail: rajeev.bhat@emu.ee

A. Ahmad
Faculty of Science and Marine Environment, Universiti Malaysia Terengganu, 21030 Kuala Nerus, Terengganu, Malaysia

S. Sharma and A. Kumar (eds.), *Lignin*, Springer Series on Polymer and Composite Materials, https://doi.org/10.1007/978-3-030-40663-9_10

Keywords Lignin · Waste valorization · Polymers · Bioactivities · Therapeutic potential

1 Introduction: Lignin-What Are They?

Lignin is a highly valued inexpensive waste, which is finding wide applications in the agri-food sector. Lignin stands next to cellulose as a natural plant-based polymer and is localized mainly in the cell walls. Chemically, lignin is a heterogeneous, biopolymeric, and polyphenolic compound providing mechanical strength and rigidity to plants. In fact, lignin is designated as a major renewable resource of carbon (and a better alternative to fossil) and aromatic molecules next to cellulose (AwungachaLekelefac et al. 2015; Bugg and Rahmanpour 2015). Naturally, as a polymer, lignin (a non-carbohydrate) is linked with cellulose and hemicellulose (carbohydrate polymers) via ester or ether or covalent bonds. As a polymer, the monomer units of lignin (rather precursor of lignin polymer) mainly comprise phenyl propane monomers (referred to as monolignol) encompassing sinapyl, p-coumaryl, and coniferyl alcohol. Syringyl–guaiacyl-type lignin is obtained from hardwoods, while it is guaiacyl type in softwoods, and with the presence of p-hydroxyphenyl lignin in both the types of woods. In Figs. 1 and 2, schematic representation of lignocellulose in biomass, its composition, precursors and basic unit in lignin molecule are provided. Lignin accounts for up to 30% (by dry mass) in plants

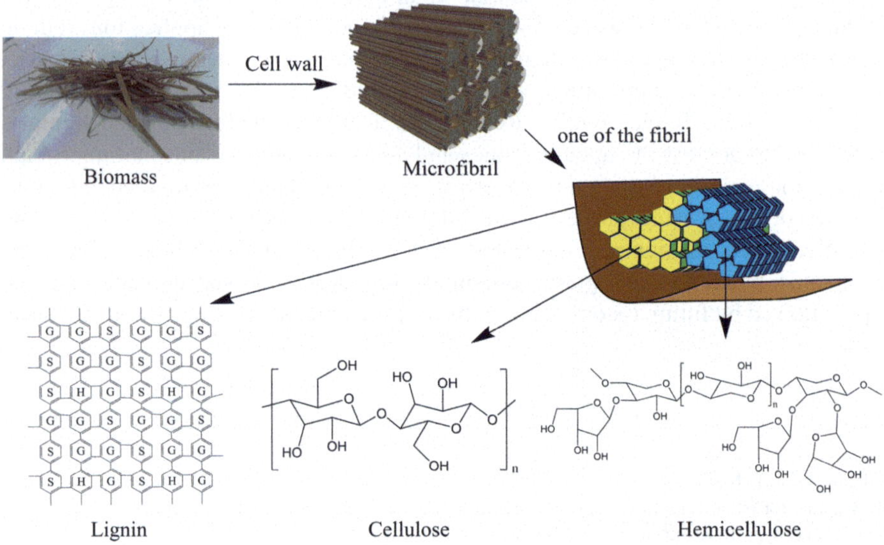

Fig. 1 Lignocellulose in biomass and its composition (reproduced with permission from Elsevier publishers, License Number 4556500933093 dated Mar 26, 2019, *Source* Chio et al. 2019, Journal: Renewable and Sustainable Energy Reviews)

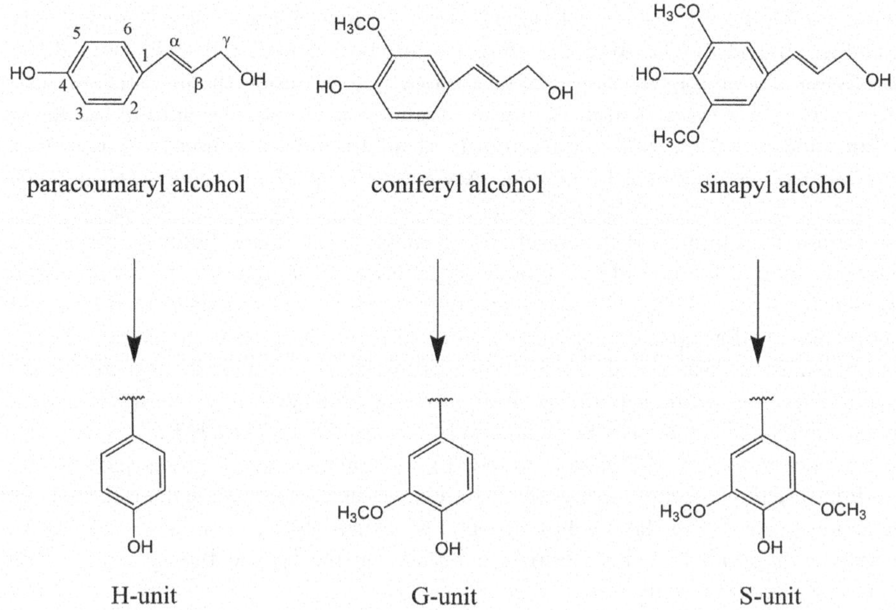

Fig. 2 Precursors and basic unit in lignin molecule (reproduced with permission from Elsevier publishers, License Number 4556511362424 dated Mar 26, 2019, *Source* Chio et al. 2019, Journal: Renewable and Sustainable Energy Reviews)

and ~40% of biomass as lignocellulose (cellulose, hemicellulose, and lignin) in agriculture residues (Lupoi et al. 2015; Ragauskas et al. 2014). Nevertheless, lignin waste is produced in abundance in pulping, wood processing, and paper industries (occurring as spent liquors). The most promising use of lignin has been generation of power (has a melting temperature of up to 160 °C) and in production of value-added biochemicals, aromatic compound (like benzene, toluene, xylene), cement additives, and biofuel in bio-refineries (Rinaldi et al. 2016; Xu et al. 2014). These potential uses have been attributed to the high carbon to oxygen ratio in lignin (Schutyser et al. 2018). Of late, lignin is explored to be used in cosmetics, agriculture, and food industry applications. Irrespective of this, huge tons of lignin waste generated contribute to serious environmental pollution-related issues. And, it is opined that ~2% of lignin to be effectively utilized for value addition (Lora and Glasser 2002).

2 Lignin Structure, Derivatives, and Modifications

Due to the heterogeneity nature, the primary chemical structure, properties, and functionality of lignin remain unidentified. The coupling or bonding, as well as arrangement between macromolecules, is also quite complicated. Nevertheless, the extraction of lignin from the lignocellulosic biomass/plant debris is technically a challenging task. The microstructure variations and isolation techniques differ for lignin isolated from various plant/agriculture residues (hemp, straws, flax, stem, etc.). Besides, isolation of lignin in pure form from agricultural wastes is tedious and expensive, and in the majority of the instances, lignin offers resistance to enzymatic or chemical treatments. According to available reports (Kim 2018; Kim et al. 2018), lignin valorization depends on efficient extraction and purification with minimal compressed structural arrangements from lignocellulose. Further, the lignin–carbohydrate complexes render the extraction of lignin to be complicated (Ragauskas et al. 2014). Of course, depolymerization of lignin via biocatalysts and finding uses in the paper, detergent, and food industries are well documented (Korbag and Saleh 2014; Wang et al. 2013, 2018). Till date, isolation, characterization, and evaluation of biosynthetic pathways of lignin have been performed mainly in silages, woods, and in selected vascular plants (Whetten and Sederoffa 1995). As extraction of lignin is expensive, attempts are made to develop engineered plants (genetically modified) with less scope for lignin accumulation or those wherein degradation can be achieved easily via chemical treatments (Mansfield 2009; Vanholme et al. 2008). Structure of lignin in lignocellulosic material and the typical linkages present in lignin are depicted in Figs. 3 and 4.

Lignin extraction (delignification) is mainly undertaken via acid hydrolysis, alkaline hydrolysis, reduction and oxidation fractionation, steam explosion, hydroxyl-methylation, acetylation, epoxidation reaction (epoxides formation), precipitation procedures, bleaching, fermentation, (Castro et al. 2019; Galkin and Samec 2016; Kim et al. 2016; Shi et al. 2019; Thomas et al. 2017) etc. Thioacidolysis coupled with the reductive cleavage method has also been explored to evaluate the cereal lignin composition (Bunzel et al. 2004; Begum et al. 2004). Kraft processing and sulfite/chemical pulping are common processes used to produce kraft lignin, lignosulfonates, and soda lignin (Figueiredo et al. 2018; Schutyser et al. 2018; Zhang et al. 2016). Further, organosolv processing (organic solvent + water + acid/base catalyst) is also recommended wherein pure lignin devoid of sulfur and hemicellulose can be obtained (Paszner and Cho 1989; Li et al. 2017; Zhang et al. 2007). Use of solvents like ethanol, methanol, acetone, and others with low boiling temperature is common in organosolv processes, and these solvents can be effortlessly recuperated via distillation method (Rodríguez and Jiménez 2008).

In obtaining aromatic compounds/aldehydes, the lignin oxidation process is recommended via the use of copper catalysts. Modification of lignin via various processes leads to cleavage in intramolecular bonds or in covalent linkages (mainly attached to the polysaccharides) leading to variations in the lignin fragments. Besides, the pulping process contributes to variations in the functional group (like sulfonate carboxyl or phenolic hydroxyl groups) of extracted lignin. The extraction process of lignin from lignocellulosic biomass is always undertaken cautiously to ensure that slow break down occurs to obtain lignin or their products with lower molecular weight (Bruijnincx and Weckhuysen 2014). However, extraction methods and original source of lignin can significantly influence the physicochemical

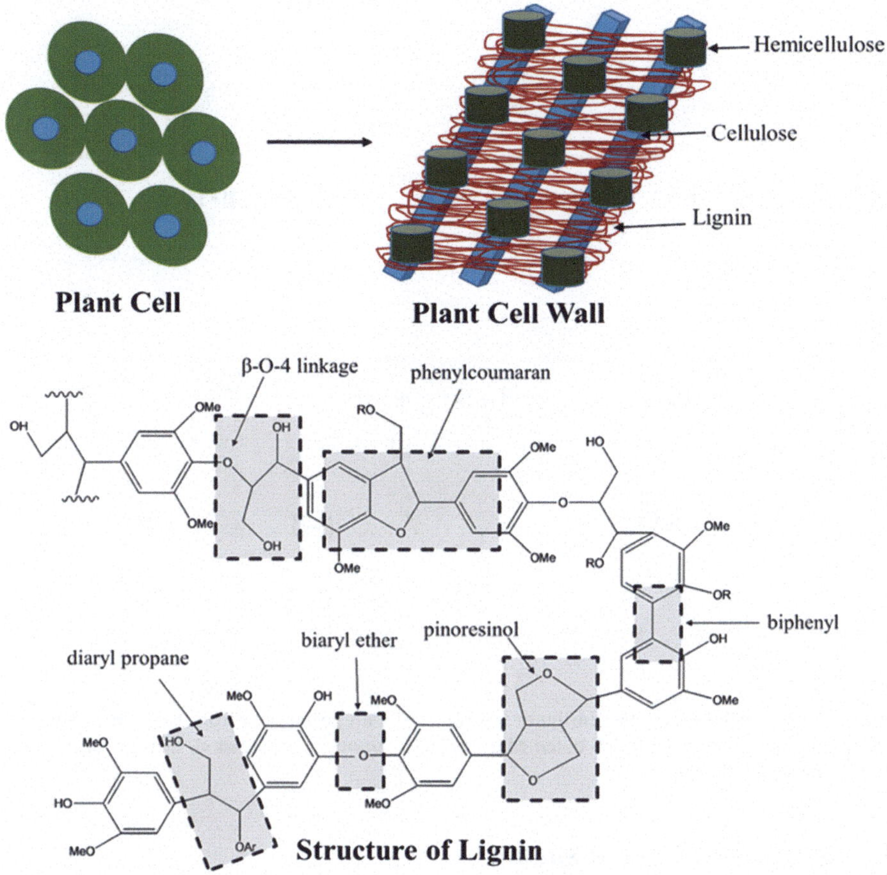

Fig. 3 Structure of lignin in lignocellulosic material (reproduced with permission from Elsevier publishers, License Number 4556530848259 dated Mar 26, 2019, *Source* Agarwal et al. 2018, Journal: Fuel Processing Technology)

properties of the extracted products (Upton and Kasko 2016). Once lignin is obtained in pure form, their characterization can be performed by various accessible techniques like NMR (31P NMR, 13C NMR), Fourier transform infrared spectroscopy, gel permeation chromatography, and other reliable analytical techniques, which are based on molecular weight determination. Some of the major components of lignin-derived extracts include p-hydroxybenzoic acid, ferulic acid, hydroxyphenyl acids, syringic acid, coumaric acid, vanillic acid, vanillin, p-hydroxybenzaldehyde, and syringaldehyde (Cruz et al. 2001).

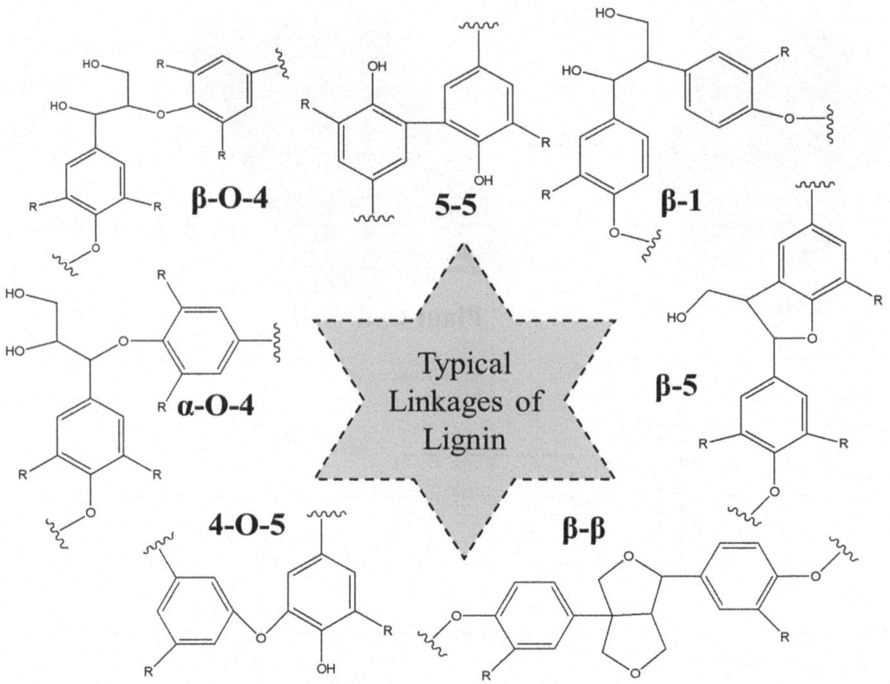

Fig. 4 Typical linkages present in lignin (reproduced with permission from Elsevier publishers, License Number 4556530848259 dated Mar 26, 2019, *Source* Agarwal et al. 2018, Journal: Fuel Processing Technology)

3 Established Uses of Lignin

The major use of lignin is their use as a fuel to generate energy. Production of biofuels is one the major area of the application of lignocellulosic materials (Crestini et al. 2010; Lou et al. 2013, 2014; Ro et al. 2019). Value addition in biorefineries via conversion into economically valued chemicals has generated much interest (Abo et al. 2019; Chen and Wan 2017). The major phenolic compounds and their derivatives obtained as intermediates during lignin biosynthesis hold much value and are used as a base material in cosmetics (e.g., moisturizing cream, sunscreen), bioplastics, resins, pharmaceuticals, food, and agriculture fields (such as herbicides). Wood adhesives and lignin–epoxy resins are prepared from lignin owed to their polyphenolic structure. Lignosulfonates also find potential usage as slurry dispersants, dyestuff dispersants, oil well dispersant, and surfactants (Qian et al. 2015; Stephen 1981; Wu et al. 2012). Promising use of lignin and their composites for developing functional materials with a wide range of applications as aerogels, nanofibers, and hydrogels has been detailed (Chen et al. 2011; Fernandes et al. 2013; Kai et al. 2015). In addition, depending on the source material, lignin can exhibit hydrophobic as well as hydrophilic nature, display good viscoelasticity and rheological features (Doherty et al. 2011). However, polyphenolic compounds in lignin and the action type of their derivatives can directly be linked with modification modes employed and on composition (Popaa et al. 2008). Polyphenolic compounds along with ketones not only provide protection against UV light but also perform the role of natural antioxidants (hence used in cosmetic industries like in sunscreens) (Qian et al. 2015; Zimniewska et al. 2012). Recently, Lee et al. (2019) reported on the synergistic effects of lignin isolated from *Miscanthus sacchariflorus,* which on addition to sunscreen significantly enhanced the protective effects against UV radiation. Besides, with oxidation leading to the formation of aromatic compounds and/or aldehydes, lignin is valued in the perfume industry too (Priefert et al. 2001; Qu et al. 2017).

Of late, in the modern agriculture system, lignin-based products are gaining popularity toward controlled/slow release of herbicides, pesticides, and fertilizers (e.g., urea granules) (Pereira et al. 2003; Mulder et al. 2011; Wang and Zhao 2013). Du et al. (2014) have reported 'layer-by-layer' self-assembly method for controlled herbicide and pesticide release via the use of microcapsules. Nevertheless, it is well established that lignin-based microcapsules do not exhibit any type of cytotoxicity and can be easily incorporated into cells (Tortora et al. 2014). Furthermore, for delivery of hydrophobic molecules, the same authors (Tortora et al. 2014) have reported on the use of ultrasound-assisted kraft lignin for microcapsulation. However, there is a word of caution, wherein controlled release to the soil can face wide challenges in the agriculture field where natural processes like bio-leaching/biodegradation/photo-degradation/volatilization, and other chemically induced degradations can occur, thus limiting the efficacy of the controlled release technique. Moreover, owing to high resistance offered to salt contamination, lignosulfonates are recommended to be used as a water–oil emulsion stabilizer in various field of applications like in asphalt stabilization, paper gluing, (Browning 1955; Cateto et al. 2014) etc. By the use of selective solvents and by precipitating lignin, nanoparticles were developed which were environmentally friendly, safe, and non-toxic (Frangville et al. 2012). These nanoparticles were recommended to find practical applications in heavy metals absorption as well as in drug delivery. Further, lignin-based complexes/nanoparticles are synthesized to explore their potential use in biomedical industrial applications (Figueiredo et al. 2017a, b). So also, several other potential applications of lignin-based nanoparticles have been reported (Chen et al. 2019; Mishra and Ekielski 2019; Zikeli et al. 2019).

Richter et al. (2016) have demonstrated the ability of lignin nanoparticles (precursors used was kraft lignin and organosolv lignin) to fulfill the purpose of producing novel biodegradable colloids with modifiable surface properties. According to the authors, the lignin-based nanoparticles are expected to perform the role of a biodegradable carrier and also had biocidal actives. Authors have opined that the new lignin-based nanoparticle formulation can be an effective base for future uses for sustainable transfer of nano-vehicles as pesticides, fungicides, antimicrobials, and drugs.

Moving ahead, lignin finds its application in the development of food packaging films, bioplastics development, polymer production, and biodegradability. Besides, lignin finds potential usage in textile and fabric industries (as protector or a barrier against intense UV light), in electrochemical uses, as heavy metals adsorbents, as a plasticizer, binder, dispersants, and much more (Bhat et al. 2013; Fierro et al. 2013; He et al. 2013; Klapiszewski et al. 2013; Lee and Lee 2014; Luo et al. 2018; Nagaraju et al. 2014; Stewart 2008; Zimniewska et al. 2008). Another interesting field of lignin application is its potential use as a basic raw material for the fabrication of cheap and economically feasible carbon fiber (Baker et al. 2012). It is expected that in the near future, this carbon fiber (nanofibers) can find wide applications as a replacement for polyacrylonitrile in industries linked with aerospace, automation, sports equipment production, capacitor, and batteries, etc.

Vanillin (4-hydroxy-3-methoxy benzaldehyde), a much appreciated and expensive flavoring compound originally isolated from vanilla beans can be produced from black liquor wastes of paper industry (like kraft lignin) or from lignosulfonates obtained from softwood of coniferous plants. Apart from food industrial use, vanillin also finds its applicability in the pharmaceutical and cosmetic industry (as an ingredient in perfume, lotion, sun protective cream, etc.).

Use of lignin as a natural biopolymer and as a filler material (cross-linking agent) in the development of novel packaging films has been studied in detail by various researchers. Owing to its stable inter and intramolecular bonding, coupled with thermoplastic nature and behavior, lignin has been a polymer in demand (Nimz 1974; Kubo and Kadla 2005; Ugartondo et al. 2008). According to Stewart (2008), chemical-based modification of lignin nanomaterials opens up the possibilities to develop novel, tailored multicomponent polymer system with desirable properties, which can be used for producing bioplastics for agricultural purposes (e.g., lignin-mulch films). Vengal and Srikumar (2005) extracted lignin from wood chips waste (obtained from the paper industry) and blended this with starch and gelatin to produce lignin-starch and lignin-gelatin biodegradable polymeric films. Calgeris et al. (2012) reported developing lignin-starch biodegradable films and studied their drug release properties. Their results revealed a faster drug release rate of ciprofloxacin (<1 h) from biofilms, with the drug release mechanism being pH-dependent.

Further, Bhat et al. (2013) have developed a low-cost, environmental friendly novel food packaging films from sago palm starch (as the film matrix) and lignin (as a reinforcing material) obtained from oil palm black liquor waste (with 30% w/w glycerol used as plasticizer). Packaging films developed exhibited good thermo-mechanical properties, barrier properties, and seal strength with improved water resistance. Besides, there are several reports focusing on the potential use of lignin in combination with their biopolymers (e.g., blending of starch/gelatin with lignin) and with chemical modifications (for polypropylene films) to produce packaging films for wider food industry applications (Baumberger et al. 1997).

Recently, Liu et al. (2019) have proposed bioconversion of kraft lignin to polyhydroxyalkanoate by using *Pandoraea* sp. B-6 (Gram-negative bacterium). Depolymerization of kraft lignin followed characterization that revealed 24.7% to be amassed in B-6 (grown in 6 g/L kraft lignin mineral medium). Further, it was observed that the monomer 3-hydroxybutyrate is to be present as a single major component, with PHA comparable to commercial bioplastics.

Lignin being a natural component of fiber, a strong bonding exists between the lignin matrix and fiber component (Naegelem et al. 2016). Natural composite films produced from this combination are always strong and display good mechanical properties. Besides, food packaging films developed from fiber/lignin-based materials can always be a good replacement for oil/petroleum-based polymers. Lignin used in different forms as a filler material (e.g., organosolv, kraft, soda lignin) is reported to exhibit better mechanical properties (Pucciariello et al. 2010). The electro-spinning of C-lignin (devoid of chemical treatments obtained from vanilla seeds) into fibers was successful and was foreseen to find wide industrial applications (Nar et al. 2016).

Hydrogels, which are either monomer or polymers suspended in aqueous solution, (water-based medium) have a wide range of medical applications (wound healing properties, providing relief against burns, drug delivery, tissue engineering, etc.) (Curvello et al. 2019; Mishra and Wimmer 2017). Of late, lignin-based hydrogels are gaining popularity owing to superior water absorption and retention properties, tissue imitating abilities, low toxicity, biocompatibility as well as controlled diffusion of desirable materials (Thakur and Thakur 2015; Zamboni et al. 2018). In agriculture fields, lignin-based hydrogels can be valuable to be utilized for reducing drought stress and to increase irrigation productivity where low water retention is a challenge. Hydrogels have been reported (Peng and Chen 2011) to be effective for slow release and movement of fertilizers too.

Lignin-based hydrogels have been prepared by graft copolymerization (El-Zawawy 2005). It was noticed that hydrogel prepared from alkaline lignin to exhibit higher swelling ratio with and slow water uptakes compared with kraft lignin, and this was assigned to the internal structural network (El-Zawawy 2005).

4 Biological Activities and Therapeutic Potential

Available research articles and databases clearly indicate plant-derived lignin to possess rich biological activities and therapeutic potential. Some of these include antioxidant properties, antimicrobial activity, anticancer/anti-tumor activity, anti-thrombosis, anti-HIV properties, and much more. The main compounds responsible for these activities are assigned to polyphenolic compounds, guaiacol and guaiacyl units, 4-hydroxy-3-methoxy-cinnamic acid, and ferulic acid (Azadfar et al. 2015; Sadeghifar and Argyropoulos 2015; Ou and Kwok 2004). However, compared to other macromolecules and bioactive compounds, research on identifying potential pharmaceutical benefits of lignin still remains in the infant stage. In Table 1, a summary on some biological properties and uses of lignin in the agri-food industry are depicted. In the preceding text, some reported activities are being discussed:

Table 1 Summary on the some biological properties and uses of lignin in agri-food industry

Active chemicals	Sources	Biological properties	References
Podophyllotoxin derivatives: etoposide, teniposide, and etopophos	Rhizome of *Podophyllum emodi* L (Berberidaceae)	Anti-cancers (colon, lung, testicle cancer) leukaemia	Gong et al. (2019)
Podophyllotoxin derivatives: Deoxypicropodophyllin, thuriferic acid deoxypodophyllotoxin	Leaves of *Juniperus thurifera* Linne (Cupressaceae)	Analgesic; anti-inflammatory	Guerrero et al. (2013)
Organosolv ethanol lignin (OEL)	Corn stalk	Tyrosinase inhibitor	Wang et al. (2019a, b, c)
Alkali and organosolv lignins	Wheat straw	Antioxidant	Gil-Chávez et al. (2019)
Lignin–carbohydrate complexes	Bamboo	Antioxidants or immune-stimulants	Huang et al. (2018)
Biocomposite lignin–chitosan	*Artocarpus heterophyllus*	Wound healing and dressing materials	Jaganathan et al. (2018)
Lignin-inspired bisguaiacol F (BGF)		Estrogenic activity	Peng et al. (2018)
Poly (lactic acid)–lignin biocomposites	Birch wood		Spiridon (2018)
Lignocellulosic	Oil palm empty fruit bunches	Antibacterial	Medina et al. (2016)
Lignosulfonic acid tannins	Cacao husk	Anti-viral (anti-HIV)	Sakagami et al. (2008)
Avermectin encapsulated with acetylated lignin and benzoylated lignin	Pulp-making black liquor	Slow release of pesticides	Zhou et al. (2019)

4.1 Antioxidant Activity

The main contributing factor for the exhibited antioxidant properties/radical scavenging activities of lignin is the presence of natural polyphenols. The effectiveness of polyphenolic components in lignin as a major contributing factor as an antioxidant (Barapatre et al. 2016; Hasnaoui et al. 2014; Zhu and Xu 2013) as well as providing protection against UV is well established in the cosmetic industry. The effective antioxidant activity of using lignin (and their linked mono- and dimeric compounds) has been tapped in dietary and medicinal products as well as in other soft materials like rubber and pulp (Barclay et al. 1997; Satoh et al. 1999).

Several available reports have indicated the potential antioxidant activities of lignin extracted from various plant resources like rice husk, hybrid poplar, apple tree, sugar cane, wheat straw, stover (Azadfar et al. 2015; Faustino et al. 2010; Pan et al. 2006; Vinardell et al. 2008; Salanti et al. 2010). Apart from being a potential free radical scavenger, lignin can stabilize various oxygen and free radical species induced reactions (Sakagami et al. 2005).

Vinardell et al. (2008) evaluated antioxidant activity in cosmetic or topical application of industrial lignin obtained from bagasse, curan, lignosulfonate, and by steam explosion. Among all, lignin obtained from bagasse was observed to be comparable with epi-catechin, a well-established antioxidant compound. Further, the authors concluded that all the lignin were safe on testing for eye and skin irritation leading the way for commercial exploitation in topical formulations and cosmetics. The antioxidant properties of lignin are also known to provide resistance against thermo-oxidation protection (Pouteau et al. 2003). Nevertheless, the efficacy of lignin antioxidants effects as stabilizers is well studied (Barana et al. 2016; Espinoza-Acosta et al. 2016). Presence of non-etherified phenolic hydroxyl groups serving as peroxyl radical scavengers is well demonstrated (García et al. 2010; Sadeghifar and Argyropoulos 2015). So also, Azadfar et al. (2015) have linked hydroxycinnamic acids and guaiacyl monomers (in pre-treated wheat straw with ozone and soaked in aqueous ammonia) with antioxidant activities and have concluded that lignin can be a good feedstock of 4-vinylguaiacol and guaiacol as antioxidants.

Lignin extracted from sugarcane has been studied for their biological activity and health benefits including that of antioxidant activity (Mitjans and Vinardell 2005). Kaur et al. (2017) via chemical modification (viz. acetylation, epoxidation, and hydroxymethylation) of sugarcane bagasse obtained lignin and reported unmodified lignin and hydroxymethyl lignin to exhibit high antioxidant activity. Antioxidant activity of lignin from 'corn stover' residues is reported to exhibit strong antioxidant activity, which was measured by employing Folin–Ciocalteureagent method and hydrophilic oxygen radical absorbance capacity (Dong et al. 2011).

Combination of gelatin and lignin for producing films is reported to enhance the oxidative stability, appearance, and of protein quality in 'salmon fillets' which were pre-exposed to high-pressure processing (Ojagh et al. 2011). Further, Núñez-Flores et al. (2013) used lignin as an active food packaging material (to tap antioxidant potentiality) for developing fish gelatin films (85% gelatin and 15% lignin: w/w) with antioxidant capacity (via the use of sulfur-free water-insoluble lignin powders L- 1000 and 2400) at non-cytotoxic concentrations. Warm water species extracted fish gelatin used with glycerol and sorbitol as a plasticizer. On visual examination at the macroscopic level, lignin incorporation led to a dark brownish color with reduced transparency and acceptable levels of miscibility. Further, at micro-level, results of the structural evaluation indicated lignin to inhibit gelatin molecule interaction. Films developed exhibited good plasticizing nature and light barrier properties, with negligible impact on water solubility. It was also reported that at non-cytotoxic concentration, lignin proves to be an effective antioxidant. The authors concluded by recommending the use of lignin-based composite films in food applications wherein UV induced lipid oxidation needs to be prevented.

Li and Ge (2012) investigated lignin extracted from sugarcane bagasse by employing ethanol and alkaline solutions. After characterizing, alkaline lignin was found to have large molecular mass with a higher number of phenolic hydroxyl and methoxyl groups offering good thermal stability than ethanol-extracted lignin. Further, their studies revealed the radical scavenging activity of alkaline lignin to be much higher than alcohol extracted lignin. Overall, in food industries, lignin's antioxidant capacity and radical scavenging ability are tapped for use in bio-packaging films as well as to enhance the compatibility of lignin with another polymer matrix. Lignin also finds application to be used as a stabilizer for foodstuff and livestock feed owed to their antioxidant and antimicrobial properties. However, due to the complexity in the structure, high molecular weight, presence of other polymers (admixture), and polydispersity, the efficacy of antioxidant and free radical scavenging activities can be significantly compromised/decreased (Dizhbite et al. 2004; Pouteau et al. 2005; Sederoff et al. 1999).

4.2 Antimicrobial Effects

Lignin has been explored for its potential antimicrobial activities against human and plant pathogens. The main compounds responsible for the activity have been assigned to the presence of polyphenols, sinapyl, and coniferyl alcohol (Zemek et al. 1979). Polyphenolic compounds have been shown to interact with the microbial cell membrane and disrupt their cell structure leading to cell component leakage (Gyawali and Ibrahim, 2014; Yang et al. 2016). Lignin obtained from pulp by-products (liquor) has been revealed to exhibit rich anti-bacterial activity against Gram-positive bacteria like *Bacillus subtilis* and *B. mycoids*. However, lignin had not exhibited any activities against Gram-negative bacteria or fungi (Nada et al. 1989). Oh-Hara et al. (1990) have shown antimicrobial activities exhibited by commercial lignin to be significantly low when compared with lignin fractions obtained via alkaline extracted pinecone.

Slavikova and Kosikova (1994) report antimicrobial activity against yeasts from lignin isolated from soft and hardwoods. Dizhbite et al. (2004) have demonstrated kraft lignin to be effective against phytopathogenic microorganisms. Dong et al. (2011) have reported antimicrobial activities of lignin extracted from corn stover residues against *Listeria monocytogenes* and *Staphylococcus aureus* (Gram-positive bacteria) and *Candida lipolytica* (yeast). However, authors reported that no activity exists for Gram-negative bacteria (bacteriophage MS2, *Escherichia coli* O157:H7 or *Salmonella enteritidis*).

Cruz et al. (2001) evaluated the antioxidant and antimicrobial activities of different lignocellulosic materials (barley bran, eucalyptus globulus wood, corncobs, corn leaves) obtained from acid hydrolysates as ethyl acetate extracts. Their results revealed eucalyptus wood extracts to have better inhibition among all of the extracts against yeasts and bacteria, with minimum inhibition concentration ranging between 10^2 and 5×10^3 μg/ml and minimum bactericide concentration being 10^3–10^5 μg/mL.

Antimicrobial activity against *Escherichia coli* was evaluated in cellulose fiber coated with chitosan and lignosulfonate (Hui and Lincai 2015). Alternated immersion technique was employed and results revealed chitosan to exhibit better antimicrobial activity compared to the lignosulfonates coated layer. Nanocomposite films produced by using poly (lactic acid) incorporated with two types of bio-based nano-fillers, viz. cellulose nanocrystals and lignin nanoparticles have been reported to exhibit antimicrobial activity against pathogenic Gram-negative bacteria like *Xanthomonas axonopodis* pv. vesicatoria and *X. arboricola*pv. pruni (Yang et al. 2016). Yearla and Padmasree (2016) by employing nanoprecipitation technique produced dioxane lignin nanoparticles and alkali lignin nanoparticles from two different sources of lignin (hardwood dioxane lignin obtained from the subabul stem and softwood alkali lignin obtained from a commercial source). Results of this study revealed that on exposure to UV at different time interval, dioxane lignin nanoparticles offer enhanced protection for *Escherichia coli* against UV. Crude lignin extract obtained from sugarcane bagasse has been testified to exhibit anti-bacterial effects against *Staphylococcus epidermidis* (DMST 15505; Minimum Inhibition Concentration being 4096 μg/mL) (Sunthornvarabhas et al. 2017). Kaur et al. (2017) on comparing unmodified and modified lignin from sugarcane bagasse reported epoxy lignin to be highly effective against pathogenic bacteria like *Bacillus* sp. and *Klebsiella* sp. Additionally, anti-bacterial activity against pathogenic *Staphylococcus aureus* (a Gram-positive bacteria) is reported from hydrotropic lignin obtained from birch wood (Gabov et al. 2017). Besides, anti-bacterial and antifungal activities of lignin capped silver nanoparticles against human pathogenic *E. coli, S. aureus*, and *A. niger* is also being reported (Marulasiddeshwara et al. 2017).

Jonglertjunya et al. (2014) coated limes in different combinations with lignin to evaluate their efficacy in the retention of quality as well as for potential antifungal activities. The formulations employed for the investigations comprised of 0.8% xanthan gum + 1.5% lignin extracted with 40% NaOH, 0.4% xanthan gum and 1.5% lignin extracted with 40% NaOH in 50% ethanol, and 0.8% xanthan gum + 1.5% lignin extracted from 40% NaOH in 50% ethanol. Results of this study showed 0.8% xanthan gum +1.5% lignin extracted in 40% NaOH to sustain weight loss and changes in color along with exhibiting good antifungal activities. Further, on the overall comparison, limes coated with lignin had better antifungal activity compared to those without coated with commercial lignin.

Additionally, several studies have reported on the anti-viral properties (e.g., anti-HIV) activity of lignin extracted from different sources. High levels of anti-HIV activities of lignin fractions. In comparison with hydrolyzable and condensed tannins have been reported by Nakashima et al. (1992). Lee et al. (2011) have established anti-viral activities against HSV 1 and HSV 2 in the lignin–carbohydrate–protein complex in aniseed plant (*Pimpinella anisum*). Matsuhisa et al. (2015) report strong inhibition of hepatitis C virus in cultured cells by alkaline hydrolysate lignin. Andrei et al. (2011) described the strong anti-HIV activity of lignin-sulfonic acid under in vivo (in HSV-2 infected mice) and in vitro conditions. On the other note, lignin application as an ultraviolet protectant against the beet armyworm nucleopolyhedrovirus and baculovirus is reported (Salamouny et al. 2002, 2009).

Baurhoo et al. (2008) in one of their review papers have discussed the effectiveness of Alcell lignin (purified lignin), which exhibited rich prebiotic effects in ‘monogastric’ animals by encouraging beneficial bacterial growth as well as enhancing the intestinal morphological structure (measured as villi height and goblet cell numbers). Further, lignin fragments have been accounted for inducing bactericidal effects and managing microbial pathogens in the intestine.

4.3 Anticancer/Anti-tumor Activity

There are a couple of studies reporting on strong anticancer/anti-tumor activities of lignin isolated from various sources. Plant-based dietary fibers (encompassing oligosaccharides, lignin, etc.) are already established to provide protection against various types of cancers. Protection against colon cancer by plant-based dietary fiber rich in lignin is reported (Meister 2007). Lu et al. (1998) have illustrated the protective effects of alkali lignin against colorectal cancer. These authors indicated the protective property against colon cancer to be dependent on the levels of lignin present in dietary fiber and on the free radical-scavenging ability of lignin. Siddiqui et al. (2018) have reported utilizing an environmentally friendly, cost-effective, biocompatible nano-particulate carrier composed of lignin formulations filled with an anticancer agent, P-gp modulator, functionalized with a ligand for CD44 receptors expressed on colon cancer cells.

Carboxylated lignin nanoparticles were used for anticancer therapy. Low water-soluble cytotoxic agents were loaded into carboxylated lignin nanoparticles which showed improved release (was pH-sensitive) and exhibited increased anti-proliferative effects in various types of cancer cells in comparison with normal endothelial cell lines (Figueiredo et al. 2017a, b).

In one of the studies by Ugartondo et al. (2008), lignin has been presented to exhibit rich antioxidant properties (at lower concentrations) along with exhibiting cytotoxic effects (at high concentrations). The authors established this via their studies on antioxidant activity and cytotoxic effects of different types of industrial lignin by using human keratinocyte HaCaTcell membrane and murine fibroblast 3T3 cells. Further, Lu et al. (1998) have attributed anticancer activity (growth and reduction in viable cancer cells) of lignin for their capability to inhibit enzyme activity, which was linked with superoxide anion radical generation. On the other note, inhibition of UV induced mutagenicity in *Euglena gracilis*(a single-celled flagellate eukaryote) is reported by Belicová et al. (2000).

4.4 Anticoagulant, Anti-emphysema, Anti-diabetic, and Anti-obesity

A wide range of research activities is being concentrated toward finding other potential therapeutic applications of lignin in the medicinal field. Controlled release concept proposed in agriculture fields (for fertilizers and herbicides) has been extended for release of various drugs used in human medicine. Some of these include evaluating the anticoagulant, anti-emphysema, anti-diabetic, and anti-obesity activities. Reports available indicate the inhibition potential of thrombin by sulfated low molecular weight lignin (Henry and Desai 2014; Mehta et al. 2016). Novel anticoagulants exhibiting dual anionic properties and hydrophobicity were produced by using sulfated low molecular weight lignin which accomplished the role of practical mimetics compared to low molecular weight heparin (Monien et al. 2006).

Lignin's anti-nociceptive and anti-inflammatory properties have been assigned to the presence of sinapyl alcohol, a lignin precursor. Anti-proliferative activity of water-soluble fractions of carbohydrate–lignin metabolites (obtained from Chaga mushroom) is reported (Song et al. 2008).

Barapatre et al. (2015) have reported anti-diabetic activity of alkali lignin by in vitro alfa-amylase inhibition. Further, Hasegawa et al. (2015) by employing in vivo (rat) and in vitro experiments (alfa-glucosidase inhibition) reported the efficacy of lignosulfonate acid to show anti-diabetic activity, wherein the mechanism of action includes inhibition of alfa-glucosidase and decreased levels of glycaemia in blood.

Anti-obesity properties of lignin-phenols have been shown to manage obesity in HepG2. Norikura et al. (2010) by employing in vitro models showed a decrease in oleate-induced apo-B secretion. Further, lignin-phenols have been linked with obesity control under in vivo conditions wherein rats which were fed with a high-fat diet showed decreased plasma triglyceride level (Sato et al. 2009).

On another note, Zhang et al. (2013) have reported on the formation of a new activator of α-amylase and lignin, which significantly enhanced α-amylase activities. Their results showed lignin to interact with α-amylase forming a 1:1 lignin/α-amylase complex. As per the authors, this result is useful to understand lignin as a dietary fiber component and was envisaged to contribute as an activator of α-amylase, in food industries.

Pancreatic enzymes activity is linked with human health as it helps in hydrolyzing fat molecules and in active digestion and adsorption of micro nutrients. Any disturbances of this enzyme can lead to diabetic conditions and adiposity. Zhang et al. (2014) have studied the effects of lignin on pancreatic lipase and reported a pancreatic lipase–lignin complex to be formed with enhanced enzyme activity. As per the authors, generated results help in understanding the activation effects and interaction between lignin and pancreatic lipase, and this is expected to find wide applications in both food and pharmaceutical industries.

5 Conclusion and Future Prospects

In the present review, we have summarized information from the currently available databases on the potentiality of plant-derived lignin. The prospective uses and application of lignin which remain significantly underutilized as a plant by-product are discussed in detail. Though many interesting research works have been initiated and have been reported in recent years, still there is a wide gap which exists relevant to tapping lignin's potentiality and valorization of lignin for agri-food industry applications. Though reports indicate highly value-added products (as biofuel, biopolymers, chemicals, aromatic monomers, cosmetic, and medicinal uses) to be produced, lignin's application remains in infancy, especially in agri-food and pharmaceutical sectors. The structural complexity of lignin isolated from differed sources has been considered to be a major hindrance for practical exploitation. Some of the common challenges incurred on purification have been overcome, and available reports indicate lignin's efficacy to be useful in drug delivery, microencapsulation for controlled release of desired materials in food industry (release of bioactive compounds in bio-packaging films) and agricultural fields (release of fertilizers or pesticides) and cosmetics applications (as in sunscreen). Lignin's use as a filler material in packaging films and as composites in hydrogels (eco-friendliness) offers much scope in pharmaceuticals and agri-food industries, and this gap needs to be filled at the earliest. Use of lignin/carbon as a bio-sorbent for in situ treatment of agriculture and food industry wastewater is another arena that needs to be explored. Nano-composites/nanoparticles/nanostructured materials from lignin as well as developing carbon nanofibers are still in initial stages. Further, proved antioxidant potential can be a boon for the food industry to use lignin as a natural preservative; whereas, proved antimicrobial activities can also be beneficial for application of lignin and their products for safe storage of foods (active food packaging material) or in agriculture field (as antimicrobial coating agent) to provide protection against plant pathogens. Owed to lignin's natural biodegradability and being devoid of cytotoxicity, they can be a good basic raw material in food packaging industries.

Future research needs to focus on developing low-cost technologies for easy conversion of lignin biomass to obtain purity and identification techniques to ensure the structural components/integrity. The scope is also there for utilizing lignin in livestock and pets food as well as to those related to pharma–cosmetic–food applications. Attempts can also be made to offer genetic modification of lignin to obtain higher amounts of biomass to be exclusively used in the agri-food industry. Even though a wide range of opportunities coexists, newer challenges can also be expected, especially considering the purity and safety issues. Finally, sustainable approaches/green technology concepts need to be adopted for tapping the potentiality of lignin for agri-food industrial applications.

Acknowledgements The theme of this chapter is based on our ongoing project— VALORTECH, which has received funding from the European Union's Horizon 2020 research and innovation program under grant agreement No 810630.

References

Abo BO, Gao M, Wang Y, Wu C, Ma H, Wang Q (2019) Lignocellulosic biomass for bioethanol: an overview on pre-treatment, hydrolysis and fermentation processes. Rev Environ Health 34(1):57–68

Agarwal A, Rana M, Park JH (2018) Advancement in technologies for the depolymerization of lignin. Fuel Process Technol 181:115–132
Andrei G, Lisco A, Vanpouille C, Introini A et al (2011) Topical tenofovir, a microbicide effective against HIV, inhibits herpes simplex virus-2 replication. Cell Host Microbe 10:379–389
AwungachaLekelefac C, Busse N, Herrenbauer M, Czermak P (2015) Photocatalytic based degradation processes of lignin derivatives. Int J Photoenergy 12:1–18
Azadfar M, Gao AH, Bule MV, Chen S (2015) Structural characterization of lignin: a potential source of antioxidants guaiacol and 4-vinylguaiacol. Int J Biol Macromol 75:58–66
Baker DA, Gallego NC, Baker FS (2012) On the characterization and spinning of organic purified lignin toward the manufacture of low-cost carbon fiber. J Appl Polym Sci 124:227–334
Barana D, Ali SD, Salanti A, Orlandi M, Castellani L, Hanel T, Zoia L (2016) Influence of lignin features on thermal stability and mechanical properties of natural rubber compounds. ACS Sustain Chem Eng 4:5258–5267
Barapatre A, Aadil KR, Tiwary BN, Jha H (2015) *In vitro* antioxidant and antidiabetic activities of biomodified lignin from *Acacia nilotica* wood. Int J Biol Macromol 75:81–89
Barapatre A, Meena AS, Mekala S, Das A, Jha H (2016) In vitro evaluation of antioxidant and cytotoxic activities of lignin fractions extracted from *Acacia nilotica*. Int J Biol Macromol 86:443–453
Barclay LRC, Xi F, Norris JQ (1997) Antioxidant properties of phenolic lignin model compounds. J Wood Chem Technol 17(1–2):73–90
Baumberger S, Lapierre C, Monties B, Lourdin D, Colonna B (1997) Preparation and properties of thermally moulded and cast lignosulfonates-starch blends. Ind Crops Prod 6:253–258
Baurhoo B, Ruiz-Feria CA, Zhaoa X (2008) Purified lignin: nutritional and health impacts on farm animals-A review. Anim Feed Sci Technol 144:175–184
Begum AN, Nicolle C, Mila I, Lapierre C, Nagano K, Fukushima K, Heinonen S-M, Adlercreutz H, Remesy C, Scalbert A (2004) Dietary lignins are precursors of mammalian lignans in rats. J Nutr 134:120–127
Belicová A, Krajcovic J, Krizková L et al (2000) Anti-UV activity of lignin biopolymers on *Euglena gracilis*. World J Microbiol Biotechnol 16:91–93
Bhat R, Abdullah N, Din RH, Tay G-S (2013) Producing novel sago starch-based food packaging films by incorporating lignin isolated from oil palm black liquor waste. J Food Eng 119:707–713
Browning WC (1955) Lignosulfonate stabilized emulsions in oil well drilling fluids. J Petrol Technol 7:9–15
Bruijnincx PCA, Weckhuysen BM (2014) Biomass conversion: lignin up for break-down. Nat Chem 6:1035–1036
Bugg TDH, Rahmanpour R (2015) Enzymatic conversion of lignin into renewable chemicals. Curr Opin Chem Biol 29:10–17
Bunzel M, Ralph J, Lu F, Hatfield RD, Steinhart H (2004) Lignins and ferulate-coniferyl alcohol cross-coupling products in cereal grains. J Agric Food Chem 52:6496–6502
Calgeris I, Cakmakci E, Ogan A, Kahraman MV, Kayaman-Apohan N (2012) Preparation and drug release properties of lignin– starch biodegradable films. Starch/Starke 64:399–407
Castro RCA, Ferreira IS, Roberto IC, Mussatto SI (2019) Isolation and physicochemical characterization of different lignin streams generated during the second-generation ethanol production process. Int J Biol Macromol 129:497–510
Cateto CA, Barreiro MF, Ottati C, Lopretti M, Rodrigues AE, Belgacem MN (2014) Lignin-based rigid polyurethane foams with improved biodegradation. J Cell Plast 50:81–95
Chen F, Xu M, Wang L, Li J (2011) Preparation and characterization of organic aerogels by the lignin-resorcinol-formaldehyde copolymer. BioResources 6:1262–1272
Chen Y, Zheng K, Niu L, Zhang Y, Liu Y, Wang C, Chu F (2019) Highly mechanical properties nanocomposite hydrogels with biorenewable lignin nanoparticles. Int J Biol Macromol 128:414–420
Chen Z, Wan C (2017) Biological valorization strategies for converting lignin into fuels and chemicals. Renew Sustain Energy Rev 73:610–621

Crestini C, Crucianelli M, Orlandi M, Saladino R (2010) Oxidative strategies in lignin chemistry: a new environmental friendly approach for the functionalisation of lignin and lignocellulosic fibers. Catal Today 156:8–22
Cruz JM, Dominguez JM, Dominguez H, Parajo JC (2001) Antioxidant and antimicrobial effects of extracts from hydrolysates of lignocellulosic materials. J Agric Food Chem 2001(49):2459–2464
Curvello R, Raghuwanshi VS, Garnier G (2019) Engineering nanocellulose hydrogels for biomedical applications. Adv Colloid Interface Sci 267:47–61
Dizhbite T, Telysheva G, Jurkjane V, Viesturs U (2004) Characterization of the radical scavenging activity of lignins-natural antioxidants. Bioresour Technol 95:309–317
Doherty WOS, Mousavioun P, Fellows CM (2011) Value-adding to cellulosic ethanol: lignin polymers. Ind Crops Prod 33:259–276
Dong X, Dong M, Lu Y, Turley A, Jin T, Wu C (2011) Antimicrobial and antioxidant activities of lignin from residue of corn stover to ethanol production. Ind Crops Prod 34:1629–1634
Du XY, Li LB, Lindström ME (2014) Modification of industrial softwood kraft lignin using Mannich reaction with and without phenolation pre-treatment. Ind Crops Prod 52:729–735
El-Zawawy WK (2005) Preparation of hydrogel from green polymer. Polym Adv Technol 16:48–54
Espinoza-Acosta JL, Torres-Chávez PI, Ramírez-Wong B et al (2016) Antioxidant, antimicrobial, and antimutagenic properties of technical lignins and their applications. BioResources 11:5452–5481
Faustino H, Gil N, Baptista C, Duarte AP (2010) Antioxidant activity of lignin phenolic compounds extracted from kraft and sulphite black liquors. Molecules 15:9308–9322
Fernandes EM, Pires RA, Mano JF, Reis RL (2013) Bionanocomposites from lignocellulosic resources: properties, applications and future trends for their use in the biomedical field. Prog Polym Sci 38:1415–1441
Fierro CM, Gorka J, Zazo JA, Rodriguez JJ, Ludwinowicz J, Jaroniec M (2013) Colloidal templating synthesis and adsorption characteristics of microporous–mesoporous carbons from Kraft lignin. Carbon 62:233–239
Figueiredo P, Ferro C, Kemell M et al (2017a) Functionalization of carboxylated lignin nanoparticles for targeted and pH-responsive delivery of anticancer drugs. Nanomedicine 12(21):2581–2596
Figueiredo P, Lintinen K, Kiriazis A, Hynninen V, Liu Z, Bauleth-Ramos T et al (2017b) *In vitro* evaluation of biodegradable lignin-based nanoparticles for drug delivery and enhanced antiproliferation effect in cancer cells. Biomaterials 121:97–108
Figueiredo P, Lintinen K, Hirvonen JT, Kostiainen MA, Santos HA (2018) Properties and chemical modifications of lignin: towards lignin-based nanomaterials for biomedical applications. Prog Mater Sci 93:233–269
Frangville C, Rutkevicius M, Richter AP, Velev OD, Stoyanov SD, Paunov VN (2012) Fabrication of environmentally biodegradable lignin nanoparticles. ChemPhysChem 13:4235–4243
Gabov K, Oja T, Deguchi T, Fallarero A, Fardim P (2017) Preparation, characterization and antimicrobial application of hybrid cellulose-lignin beads. Cellulose 24:641–658
Galkin MV, Samec JS (2016) Lignin valorization through catalytic lignocellulose fractionation: a fundamental platform for the future biorefinery. Chemsuschem 9:1544–1558
García A, Toledano A, Andrés MÁ, Labidi J (2010) Study of the antioxidant capacity of *Miscanthus sinensis* lignins. Process Biochem 45:935–940
Gil-Chávez GJ, Padhi SSP, Pereira CV et al (2019) Cytotoxicity and biological capacity of sulfur-free lignins obtained in novel biorefining process. Int J Biol Macromol 136:697–703
Gong N, Yang D, Jin G, Liu S, Du G, Lu Y (2019) Structure, characterization, solubility and stability of podophyllotoxin polymorphs. J Mol Struct 1195:323–330
Guerrero E, Abad A, Montenegro G, del Olmo E, López-Pérez JS, Feliciano AS (2013) Analgesic and anti-inflammatory activity of podophyllotoxin derivatives. J Pharm Biol 51(5):566–572
Gyawali R, Ibrahim SA (2014) Natural products as antimicrobial agents. Food Control 46:412–429
Hasegawa Y, Kadota Y, Hasegawa C, Kawiminami S (2015) Lignosulfonic acid-induced inhibition of intestinal glucose absorption. J Nutr Sci Vitaminol 61:449–454

Hasnaoui N, Wathelet B, Jiménez-Araujo A (2014) Valorization of pomegranate peel from 12 cultivars: dietary fibre composition, antioxidant capacity and functional properties. Food Chem 160:196–203

He ZW, He LH, Yang J, Lu QF (2013) Removal and recovery of Au (iii) from aqueous solution using a low-cost lignin-based biosorbent. Ind Eng Chem Res 52:4103–4138

Henry BL, Desai UR (2014) Sulfated low molecular weight lignins, Allosteric inhibitors of coagulation proteinases via the heparin binding site, significantly alter the active site of thrombin and factor xa compared to heparin. Thromb Res 134:1123–1129

Huang C, Tang S, Zhang W, Tao Y, Lai C, Li X, Yong Q (2018) unveiling the structural properties of lignin–carbohydrate complexes in bamboo residues and its functionality as antioxidants and immunostimulants. ACS Sustain Chem Eng 69:12522–12531

Hui L, Lincai P (2015) Antimicrobial and antioxidant surface modification of cellulose fibers using layer-by-layer deposition on chitosan and lignosulfonates. Carbohydr Polym 124:35–42

Jaganathan G, Manivannan K, Lakshmanan S, Sithique MA (2018) Fabrication and characterization of *Artocarpus heterophyllus* waste derived lignin added chitosan biocomposites for wound dressing application. Sustain Chem Pharm 10:27–32

Jonglertjunya W, Juntong T, Pakkang N, Srimarut N, Sakdaronnarong C (2014) Properties of lignin extracted from sugarcane bagasse and its efficacy in maintaining postharvest quality of limes during storage. LWT Food Sci Technol 57:116–125

Kai D, Jiang S, Low ZW, Loh XJ (2015) Engineering highly stretchable lignin-based electrospun nanofibers for potential biomedical applications. J Mater Chem B 3:6194–6204

Kaur R, Uppal SK, Sharma P (2017) Antioxidant and antibacterial activities of sugarcane bagasse lignin and chemically modified lignin. Sugar Technol 19:675–680

Kim D (2018) Physico-chemical conversion of lignocellulose: inhibitor effects and detoxification strategies: a mini review. Mol (Basel, Switzerland), 23(2):309. https://doi.org/10.3390/molecules23020309

Kim JS, Lee YY, Kim TH (2016) A review on alkaline pretreatment technology for bioconversion of lignocellulosic biomass. Bioresour Technol 199:42–48

Kim KH, Dutta T, Sun J, Simmons B, Singh S (2018) Biomass pretreatment using deep eutectic solvents from lignin derived phenols. Green Chem 20:809–815

Klapiszewski L, Nowacka M, Milczarek G, Jesionowski T (2013) Physicochemical and electrokinetic properties of silica/lignin biocomposites. Carbohydr Polym 94:345–355

Korbag I, Saleh SM (2014) Extraction of lignin from paper industry waste. Int J Appl Eng Res 9:19421–19428

Kubo S, Kadla JF (2005) Hydrogen bonding in lignin: a fourier transform infrared model compound study. Biomacromol 6:2815–2821

Lee E, Lee S (2014) Fabrication of lignin nanofibers using electrospinning. J Korean Soc Clothing Text 38:372–385

Lee JB, Yamagishi C, Hayashi K, Hayashi T (2011) Antiviral and immunostimulating effects of lignin-carbohydrate-protein complexes from *Pimpinellaanisum*. Biosci Biotechnol Biochem 75(3):459–465

Lee SC, Thu Tran TM, Choi JW, Won K (2019) Lignin for white natural sunscreens. Int J Biol Macromol 122:549–554

Li H-L, She D, Peng P, Xu Q, Liu J-K, Zhang X-M, Geng Z-C (2017) Optimizing extraction and structural characterization of organosolv lignin from wheat straw. Cellul Chem Technol 51:433–445

Li Z, Ge Y (2012) Antioxidant activities of lignin extracted from sugarcane bagasse via different chemical procedures. Int J Biol Macromol 51:1116–1120

Liu D, Yan X, Si M, Deng X, Min X, Shi Y, Chai L (2019) Bioconversion of lignin into bioplastics by Pandoraea sp. B-6: molecular mechanism. Environ Sci Pollut Res 3:2761–2770

Lora JH, Glasser WG (2002) Recent industrial applications of lignin: a sustainable alternative to non-renewable materials. J Polym Environ 10:39–48

Lou HM, Zhou HF, Li XL, Wang MX, Zhu JY, Qiu XQ (2014) Understanding the effects of lignosulfonate on enzymatic saccharification of pure cellulose. Cellulose 21:1352–1359
Lou HM, Zhu JY, Lan TQ, Lai HR, Qiu XQ (2013) pH-induced lignin surface modification to reduce nonspecific cellulase binding and enhance enzymatic saccharification of lignocelluloses. Chemsuschem 6:919–927
Lu FJ, Chu LH, Gau RJ (1998) Free radical-scavenging properties of lignin. Nutr Cancer 30:31–38
Luo X, Xiao Y, Wu Q, Zeng J (2018) Development of high-performance biodegradable rigid polyurethane foams using all bioresource-based polyols: lignin and soy oil derived polyols. Int J Biol Macromol 115:786–791
Lupoi JS, Singh S, Parthasarathi R, Simmons BA, Henry RJ (2015) Recent innovations in analytical methods for the qualitative and quantitative assessment of lignin. Renew Sustain Energy Rev 49:871–906
Mansfield SD (2009) Solutions for dissolution-engineering cell walls for deconstruction. Curr Opin Biotechnol 20:286–294
Marulasiddeshwara MB, Dakshayani SS, Sharath Kumar MN et al (2017) Facile-one pot-green synthesis, antibacterial, antifungal, antioxidant and antiplatelet activities of lignin capped silver nanoparticles: a promising therapeutic agent. Mater Sci Eng, C 81:182–190
Matsuhisa K, Yamane S, Okamoto T, Watari A, Kondoh M et al (2015) Anti-HCV effect of Lentinula edodes mycelia solid culture extracts and low-molecular-weight lignin. Biochem Biophys Res Commun 462:52–57
Medina JDC, Woiciechowski AL, Filho AZ et al (2016) Biological activities and thermal behavior of lignin from oil palm empty fruit bunches as potential source of chemicals of added value. Ind Crops Prod 94:630–637
Mehta AY, Mohammed BM, Martin EJ, Brophy DF, Gailani D, Desai UR (2016) Allosterism-based simultaneous, dual anticoagulant and antiplatelet action. Allosteric inhibitor targeting the glycoprotein ibα and heparin-binding site of thrombin. J Thromb Haemost 14:828–838
Meister JJ (2007) Modification of lignin. J Macromol Sci Part C: Polym Rev 42(2):235–289
Mishra PK, Ekielski A (2019) The self-assembly of lignin and its application in nanoparticle synthesis: a short review. Nanomaterials (Basel) 9(2). https://doi.org/10.3390/nano9020243 (in press)
Mishra PK, Wimmer R (2017) Aerosol assisted self-assembly as a route to synthesize solid and hollow spherical lignin colloids and its utilization in layer by layer deposition. Ultrason Sonochem 35:45–50
Mitjans M, Vinardell MP (2005) Biological activity and health benefits of lignans and lignins. Trends Comp Biochem Physiol 11:55–62
Monien BH, Henry BL, Raghuraman A, Hindle M, Desai UR (2006) Novel chemo-enzymatic oligomers of cinnamic acids as direct and indirect inhibitors of coagulation proteinases. Bioorg Med Chem 14:7988–7998
Mulder WJ, Gosselink RJA, Vingerhoeds MH, Harmsen PFH, Eastham D (2011) Lignin based controlled release coatings. Ind Crops Prod 34:915–920
Nada AMA, El-Diwanya AI, Elshafei AM (1989) Infrared and antimicrobial studies on different lignins. Acta Biotechnol 9(3):295–298
Naegelem H, Pfitzer J, Zieglerm L, Inone-Kauffmann ER, Eisenreich N (2016) Applications of lignin materials and their composites. In: Faruk O, Sain M (eds) Lignin in polymer composites. Elsevier Inc., Amsterdam, pp 233–244
Nagaraju DH, Rebis T, Gabrielsson R, Elfwing A, Milczarek G, Inganäs O (2014) Charge storage capacity of renewable biopolymer/conjugated polymer interpenetrating networks enhanced by electroactive dopants. Adv Energy Mater 4:1–7
Nakashima H, Murakami T, Yamamoto N, Sakagami H, Tanuma S et al (1992) Inhibition of human immunodeficiency viral replication by tannins and related compounds. Antiviral Res 18:91–103
Nar M, Rizvi HR, Dixon RA, Chen F, Kovalcik A, D'Souza N (2016) Superior plant based carbon fibers from electrospun poly-(caffeyl alcohol) lignin. Carbon 103:372–383
Nimz H (1974) Beech lignin-proposal of a constitutional scheme. Angew Chem Int Edn 13:313–321

Norikura T, Mukai Y, Fujita S, Mikame K, Funaoka M, Sato S (2010) Lignophenols decrease oleate-induced apolipoprotein-B secretion in HepG2 cells. Basic Clin Pharmacol Toxicol 107:813–817

Núñez-Flores R, Giménez B, Fernández-Martín F, López-Caballero ME, Montero MP, Gómez-Guillén MC (2013) Physical and functional characterization of active fish gelatin films incorporated with lignin. Food Hydrocolloids 30:163–172

Oh-Hara T, Sakagami H, Kawazoe Y, Kaiya T, Komatsu N, Ohsawa N et al (1990) Antimicrobial spectrum of lignin-related pine cone extracts of *Pinus parviflora* Sieb. et Zucc. In Vivo 4:7–12

Ojagh SM, Núñez-Flores R, López-Caballero ME, Montero MP, Gómez-Guillén MC (2011) Lessening of high-pressure-induced changes in Atlantic salmon muscle by the combined use of a fish gelatine lignin film. Food Chem 125(2):595–606

Ou S, Kwok KC (2004) Ferulic acid: pharmaceutical functions, preparation and applications in foods. J Sci Food Agric 84:1261–1269

Pan X, Kadla JF, Ehara K, Gilkes N, Saddler JN (2006) Organosolv ethanol lignin from hybrid poplar as a radical scavenger: relationship between lignin structure, extraction conditions, and antioxidant activity. J Agric Food Chem 54:5806–5813

Paszner L, Cho HJ (1989) Organosolv pulping: acidic catalysis options and their effect on fiber quality and delignification. Tappi J 72:135–142

Peng Z, Chen F (2011) Synthesis and properties of lignin-based polyurethane hydrogels. Int J Polym Mater 60:674–683

Peng Y, Nicastro KH, Epps TH, Wu C (2018) Evaluation of estrogenic activity of novel bisphenol a alternatives, four bioinspired bisguaiacol of specimens, by in vitro assays. J Agric Food Chem 6644:11775–11783

Pereira FM, Gonçalves AR, Ferraz A, Silva FT, Oliveira SC (2003) Estimation of solubility effect on the herbicide controlled-release kinetics from lignin-based formulations. Appl Biochem Biotechnol 108:913–919

Popaa VI, Dumitrua M, Volf I, Anghelb N (2008) Lignin and polyphenols as allelochemicals. Ind Crops Prod 27:144–149

Pouteau C, Cathala B, Dole P, Kurek B, Monties B (2005) Structural modification of Kraft lignin after acid treatment: characterization of the apolar extracts and the influence on the antioxidant properties in polypropylene. Ind Crops Prod 21:101–108

Pouteau C, Dole P, Cathala B, Averous L, Boquillon N (2003) Antioxidant properties of lignin in polypropylene. Polym Degrad Stab 81:9–18

Priefert H, Babenhorst J, Steinbuchel A (2001) Biotechnological production of vanillin. Appl Microbiol Biotechnol 56(3–4):296–314

Pucciariello R, D'Auria M, Villani V, Shulga G (2010) Lignin/poly (ε-caprolactone) blends with tuneable mechanical properties prepared by high-energyball-milling. J Environ Polym Degrad 18:326–334

Qian Y, Qiu X, Zhu S (2015) Lignin: a nature-inspired sun blocker for broad spectrum sunscreens. Green Chem 17:320–324

Qu L, Chen JB, Zhang GJ, Sun SQ, Zheng J (2017) Chemical profiling and adulteration screening of Aquilariae Lignum Resinatum by Fourier transform infrared (FT-IR) spectroscopy and two-dimensional correlation infrared (2D-IR) spectroscopy. Spectrochim Acta Part A Mol Biomol Spectrosc 174:177–182

Ragauskas AJ, Beckham GT, Biddy MJ, Chandra R et al (2014) Lignin valorization: improving lignin processing in the biorefinery. Science 344:709–720

Richter AP, Bharti B, Armstrong HB, Brown JS, Plemmons D, Paunov VN, Stoyanov SD, Velev OD (2016) Synthesis and characterization of biodegradable lignin nanoparticles with tunable surface properties. Langmuir 32(25):6468–6477

Rinaldi R, Jastrzebski R, Clough MT et al (2016) Paving the way for lignin valorisation: recent advances in bioengineering, biorefining and catalysis. Angew Chem Int Edn 55:8164–8215

Ro D, Shafaghat H, Jang SH et al (2019) Production of an upgraded lignin-derived bio-oil using the clay catalysts of bentonite and olivine and the spent FCC in a bench-scale fixed bed pyrolyzer. Environ Res 172:658–664

Rodríguez A, Jiménez L (2008) Pulping with organic solvents other than alcohols. Afinidad 65:88–196
Sadeghifar H, Argyropoulos DS (2015) Correlations of the antioxidant properties of softwood kraft lignin fractions with the thermal stability of its blends with polyethylene. ACS Sustain Chem Eng 3:349–356
Sakagami H, Hashimoto K, Suzuki F, Ogiwara T, Satoh K et al (2005) Molecular requirements of lignin-carbohydrate complexes for expression of unique biological activities. Phytochemistry 66:2108–2120
Sakagami H, Satoh K, Fukamachi H, Ikarashi T, Shimizu A et al (2008) Anti-HIV and vitamin C-synergized radical scavenging activity of cacao husk lignin fraction. Vivo 22(3):327–332
Salamouny SE, Herz A, Huber J (2002) Suitability of three lignin products as UV protectants to baculovirus. Bull Entomol Soc Egypt 28:103–111
Salamouny SE, Shapiro M, Ling KS et al (2009) Black tea and lignin as ultraviolet protectants for the beet armyworm nucleopolyhedrovirus. J Entomol Sci 44:50–58
Salanti A, Zoia L, Orlandi M, Zanini F, Elegir G (2010) Structural characterization and antioxidant activity evaluation of lignins from rice husk. J Agric Food Chem 58:10049–10055
Sato S, Mukai Y, Yamate J, Norikura T, Morinaga Y, Mikame K, Funaoka M, Fujita S (2009) Lignin-derived lignophenols attenuate oxidative and inflammatory damage to the kidney in streptozotocin-induced diabetic rats. Free Radical Res 43:1205–1213
Satoh K, Kihara T, Ida Y et al (1999) Radical modulation activity of pine cone extracts of *Pinuselliottii* var. Elliottii. Anticancer Res 19:357–364
Schutyser W, Renders T, Van den Bosch S, Koelewijn SF, Beckham G, Sels B (2018) Chemicals from lignin: an interplay of lignocellulose fractionation, depolymerisation, and upgrading. Chem Soc Rev 47:852–908
Sederoff RR, MacKay JJ, Ralph J, Hatfield RD (1999) Unexpected variation of lignin. Curr Opin Plant Biol 2:145–152
Shi L, Ge J, Nie S, Qin C, Yao S (2019) Effect of lignin structure on adsorbable organic halogens formation in chlorine dioxide bleaching. Royalty Soc Open Sci 6(2):182024
Siddiqui L, Mishra H, Mishra PK, Iqbal Z, Talegaonkar S (2018) Novel 4-in-1 strategy to combat colon cancer, drug resistance and cancer relapse utilizing functionalized bioinspiring lignin nanoparticle. Med Hypotheses 121:10–14
Slavikova E, Kosikova B (1994) Current awareness on yeast. Yeast 11(3):293–300
Song Y, Hui J, Kou W, Xin R, Jia F, Wang N, Hu F, Zhang H, Liu H (2008) Identification of *Inonotus obliquus* and analysis of antioxidation and antitumor activities of polysaccharides. Curr Microbiol 57:454–462
Spiridon I (2018) Biological and pharmaceutical applications of lignin and its derivatives: a mini-review. Cellul Chem Technol 52(7–8):543–550
Stephen LY (1981) New lignosulfonate dispersant for dyes. Text Chem Colorist 13:24–28
Stewart D (2008) Lignin as a base material for materials applications: chemistry, application and economics. Ind Crops Prod 27:202–207
Sunthornvarabhas J, Liengprayoon S, Suwonsichon T (2017) Antimicrobial kinetic activities of lignin from sugarcane bagasse for textile product. Ind Crops Prod 109:857–861
Thakur VK, Thakur MK (2015) Recent advances in green hydrogels from lignin: a review. Int J Biol Macromol 72:834–847
Thomas VA, Donohoe BS, Li M, Pu Y, Ragauskas AJ, Kumar R, Nguyen TY, Cai CM, Wyman CE (2017) Adding tetrahydrofuran to dilute acid pretreatment provides new insights into substrate changes that greatly enhance biomass deconstruction by *Clostridium thermocellum* and fungal enzymes. Biotechnol Biofuels 10:252 (1-13)
Tortora M, Cavalieri F, Mosesso P et al (2014) Ultrasound driven assembly of lignin into microcapsules for storage and delivery of hydrophobic molecules. Biomacromol 15:1634–1643
Ugartondo V, Mitjans M, Vinardell MP (2008) Comparative antioxidant and cytotoxic effects of lignins from different sources. Bioresour Technol 99(14):6683–6687

Upton BM, Kasko AM (2016) Strategies for the conversion of lignin to high-value polymeric materials: review and perspective. Chem Rev 116:2275–2306

Vanholme R, Morreel K, Ralph J, Boerjan W (2008) Lignin engineering. Curr Opin Biotechnol 11:278–285

Vengal JC, Srikumar M (2005) Processing and study of novel lignin-starch and lignin-gelatin biodegradable polymeric films. Trends Biomater Artif Organs 18(2):237–241

Vinardell MP, Ugartondo V, Mitjans M (2008) Potential applications of antioxidant lignins from different sources. Ind Crops Prod 27:220–223

Wang J, Cao F, Su E, Wu C, Zhao L, Ying R (2013) Improving flavonoid extraction from *Ginkgo biloba* leaves by pre-fermentation processing. J Agric Food Chem 61:5783–5791

Wang J, Cao F, Su E, Zhao L, Qin W (2018) Improvement of animal feed additives of Ginkgo leaves through solid-state fermentation using *Aspergillus niger*. Int J Biol Sci 14:736–747

Wang H, Pu Y, Ragauskas A, Yang B (2019a) From lignin to valuable products-strategies, challenges, and prospects. Bioresour Technol 271:449–461

Wang Q, Zuoa Z, Chucky Cheung CK, Yee Leung SS (2019b) Updates on thermosensitive hydrogel for nasal, ocular and cutaneous delivery. Int J Pharm 559:86–101

Wang R, Wang G, Xia Y, Sui W, Si C (2019c) Functionality study of lignin as a tyrosinase inhibitor: influence of lignin heterogeneity on anti-tyrosinase activity. Int J Biol Macromol 128:107–113

Wang X, Zhao J (2013) Encapsulation of the herbicide picloram by using polyelectrolyte biopolymers as layer-by-layer materials. J Agric Food Chem 61:3789–3796

Whetten R, Sederoffa R (1995) Lignin biosynthesis. Plant Cell 7:1001–1013

Wu H, Chen F, Feng Q, Yue X (2012) Oxidation and sulfomethylation of alkali-extracted lignin from corn stalk. BioResources 7(3):2742–2751

Xu CP, Arancon RAD, Labidi J, Luque R (2014) Lignin depolymerisation strategies: towards valuable chemicals and fuels. Chem Soc Rev 43:7485–7500

Yang W, Fortunati E, Dominici F et al (2016) Effect of cellulose and lignin on disintegration, antimicrobial and antioxidant properties of PLA active. Int J Biol Macromol 89:360–368

Yearla SR, Padmasree K (2016) Preparation and characterisation of lignin nanoparticles: evaluation of their potential as antioxidants and UV protectants. J Exp Nanosci 11:289–302

Zamboni F, Vieira S, Reis RL, Oliveira JM, Collins MN (2018) The potential of hyaluronic acid in immunoprotection and immunomodulation: chemistry, processing and function. Prog Mater Sci 97:97–122

Zemek J, Kosíková B, Augustín J, Joniak D (1979) Antibiotic properties of lignin components. Folia Microbiol 24:483–486

Zhang F, Lin J, Zhao G (2016) Preparation and characterization of modified soda lignin with polyethylene glycol. Materials (Basel) 9(10):822

Zhang J, Cui J-H, Yin T, Sun L, Li G (2013) Activated effect of lignin on α-amylase. Food Chem 141:2229–2237

Zhang J, Xiao L, Yang Y, Wang Z, Li G (2014) Lignin binding to pancreatic lipase and its influence on enzymatic activity. Food Chem 149:99–106

Zhang MY, Xu YJ, Li KC (2007) Removal of residual lignin of ethanol-based organosolv pulp by an alkali extraction process. J Appl Polym Sci 106(1):630–636

Zhou Y, Wang D, Yang D, Qiu X, Li Y (2019) Avermectin loaded nanosphere prepared from acylated alkali lignin showed anti-photolysis property and controlled release performance. Ind Crops Prod 137:453–459

Zhu L, Xu X (2013) Stimulatory effect of different lignocellulosic materials for phenolic compound production and antioxidant activity from *Inonotus obliquus* in submerged fermentation. Biotechnol Appl Biochem 169(7):2138–2152

Zikeli F, Vinciguerra V, D'Annibale A, Capitani D, Romagnoli M, Scarascia Mugnozza G (2019) Preparation of lignin nanoparticles from wood waste for wood surface treatment. Nanomaterials (Basel) 9(2): pii: E281. https://doi.org/10.3390/nano9020281 (in press)

Zimniewska M, Batog J, Bogacz E, Romanowska B (2012) Functionalization of natural fibres textiles by improvement of nanoparticles fixation on their surface. J Fiber Bioeng Inform 5:321–339
Zimniewska M, Kozlowski R, Batog J (2008) Nanolignin modified linen fabric as a multifunctional product. Mol Cryst Liq Cryst 484:409–416

The manufacturer's authorised representative in the EU is Springer Nature Customer Service Centre GmbH, Europaplatz 3, 69115 Heidelberg, Germany. If you have any concerns regarding our products, please contact ProductSafety@springernature.com

Printed and bound by CPI Group (UK) Ltd, Croydon, CR0 4YY
05/10/2026
02228120-0008